# Locksmithing and Electronic Security Wiring Diagrams

## John L. Schum

**McGraw-Hill**
New York   Chicago   San Francisco   Lisbon   London   Madrid
Mexico City   Milan   New Delhi   San Juan   Seoul
Singapore   Sydney   Toronto

**Cataloging-in-Publication Data is on file with the Library of Congress**

*McGraw-Hill*

*A Division of The McGraw·Hill Companies*

1 2 3 4 5 6 7 8 9 0 DOC/DOC 0 9 8 7 6 5 4 3 2

ISBN 0-07-139305-6

*The sponsoring editor for this book was Shelley Carr, the editing supervisor was Daina Penikas, and the production supervisor was Pamela A. Pelton. It was set in Century Schoolbook per the MHT design by Kim Sheran of McGraw-Hill Professional's Hightstown, N.J., composition unit.*

*Printed and bound by R. R. Donnelley & Sons Company.*

McGraw-Hill books are available at special quantity discounts to use as premiums and sales promotions, or for use in corporate training programs. For more information, please write to the Director of Special Sales, Professional Publishing, McGraw-Hill, Two Penn Plaza, New York, NY 10121-2298. Or contact your local bookstore.

 This book is printed on recycled, acid-free paper containing a minimum of 50% recycled de-inked fiber.

*To all struggling technical writers.*

# Contents

# Acknowledgments

I am grateful to the many students who, over the years, provided me with lots of material for this book. They inadvertently taught me how to teach this subject. I also wish to thank Sue Robinson and my daughter Dawn Speer for their typing skills; and John Sanchirico, Jr., and Glen Girard for their expertise in putting together the illustrations.

Special thanks to my editors Shelley Carr and Daina Penikas, who guided me along the way.

# Introduction

The intent of this book is to equip members of the locksmith, hardware, and security industries with information they need to profit in the rapidly growing electronic security market. Its purpose is to educate (and serve as a basic reference manual for) those who desire increased knowledge of today's sophisticated electronic security field.

The creation of wiring diagrams for security systems crosses many fields—locksmithing, builders' hardware, and alarm and access trades. In today's competitive market, one must be aware of all the elements that are required to create a total security system. Whether you are involved in the sales, installation, design, or service of security systems, this book will broaden your knowledge of a very lucrative field.

Until recently the wiring diagram was thought of as the exclusive province of specialists of one particular trade. Those brave and energetic enough to go in search of knowledge of this subject are often reluctant to expose themselves by asking even the basic questions. This manual answers all those questions, arranged step by step, from very basic information to complex concepts. It is written in concise and matter-of-fact language to help the average person better understand wiring diagrams.

*Wiring Diagrams* grew as an idea from my many years of teaching electronic security to tradespeople. Many of my classes included hands-on laboratories that required following simple wiring diagrams. Very early on, while explaining hookup diagrams, I always noticed a certain discomfort among the students. When it came time for actual hands-on exercises, many students were totally lost!

It finally struck me that they were unable to read even a basic wiring diagram. It wasn't their fault! Many were working with a wiring diagram for the first time. Finally I added what I call a "basic drafting" section to all my basic electronics classes. That has proved very effective. I then added other sections on symbols and power flow to help the student to better understand wiring diagrams. This manual is an expanded version of those classes. It is aimed primarily toward hardware and locksmith personnel who haven't had much exposure to electrical drawings. For all others it should serve as a "tune-up." At the very least, reading this manual will help tradespeople understand the basics of an electrical drawing, essential to the electronic security trade.

The original idea for this manual was to present a simple method for drawing common wiring diagrams, including some insight into their application. This inspiration was prompted by the many questions that arose during years of training industry members.

Realizing that other subject matter came into play, I began to expand the manual. Drafting basics became Chapter 1, followed by the use of standard electronic symbols and descriptions of the different types of electrical drawings in Chapters 2 and 3. In Chapter 4 I decided to show the reader step by step how to build a typical wiring diagram. Chapter 5 explains how to gather information necessary to creating a wiring diagram, followed by a discourse on controlling and routing power throughout a system. By now I thought the readers would like to try to create a wiring diagram on their own. At the end of Chapter 5 is an exercise to "test" the readers' new knowledge.

What began as a training manual was becoming a workbook! I decided to go another step further. Chapter 6 covers the use of junction boxes, fire panel tie-in, and override circuits—all items that are used on wiring diagrams for larger systems. The next logical addition to a system is monitoring, an important, often overlooked feature thoroughly covered in Chapter 7. More sophisticated systems require the use of relays. Chapter 8 covers relays in depth followed by their use in access/egress devices. Chapter 10 is a series of skill exercises to evaluate your progress. The "grand finale" of wiring diagrams is multidoor interlocks, presented in Chapter 11. Understanding and creating wiring diagrams certainly enables you to identify and correct basic problems in an installed system. A natural addition to this manual is Chapter 12 on troubleshooting.

As a conclusion, Chapter 13 presents a collection of basic applications for common security systems.

Start at the beginning. As you work through each chapter, you will begin to "see" how a wiring diagram originates. Each chapter will lead you to the next, until you actually create your own wiring diagram. The final chapters will take you into more complex designs, which you will now understand.

The ending chapters were added to show that your increased knowledge can be used in other aspects of electronic security, primarily troubleshooting and system design.

To my knowledge there is no other manual available that offers in-depth instruction on creating wiring diagrams. I am sure you will find this material useful in furthering your careers. I highly recommend increasing your knowledge by taking advantage of any educational opportunities that present themselves. Most electronic security hardware manufacturers offer classes on their products. The hardware, locksmith, and alarm industries all have trade organizations that offer scheduled classes relevant to the subject matter in this manual.

I believe this workbook will provide the groundwork that will enable you not only to create wiring diagrams but also to actually design security systems.

I sincerely hope you find this manual helpful.

John L. Schum

# 1

# Drawing Basics

In today's computer-oriented world a lot of drawings are done on computer-aided design (CAD) systems. These drawings are usually for large or complex projects. Making these drawings also requires computer hardware, software, and a certain expertise in using CAD.

I do believe, however, that a great many drawings covering small systems are done the "old-fashioned" way—by hand. Knowing a little about drawing technique and symbolism and being able to draw a straight line should enable anyone to create a viable drawing.

Let's look first at some of the equipment we might need.

## Equipment

The first thing we learned about drafting in technical school was what basic equipment we should have. This amounted to a pretty good investment for a full-fledged drafting course. The good news is, you can get started for under $15! You won't need elaborate equipment for sketch work or basic trade drawings.

The starting point is the drawing instrument, better known as a pencil! Pencils, and lead for mechanical pencils, are available in different *hardnesses*. The hardness of the lead controls the thickness and darkness of the lines and characters you are drawing. Lead hardness is designated by letters and numbers, with 6B being very soft, HB and F being medium, and 2H to 9H being hard.

A soft lead will make thicker and darker lines; a hard lead will make thinner, lighter lines. Figure 1-1 shows some line "weights" and techniques. In all honesty, you can get by with a number 2 pencil for heavy lines and a number 3 for lighter lines. You will see the value of line weight on drawings a little later.

The next thing we learned was how to draw a straight line! Now there are all kinds of instruments available to help you do this. I recommend that you keep a small, inexpensive plastic triangle at hand for this purpose. Finding one similar to the one shown in Figure 1-2 will also help in drawing small circles, rectangles, and squares.

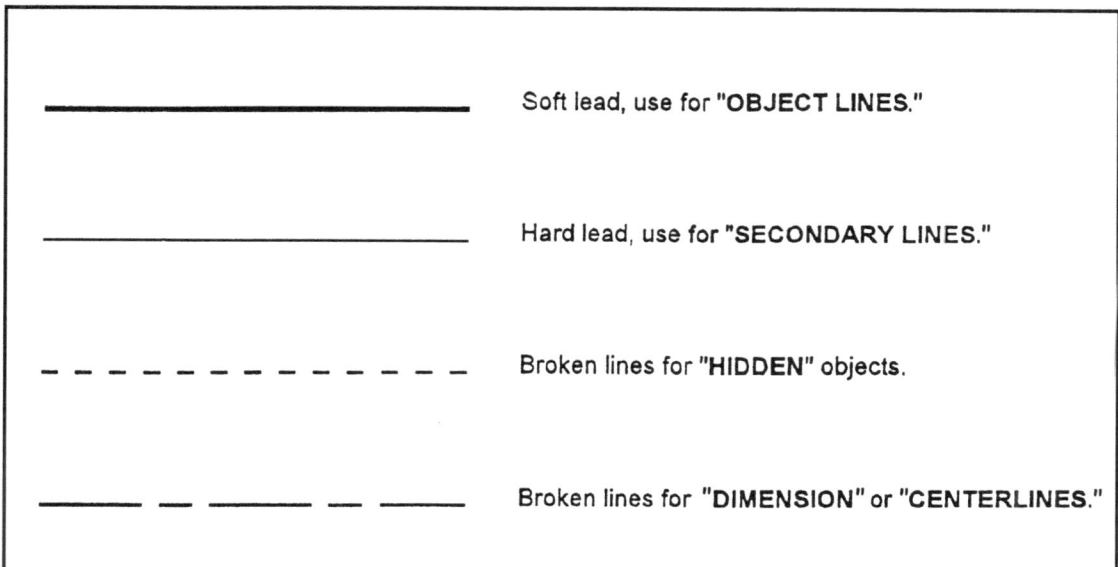

**Figure 1-1**   Line techniques.

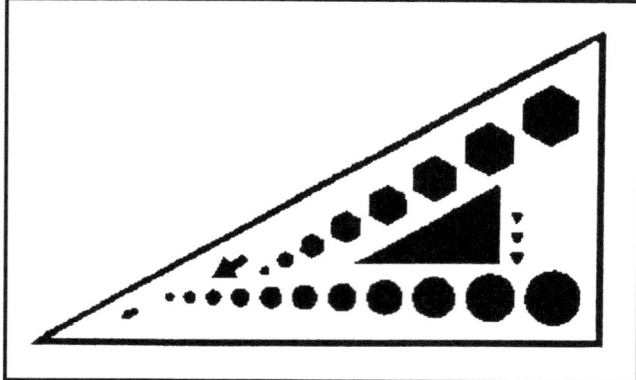

**Figure 1-2**   Triangle with Template cutouts.

In a pinch I will pick up anything—a business card or matchbook cover—to use for drawing a straight line.

By now you may think, "This guy is hung up on straight lines!" Let me show you the value of straight lines. Figure 1-3 is an actual drawing handed in for a class exercise. This was someone's attempt at a wiring diagram. I call it "the spaghetti drawing."

Now I'm not faulting the student, because he or she had never made a wiring diagram. At this point I made sure that every student thereafter had a graph type of pad for these classes.

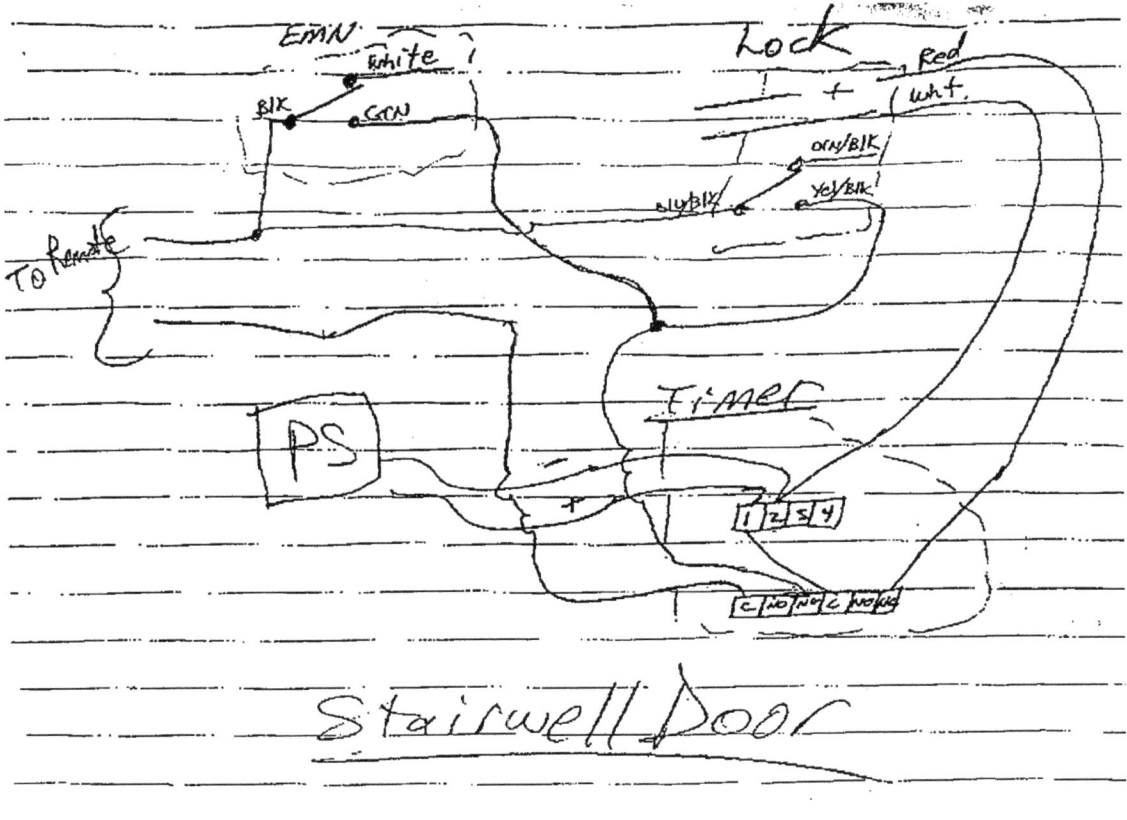

**Figure 1-3** "The spaghetti drawing."

This leads me to the subject of paper. Once we have picked a favorite pencil, we need something to draw on. I have used just about anything—paper napkins, matchbook covers, flaps from cardboard cartons—whatever was handy. I probably use the standard lined writing pad the most, only because there is always one handy. But the graph-type pad is the easiest to use for sketches or simple drawings. At this point I should mention that a true graph pad is a little different from what we should be using. Perhaps a better term would be *grid pad,* commonly sold as a "quadrille" pad.

These pads come in three basic sizes, which are identified in the drafting field by letters, as shown in Figure 1-4. These pads are also available in different size *grids.* The grid pattern is simply very pale blue lines that run horizontally and vertically over the pad. Grid size is the size of the squares formed by the lines, as shown in Figure 1-5.

The grid helps in drawing vertical and horizontal lines *freehand.* It also assists in *sizing* objects to scale when you want to keep some real proportion to the drawing.

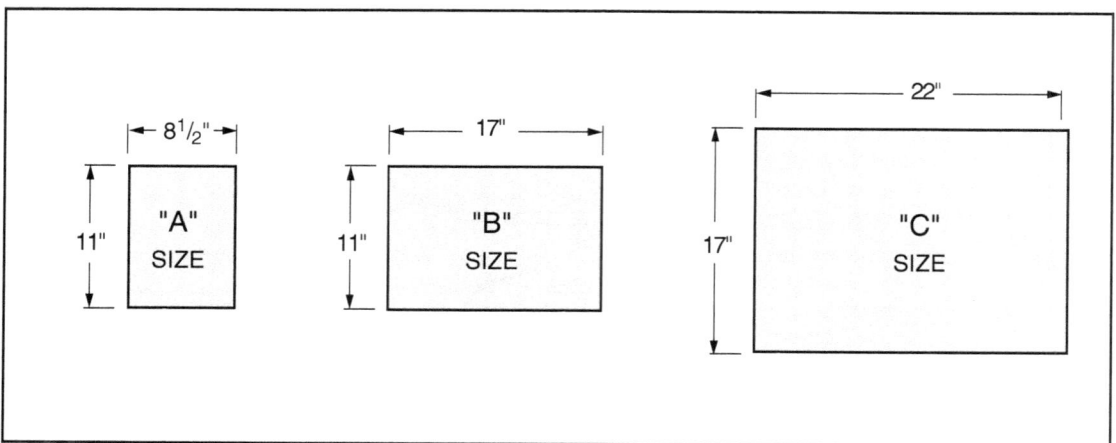

**Figure 1-4**   Basic drawing pad sizes.

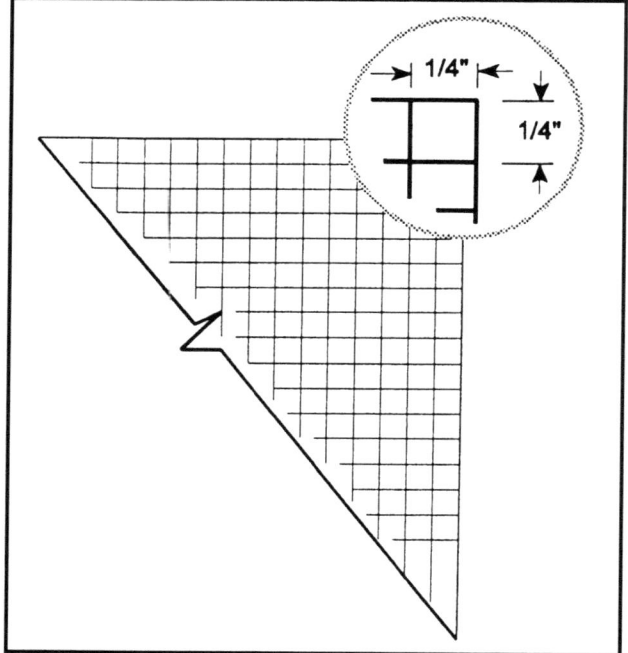

**Figure 1-5**   Grid size. $\frac{1}{4}$" grid shown.

As a simple example, suppose you wanted to draw a $3\frac{1}{2}$-inch $\times$ $2\frac{1}{2}$-inch box using a $\frac{1}{4}$-inch grid paper. You simply run your pencil horizontally 14 squares and vertically 10 squares to establish the sides of the box, as shown in Figure 1-6. Also, you could draw a much bigger object to some proportional scale. For instance, if you want to represent a 4-foot square box, you designate each

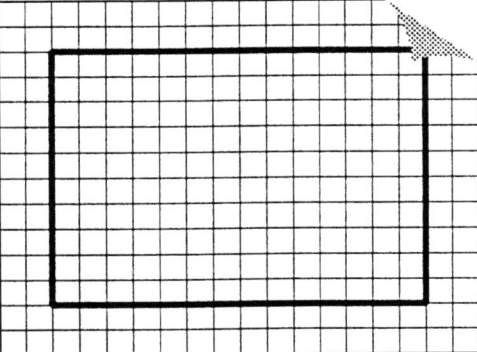

**Figure 1-6** Using $1/4$" grid layout.

$1/4$-inch grid mark to equal 1 foot. Of course, anything else drawn on the same sheet would follow the same scale: $1/4$ inch = 12 inches.

To make my point about straightness, I redrew, by hand on a grid pad, "the spaghetti drawing." Figure 1-7 shows how this drawing should look, given a little care. The drawing technically may not be accurate, but neatness counts! Somebody has to be able to easily read these drawings.

As a final note, when you photocopy grid paper on a normal setting, the blue grid lines do not reproduce. This leaves you with a nice, clean-looking drawing. It also provides a faxable document.

With pencil, paper, and tools in hand, we need to look at one more item. I call this next part *line technique*.

## Line Technique

Practicing with different lead "weights" will result in drawings that are easy to read. You can also emphasize the message you are trying to deliver. Notice the drawing in Figure 1-8. This drawing was made with a single pencil, or single type of lead.

The drawing isn't bad, the lines are straight, but the message is sort of bland. It is like a speaker droning on in a monotone voice.

Now look at the drawing in Figure 1-9.

It contains the same message, but like a good speaker, the writer has emphasized particular areas. This is the benefit of varying the line density.

Figure 1-9 was drawn with three different weights of lead. The purpose is to accentuate the parts of the drawing that are most important. The reader's eyes are first drawn to the hookup terminals. A lighter-weight line emphasizes the wiring. Even lighter lines could be used for secondary information. In this manner less important information doesn't interfere with the real thrust of the drawing—the hookup information.

If you happen to be stuck with one type of pencil or lead, you can still create this effect. For darker lines use a blunt lead and bear down heavily. For lighter

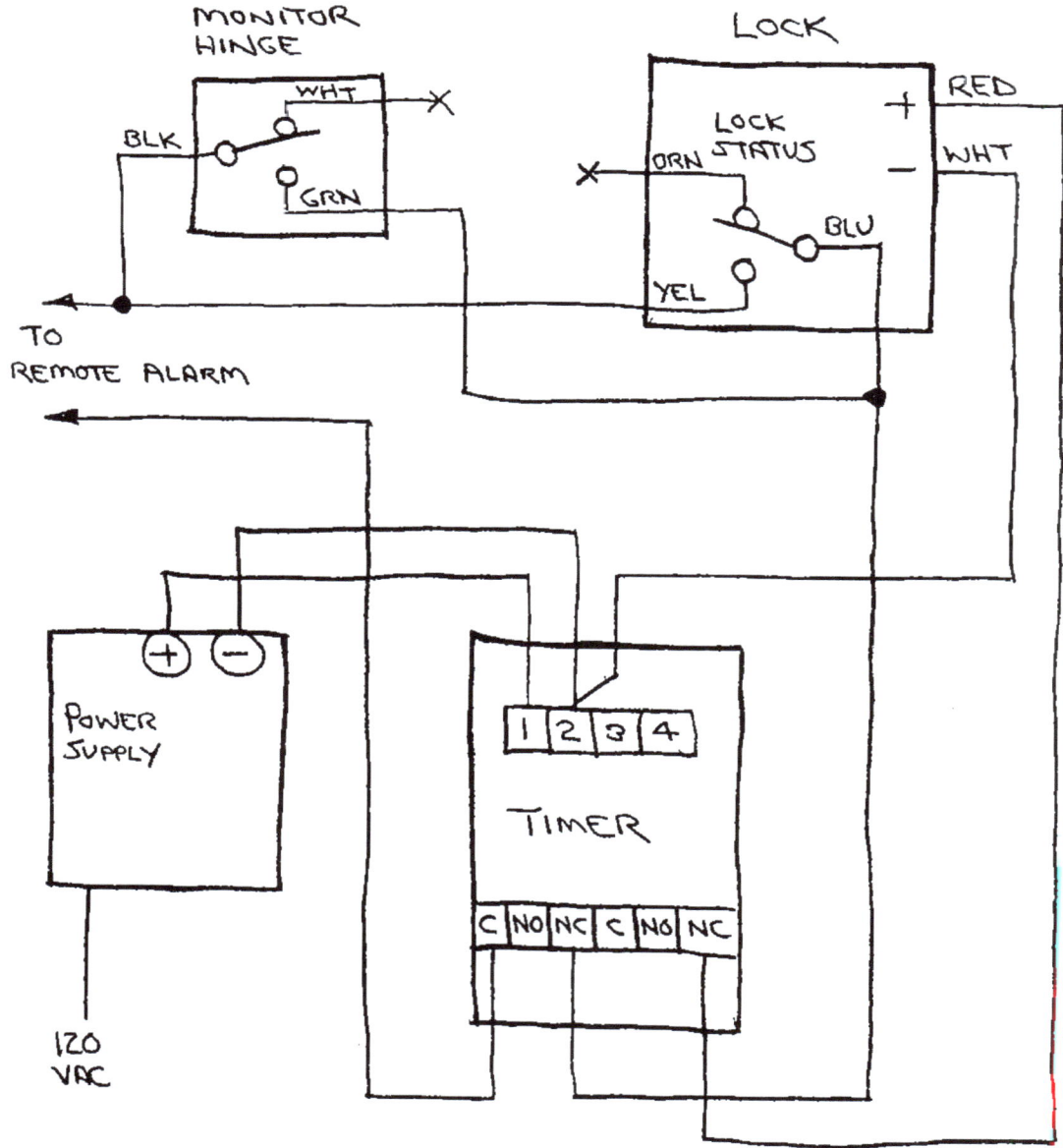

**Figure 1-7**   Hand-drawn wiring diagram.

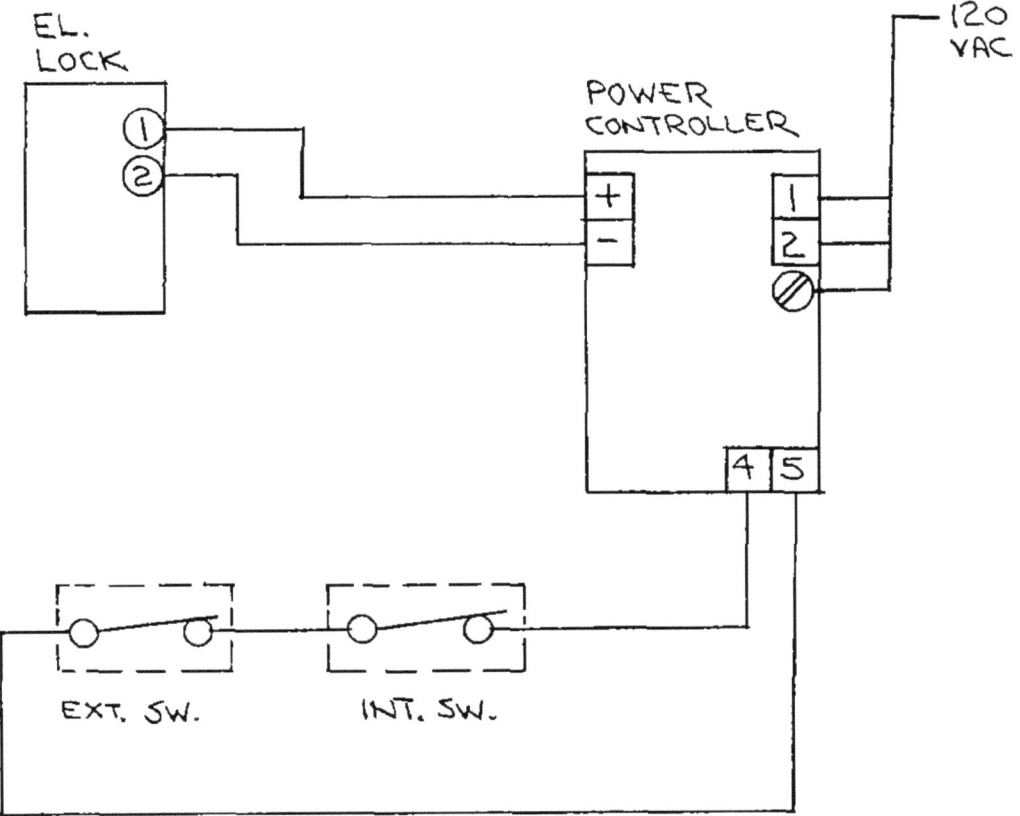

**Figure 1-8**   Drawing made with single "weight" lead.

lines use a sharper lead and less pressure. Simply keeping these ideas in mind should improve your drawing style.

On another note, printing is the norm for notes, letters, and numbers. Lettering guides, like the one shown in Figure 1-10, are available.

I find that with care and practice one can "letter" quite legibly by hand. As a drafter in my earlier years, I had difficulty with my freehand lettering. Being left-handed, I tended to slant every other letter in a different direction. A tip I was given was to purposely slant every letter in one direction. This helped considerably to "straighten out" my printing.

Next we look at some of the symbols and abbreviations that are used in the industry for electrical drawings

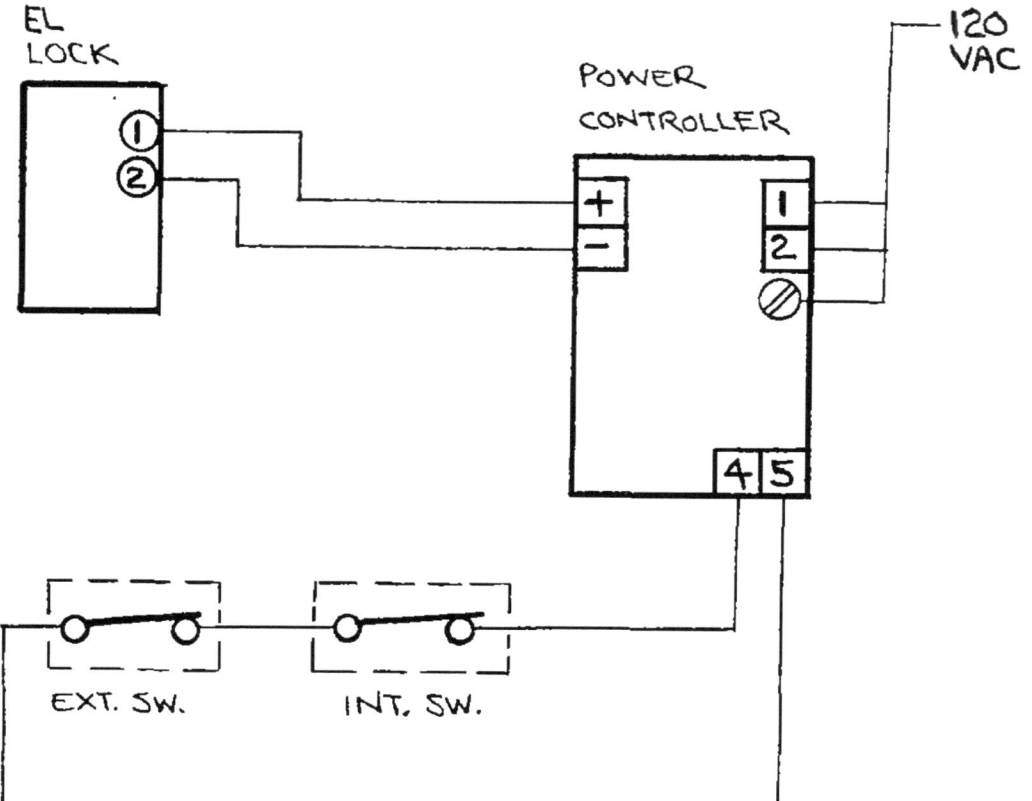

**Figure 1-9**   Drawing made with different "weight" leads.

**Figure 1-10**   Typical lettering guide.

# 2

# Symbols and Abbreviations

Use of industry standards for drawings helps in conveying information to the experienced reader. The use of nonstandard symbols is sometimes confusing and often takes on an "artform" of its own. Using standards can also cut down on your drawing time.

In the electronic security industry there is a lack of drawing symbols, or at least a lack of coordination of symbols used between trades. What I present in this chapter are symbols commonly used on security industry drawings. Some of these symbols are borrowed from several trades, for example, the architectural, electronic, and alarm trades.

In December 1995, a group called the Security Industry Association (SIA) published a standard that "defines drawing symbols for system layout and design." This group is associated with the alarm industry, and it has made this effort to offer some standardization. We discuss this further at the end of this chapter.

## Symbols

During classes I've sometimes equated reading an electrical drawing to reading a road map. There are paths to follow, obstacles to go around, and bridges to cross.

Occasionally I'll use some road map analogies. I think this will help you to understand electrical drawings, which we will cover in depth later.

The first item we will cover could be better called *drawing conventions* than symbols; but it is an important starting point.

Figure 2-1 shows two methods of drawing insulated wires that cross. Electricity can flow uninterrupted in either wire. There is no connection of the two paths. You could have 12 volts (abbreviated V) flowing in one path and 24 volts flowing in the other. Just think of this as two highways, one built as an overpass over the other. Traffic flows on one highway without interfering with traffic on the other one.

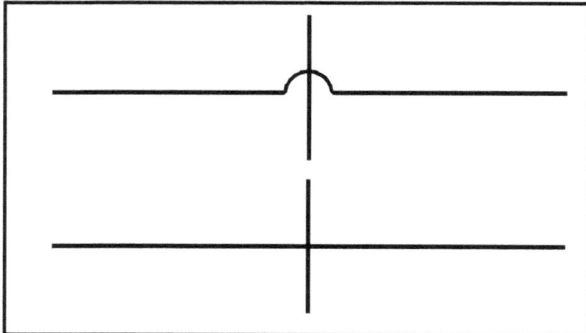

**Figure 2-1**  Wires crossing each other. No physical
connection. Electricity flows in each wire independently.

Figure 2-2 shows wires that are physically connected to each other. Electricity will flow in all directions simultaneously. Think of this as an uncontrolled road crossing—everybody on any road can drive anywhere the roads lead! It is pretty confusing if you're driving, but okay for electricity.

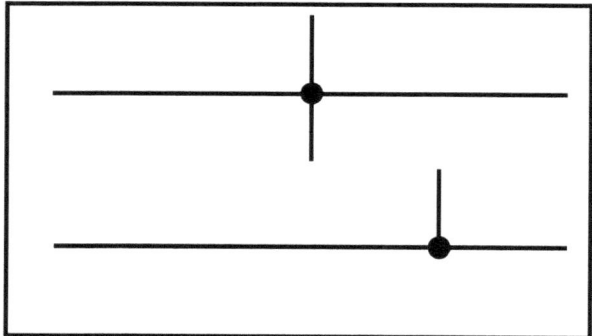

**Figure 2-2**  Wire junction. This is a physical junction of the
metal conductor of the wire. It can be a solder, wire nut, or
other type connection. Electricity can flow in all directions
simultaneously.

Some drawings will show physical wire connections, such as Figure 2-3. These are also physical wire connections. The symbols labeled TB4 usually indicate a screw terminal. TB4 stands for terminal block (or terminal board) number 4. So in this example, two wires are connected under the second terminal of the terminal block labeled TB4.

The connection marked P19 and J19 indicates wires in some kind of socket-and-plug connector. The letters J and P are commonly used to identify this type of connector; J indicates the jack and P indicates the plug.

Another area that causes some confusion is the symbol used to show a switch. Figure 2-4 shows a typical switch symbol and defines its components.

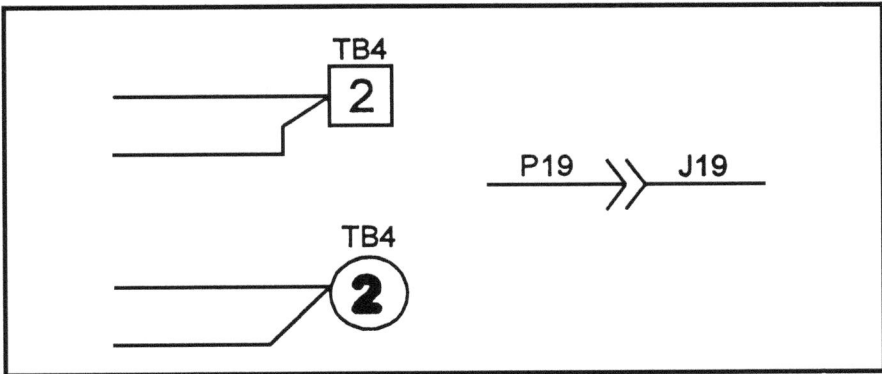

**Figure 2-3**  Symbols for physical wire connections.

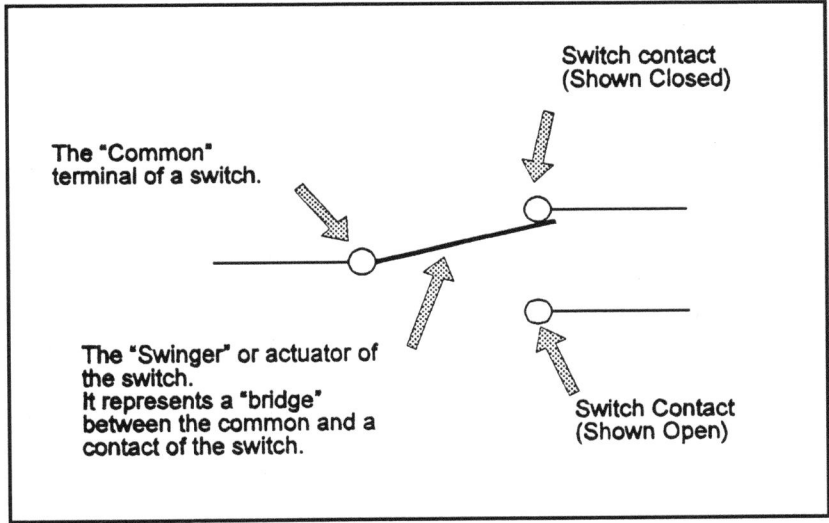

**Figure 2-4**  Identification of "parts" of a switch symbol.

   The parts of the switch are identified only to clarify the information that the symbol is conveying. Normally the same switch symbol on a drawing would be identified as shown in Figure 2-5.
   Usually switch contacts are labeled:

C = common

NO = normally open

NC = normally closed

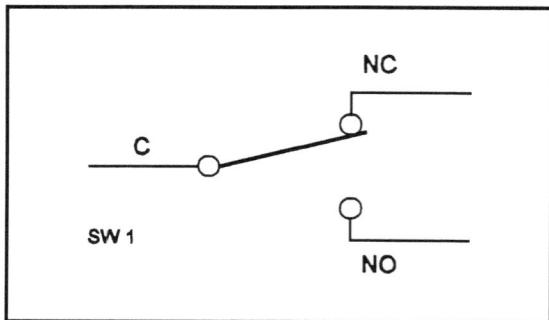

**Figure 2-5**   Typical switch symbol.

By *normal* we mean the switch is in its normal condition, that is, under no external influence. It is just the way it comes out of the box—a closed contact and an open contact.

Now at times you will see a switch on a drawing labeled as shown in Figure 2-6A or B.

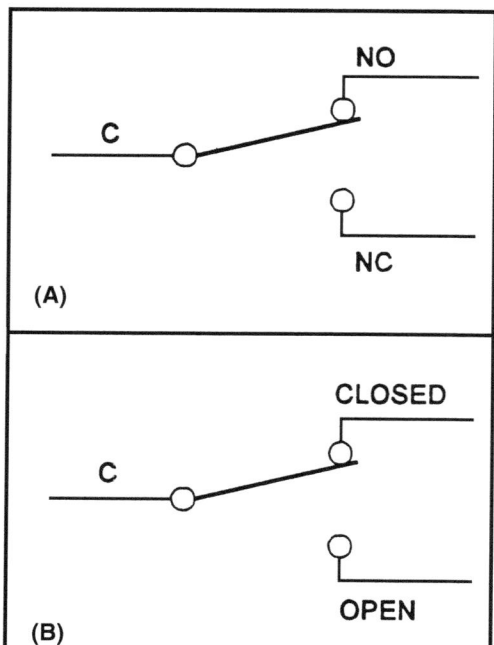

**Figure 2-6**   Switch contact identification for an "influenced" switch.

This is common in the alarm industry, especially for door or window contacts. In Figure 2-6A notice that the normally open contact is shown closed. The swinger of the switch is shown touching the open contact. This is common when the switch, in its "normal" usage in the circuit, is physically being held in its opposite to normal position. Figure 2-6B is another way this is shown, simply eliminating the word *normally* from the contact identification. Sometimes the identification will note "held closed" and "held open" to identify this condition.

No matter how the switch contacts are identified, the important thing is to show how you want the switch wired into the circuit. If there were no identification at all, the installer could still see how the switch was to be wired.

An example of this condition is a door status switch in an alarm or monitoring circuit. The switch shown in Figure 2-7 is purchased as a normally open switch.

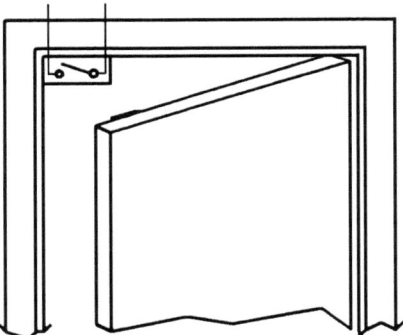

**Figure 2-7**  Surface mounted, normally open door status switch.

This switch is in its normal state when the door is open. The trouble is, the normal condition of the system is with the door closed.

When the door is closed, the permanent magnet mounted on the door pulls the contacts into the closed position. Now the switch contacts can be described as "normally open—held closed." I have often seen this type of switch identified in a monitoring circuit as shown in Figure 2-8. This method really tells it all.

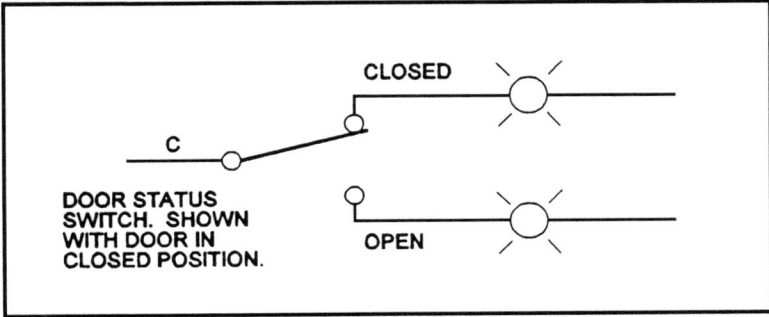

**Figure 2-8**  Door status switch in circuit.

Figure 2-9 shows some switch symbols commonly used on drawings. The symbol actually tries to impart to the reader the type of switch being used. In Figure 2-9A and B, it actually looks as though you have to push the actuator to open or close the contacts. In Figure 2-9C, D, and E, it looks as though you have to turn or "toggle" the actuator to open or close the switch contacts.

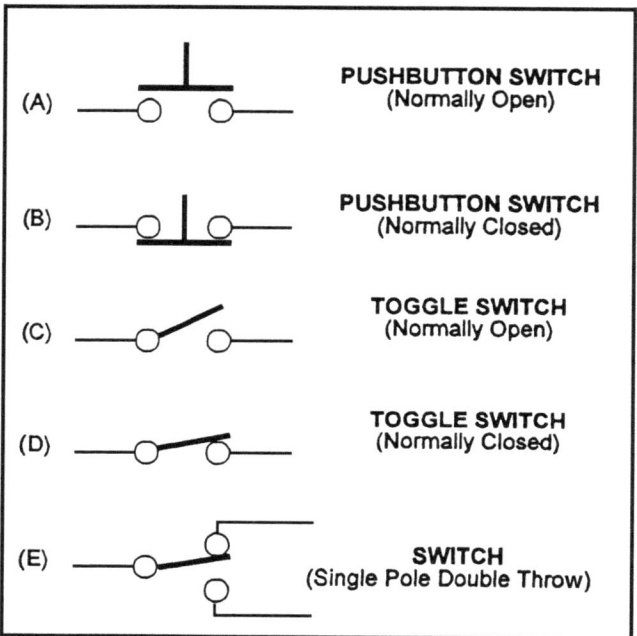

**Figure 2-9**  Common switch symbols.

In reality, it would not be totally incorrect to show any of the symbols for any type of switch. The important thing is to show the correct contact configuration—open or closed.

Other symbols for electronic devices are shown in Figure 2-10. Most of these symbols look somewhat like the items they represent.

Note that several of the symbols have notations included. It is not necessary to clutter up a drawing with a lot of writing, but important information should be briefly stated as shown.

An audible, an indicator, and a relay all use power. Good information for items like these is the voltage they operate on and the amount of current they draw. A circuit breaker and fuse should be labeled with their amperage rating. The battery is a power source and should be labeled with the voltage and amperage rating.

Other information should include the color of indicators and terminal designations as shown on the rectifier and relay. Note also that each item has an

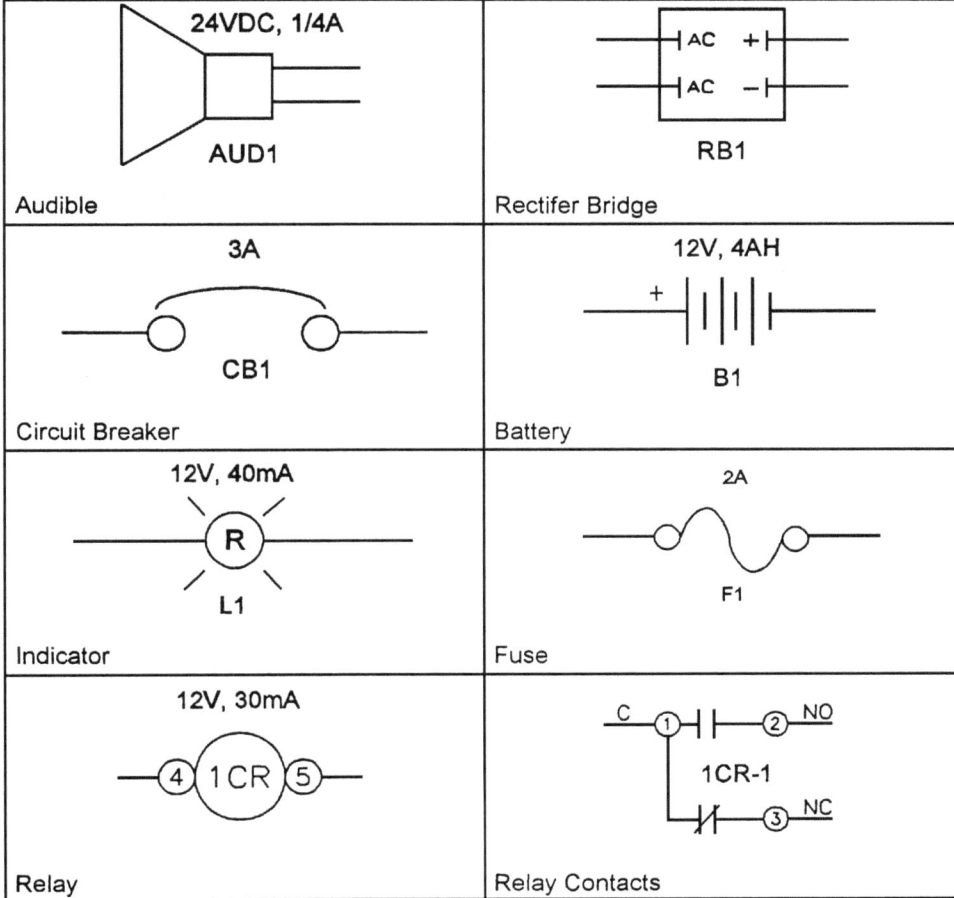

**Figure 2-10**  Symbols for common electronic devices.

identification, for example, CB1 for the circuit breaker, F1 for the fuse, and 1CR for the relay. This identification helps to locate specific items on a wiring diagram. All this information helps not only the installer but also anyone troubleshooting a system after installation.

Larger devices, for example, locks, power supplies, and access controls, don't have any special symbols. I have had wiring diagrams turned in from classroom exercises with some pretty elaborate drawings on them! Simple boxes and rectangles with some sort of identification will do just fine. Figure 2-11 shows some acceptable symbolism for these types of items.

There are no real industry standard symbols for the types of devices shown in Figure 2-11. As you can see, you can pretty much draw what you want. Just remember to keep it simple and brief.

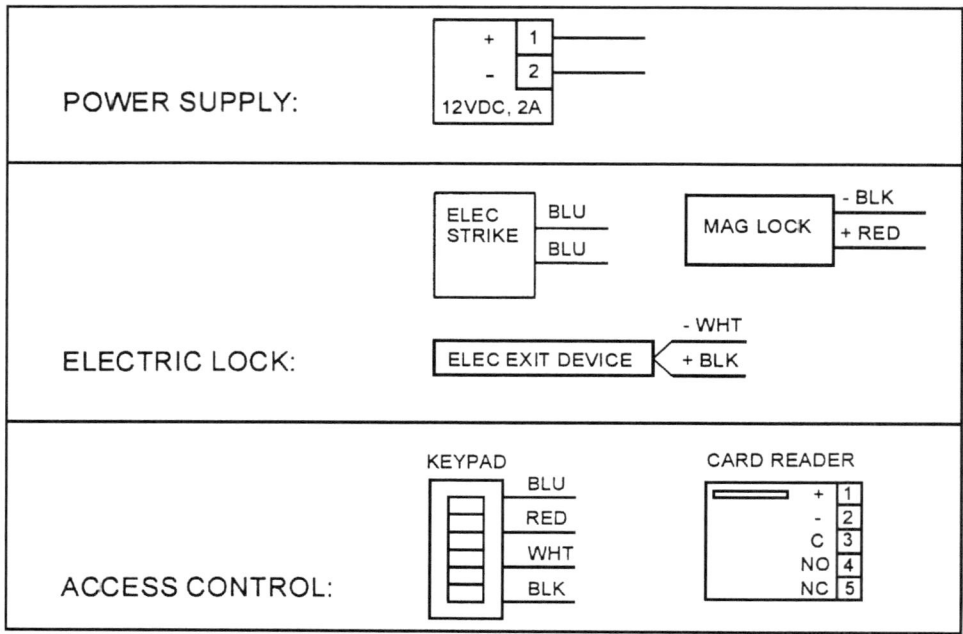

**Figure 2-11**  Symbols for large electronic devices.

## Symbol Standardization

As I mentioned earlier, an effort is being made by certain industry groups to standardize drawing symbols. The Security Industry Association (SIA) offers a diskette and printout of a standard, defining drawing symbols. It is called *Architectural Graphics—CAD Symbols* and is available in AutoCad and DXF format.

The SIA serves the alarm and access control industry. This standard is a little expensive; but if you are getting heavily involved in this type of work, it should be very helpful. The SIA can be contacted as follows:

SIA
635 Slaters Lane, Suite 110
Alexandria, VA 22314
Phone: 703/683-2075
Fax: 703/683-2469

Every so often trade magazines will publish some of the symbols used in the alarm and access control industry. Figure 2-12 shows a small collection of some of the symbols used for electronic hardware.

## Abbreviations

As you may have noted, many of the abbreviations used in the electronic security industry are just commonsense shorthand; for example, DSS = door status

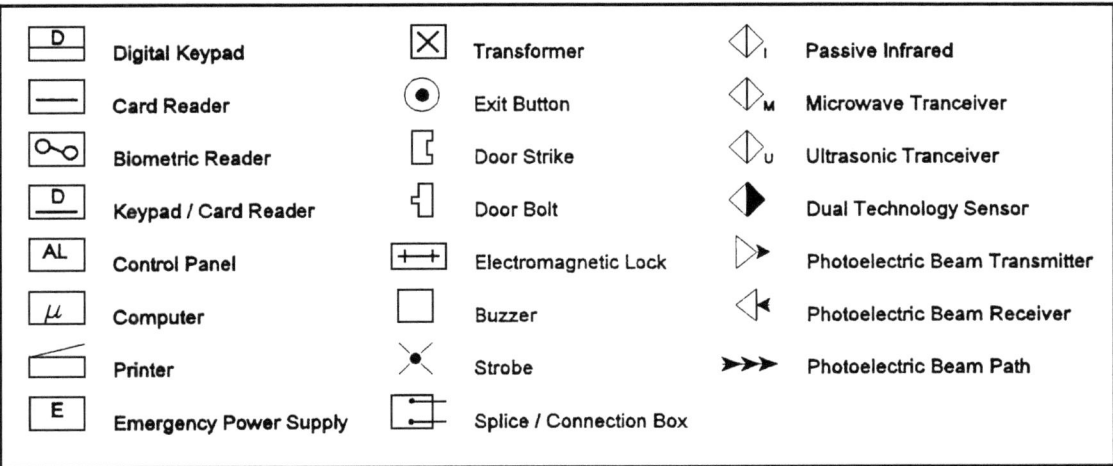

**Figure 2-12**   Architectural drawing symbols used to represent electrical hardware.

switch, T = transformer. Once again, there is no real industry standard that I am aware of. The following list comprises terms I have seen used on electrical documents. For the most part, these terms are all pretty well accepted as a de facto standard.

| | |
|---|---|
| A | amperes |
| AA | alternate action |
| AC | alternating current |
| AH | ampere-hours |
| AMP | amperes |
| ATD | adjustable time delay |
| AUD | audible |
| AWG | American Wire Gauge |
| B | battery |
| BPS | bolt position switch |
| C | common |
| C | capacitor |
| CB | circuit breaker |
| CONT | continuous duty |
| CR | card reader |
| CR | control relay |
| D | diode |
| DA | delayed action |
| dB | decibel |
| DC | direct current |

| | |
|---|---|
| DPDT | double pole, double throw |
| DPS | door position switch |
| DPST | double pole, single throw |
| DSM | door status monitor |
| DSS | door status switch |
| EH | electric hinge |
| EL | electric lock |
| EM | electromagnet |
| ER | emergency release |
| ES | electric strike |
| F | fuse |
| FA | fire alarm |
| GND | ground |
| Hz | hertz (cycles per second) |
| INT | intermittent duty |
| J | jack |
| JB | junction box |
| KA | keyed alike |
| KD | keyed different |
| KP | keypad |
| KS | key switch |
| L | lamp or light |
| LC | line cord |
| LED | light-emitting diode |
| M | magnet |
| MAINT | maintained contact |
| MOM | momentary contact |
| MR | manual release |
| MSC | manual station control |
| NC | normally closed |
| NO | normally open |
| PB | pushbutton |
| PS | power supply |
| PWR | power |
| RB | rectifier bridge |
| REX | request to exit |
| SO | silent operation |
| SPDT | single pole, double throw |
| SPNC | single pole, normally closed |

| | |
|---|---|
| SPNO | single pole, normally open |
| SPST | single pole, single throw |
| SW | switch |
| T | transformer |
| TB | terminal board or block |
| UPS | uninterruptible power system |
| V | volt |
| VA | volt-ampere |
| W | watt |

# 3

# Types of Drawings

Before we proceed to wiring diagrams, I thought it best to define all the types of trade drawings you may encounter. Documenting a system from start to finish may include all the following drawings, or possibly only one or two of them.

Pay particular attention to the elevation diagram and the wiring diagram. These are the drawings you will most likely have to create or account for.

## Schematic Diagram

Figure 3-1 shows a drawing that really should be called a *schematic* or *schematic diagram,* as noted. Many times I have heard it called a *wiring diagram,* which is okay as long as everyone involved knows what type of drawing is truly being referred to. *Wiring diagram* is actually the name of another type of drawing that we will cover a little later.

A schematic is the type of drawing you will find in the back of your television set or radio. This type of drawing is made by design engineers for production personnel to use when building electronic equipment. This same drawing could be used by service technicians for the troubleshooting and repair of electronic equipment.

## Block Diagram

In all probability you will not often come across the type of drawing shown in Figure 3-2. It is usually used in presentations to represent a very broad overview of a proposed system. The only time I can imagine using a block diagram might be to provide a rough layout of a system for planning purposes. This type of drawing might also be used in a submittal for approval of a concept or for informational purposes in a specification.

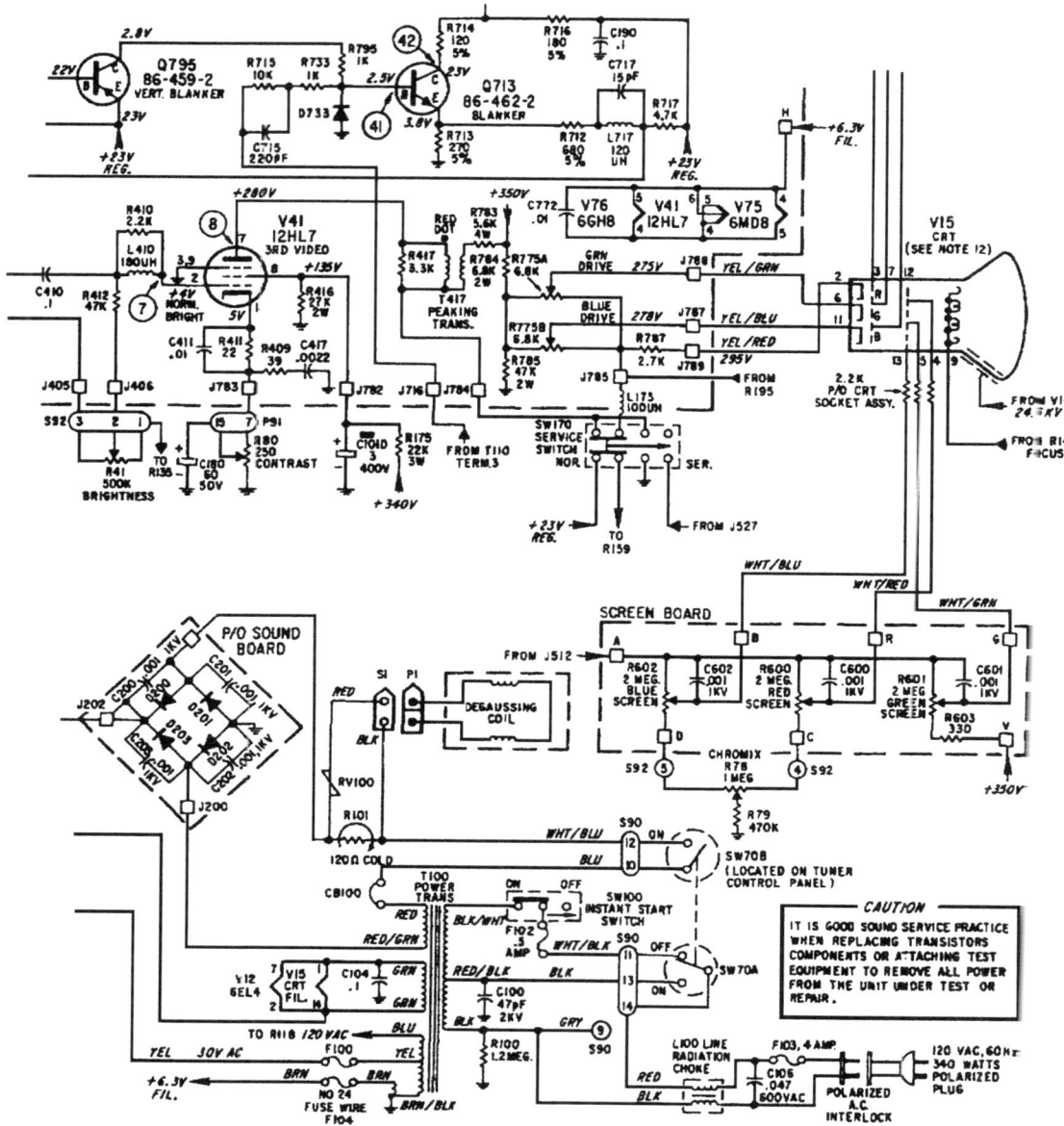

**Figure 3-1**   Schematic diagram. A drawing used for building and troubleshooting a piece of electronic equipment. It shows the complete circuitry of the equipment, using symbols to display individual components, without regard to the actual physical size, shape, or locations of the components parts.

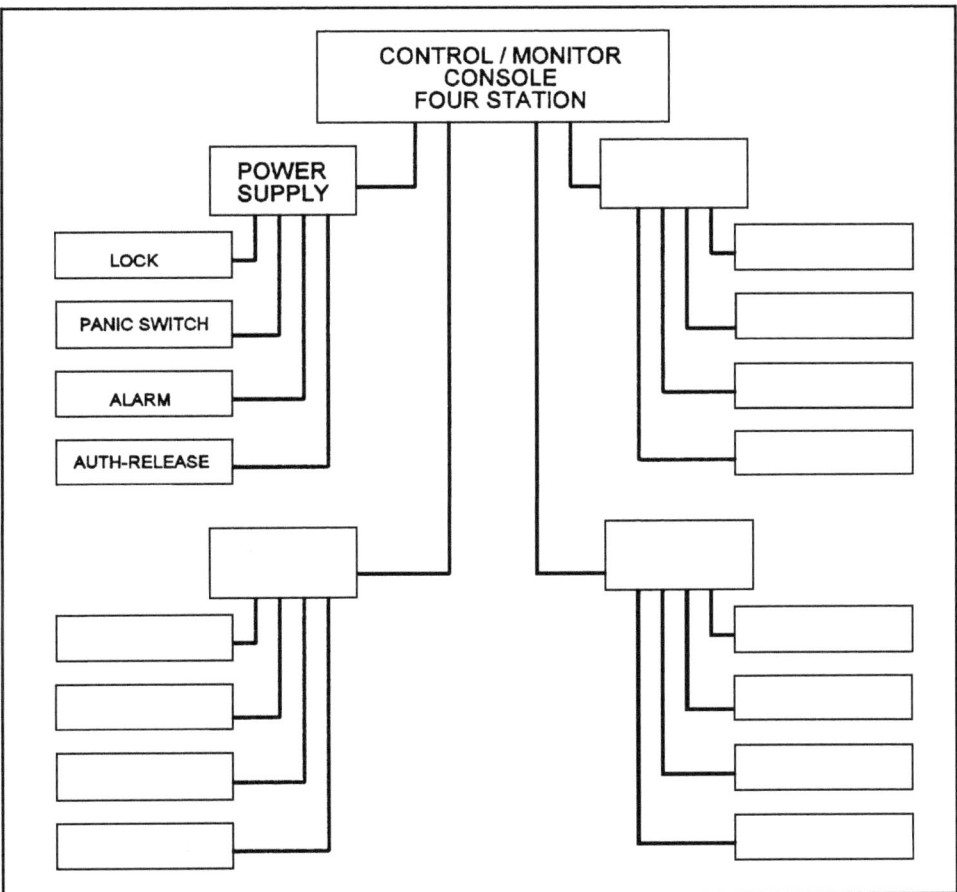

**Figure 3-2** Block diagram. A drawing of a system in which the components are represented by annotated geometrical figures. It represents the basic flow of the system and functional relationship of the components.

## Elevation Diagram

The drawing shown in Figure 3-3 is sort of a spinoff from architectural drawings. It represents the layout of equipment in a specific area. The drawing isn't necessarily to dimensional scale, but usually shows some proportional realism of the items shown. The elevation diagram does not normally include dimensional information, but dimensions locating equipment for special requirements can be added.

This type of drawing can be for a single opening, as shown in Figure 3-3, for multiple openings, or even for other related equipment. Ideally this drawing would also include a brief *description of operation* to let the reader know how the system is to work. This information is necessary for the following reasons:

- It helps the hardware supplier select the proper equipment to do the job
- It helps the designer to create the wiring diagram.
- It helps the installer to test the system.
- It helps in troubleshooting later problems.

As you can see, the elevation diagram can serve several purposes. It should be included as a specification document if a specification is being provided. It can also serve as an approval drawing between the system designer and architect or end-user. It could likewise be included in a proposal or request for quotation. At any rate, somebody along the line is going to be looking for this type of information.

Often, on simple systems, I have seen *riser diagram* information added to the elevation diagram. Many times I have copied the elevation diagram, eliminated the description of operation, and used the copy to create a riser diagram.

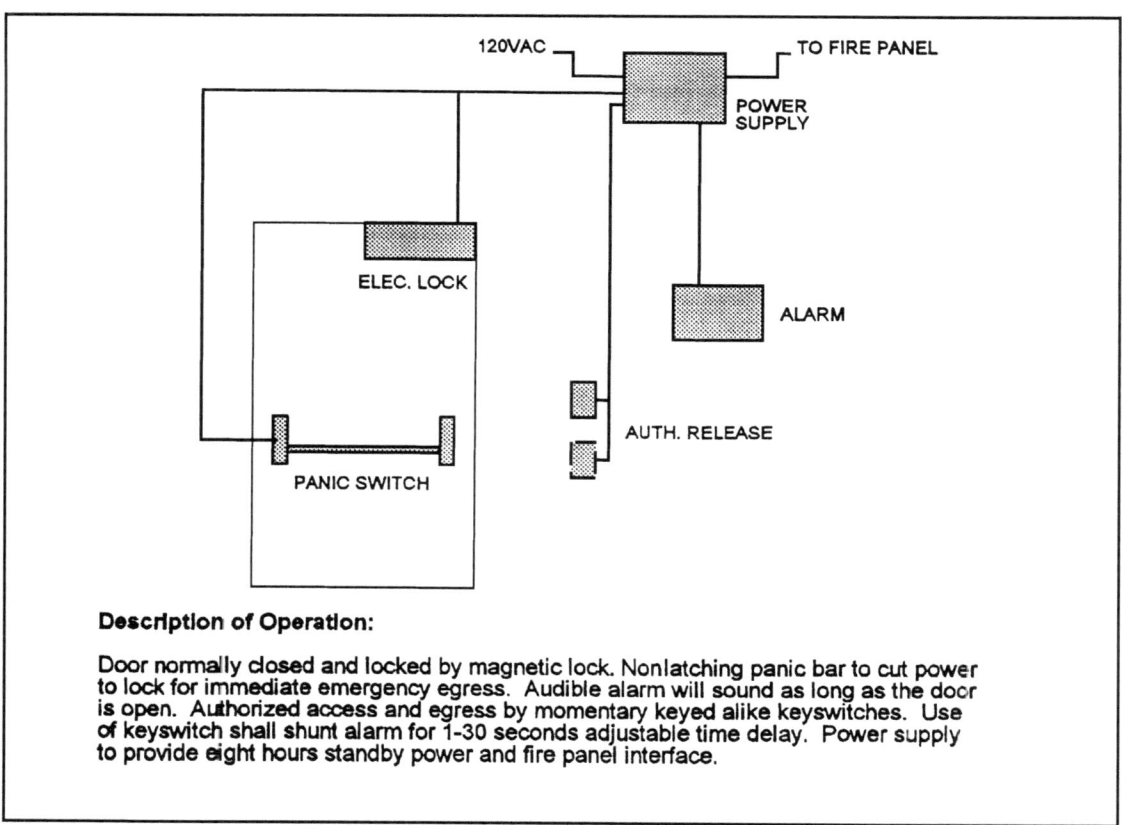

**Description of Operation:**

Door normally closed and locked by magnetic lock. Nonlatching panic bar to cut power to lock for immediate emergency egress. Audible alarm will sound as long as the door is open. Authorized access and egress by momentary keyed alike keyswitches. Use of keyswitch shall shunt alarm for 1–30 seconds adjustable time delay. Power supply to provide eight hours standby power and fire panel interface.

**Figure 3-3** Elevation diagram. A drawing used to locate the electric hardware and components about an opening or related area. This drawing may also double as a riser or cable diagram for simple systems.

## Riser Diagram

This drawing always leads to a lively discussion as to what information it should include, where the drawing should originate, and whether it should have a price tag!

The purpose of the drawing is to show the electricians the number of wires needed at specific locations. In specification work it should be included as part of the specification. We ask that an elevation diagram be included to help someone select and bid the hardware; why not the riser diagram to help in selecting and bidding wire?

The problem is, many of the people who provide elevation diagrams have trouble creating a riser diagram. Hopefully we can rectify that problem in the next chapter.

Others believe the riser diagram should be a "supply" document and be priced accordingly. The trouble with this theory is that the electricians may be looking to pull wire long before a wiring documentation package is ready. Most of the time a manufacturer is asked to provide the wiring information. When the system equipment is supplied by multiple manufacturers, there is a question as to who is going to produce any of the wiring documentation.

In another scenario, the low-voltage contractor has her or his engineering staff produce the riser and wiring diagrams.

As a result of different types of personnel supplying a riser diagram you may find varying degrees of information. Basically a riser simply shows the number of conductors (wires) to pull, as seen in Figure 3-4. Some risers will show the size of wire (wire gauge) also. This does require special expertise, as you would have to know distances between power sources and loads. Sizing the wire would also mean calculating voltage drops based on distances and current draw.

## Wiring or Hookup Diagram

The final drawing in the progression of documentation is the point-to-point wiring diagram, shown in Figure 3-5. This is the diagram that the installer uses to wire all the system components together. The drawing should be clean and concise, with all terminations clearly labeled. Theoretically, if it is properly drawn, a person with no knowledge of the equipment could wire the system— and it would work!

This type of drawing takes a little research, but could be drawn by anyone. Many manufacturers of security equipment offer "canned" wiring diagrams for repetitive systems at no charge. Most will be happy to quote wiring diagrams for custom or special systems.

Notice that in Figure 3-5 all terminations are clearly marked with numbers, letters, or wire colors. Wire runs that do not have colors are field-wired. It is good practice to leave enough room between these wires for the installer to fill in the colors used. Then the drawing can be updated to "as built," along with any other changes. A copy of the final drawing should be turned over to the end-user for the files. The installer should keep a copy in case of a callback for troubleshooting.

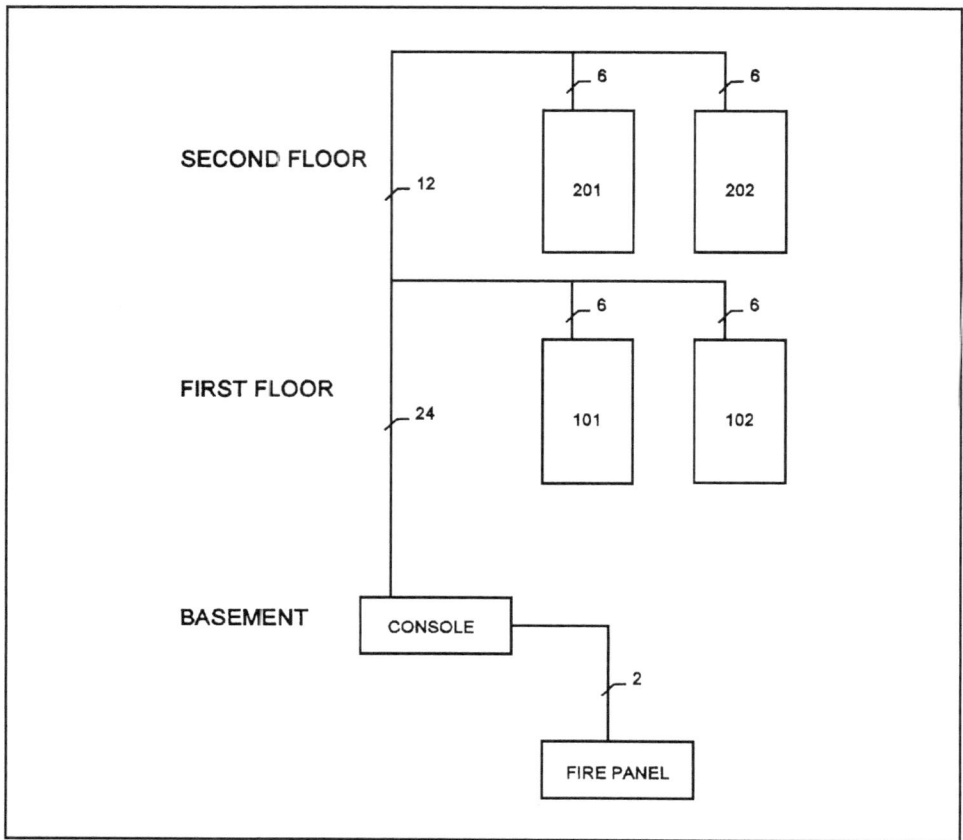

**Figure 3-4** Riser diagram. A drawing used to identify the number of wires (conductors) to be run between the major components in a system. It shows the general route of the wire runs and may include the size of wire to be used. Also may be referred to as a cable diagram.

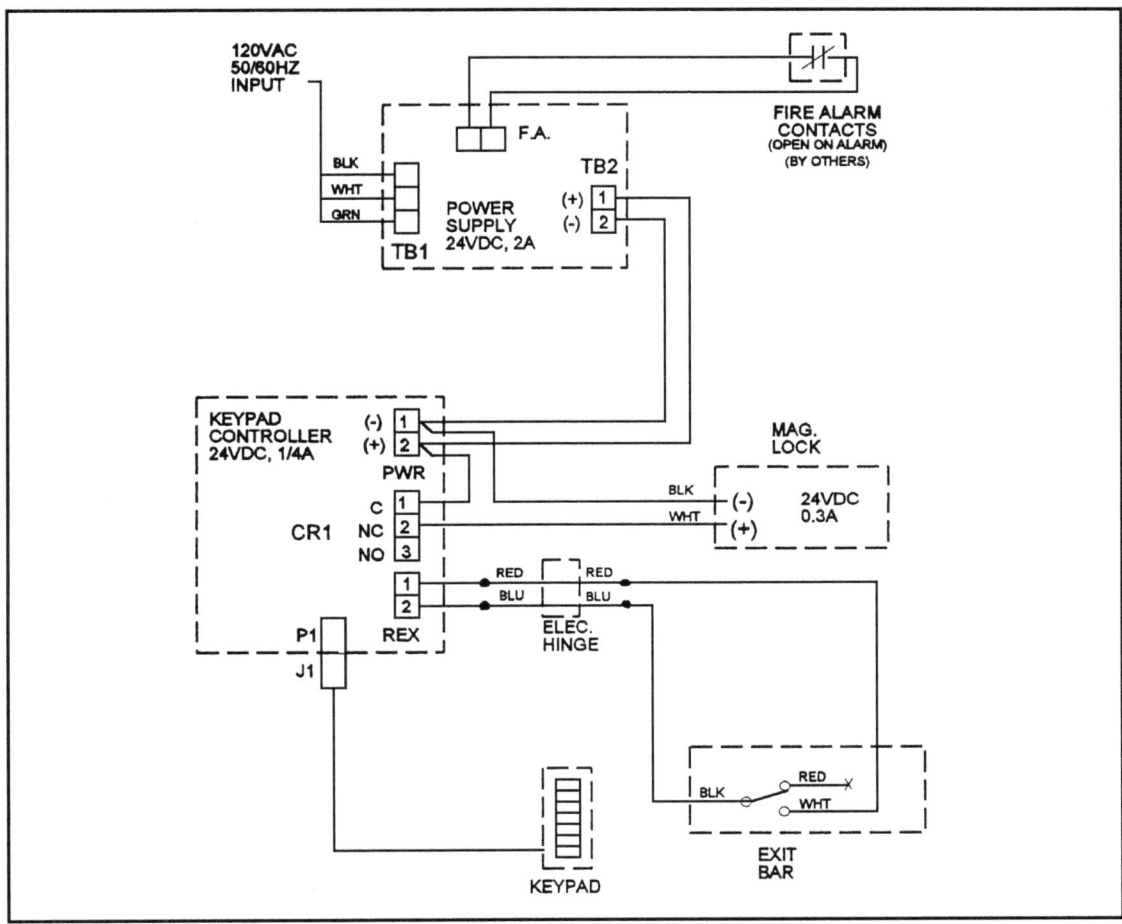

**Figure 3-5** Wiring diagram. A drawing used for point-to-point hook-up of each electrical device in a system. It shows the termination of all the interconnecting wire in the system, but does not necessarily represent the actual paths the physical wire or cable runs may take. In large systems a wire list may be provided in lieu of a diagram.

# Anatomy of a Wiring Diagram

Before we make a wiring diagram, let's first analyze how a diagram is "built." Seeing what makes up a diagram before trying to create one will help a lot.

This section will take you step by step through the creation of a simple wiring diagram. The completed diagram is shown at the end of the chapter.

Figure 4-1 is representative of what we could call the *skeleton* of a wiring diagram. This drawing is the framework of a wiring diagram before any identification or other labeling is added. It shows a system consisting of two electric locks, two switches, a power supply, and an indicator light.

The details to observe in this drawing are the line technique and drawing format. Note that the components are drawn a little heavier than the wiring lines. This adds some contrast, presenting a clear, easy-to-read drawing. There is also adequate spacing between wire lines, which will come in handy later when information is added to the drawing.

Another item to notice is the drawing format. Good practice dictates that all lines be drawn vertically or horizontally. Very seldom will you see electrical drawings with long diagonal lines. Occasionally, you may have to use a short diagonal line for the sake of clarity. Additionally, you should never use curved lines, as we saw in the "spaghetti" drawing in Chapter 1.

It is also good practice to draw as few crossover lines as possible. Crossover junctions can be mistaken for physical connections. Too many crossovers also make a drawing confusing to read.

Let's see what we can add to this drawing next to "build" it into a complete wiring diagram.

## Identifications

Notice in Figure 4-2 that each component has been given a brief identification. This makes it a lot easier to locate specific items on a drawing later, especially

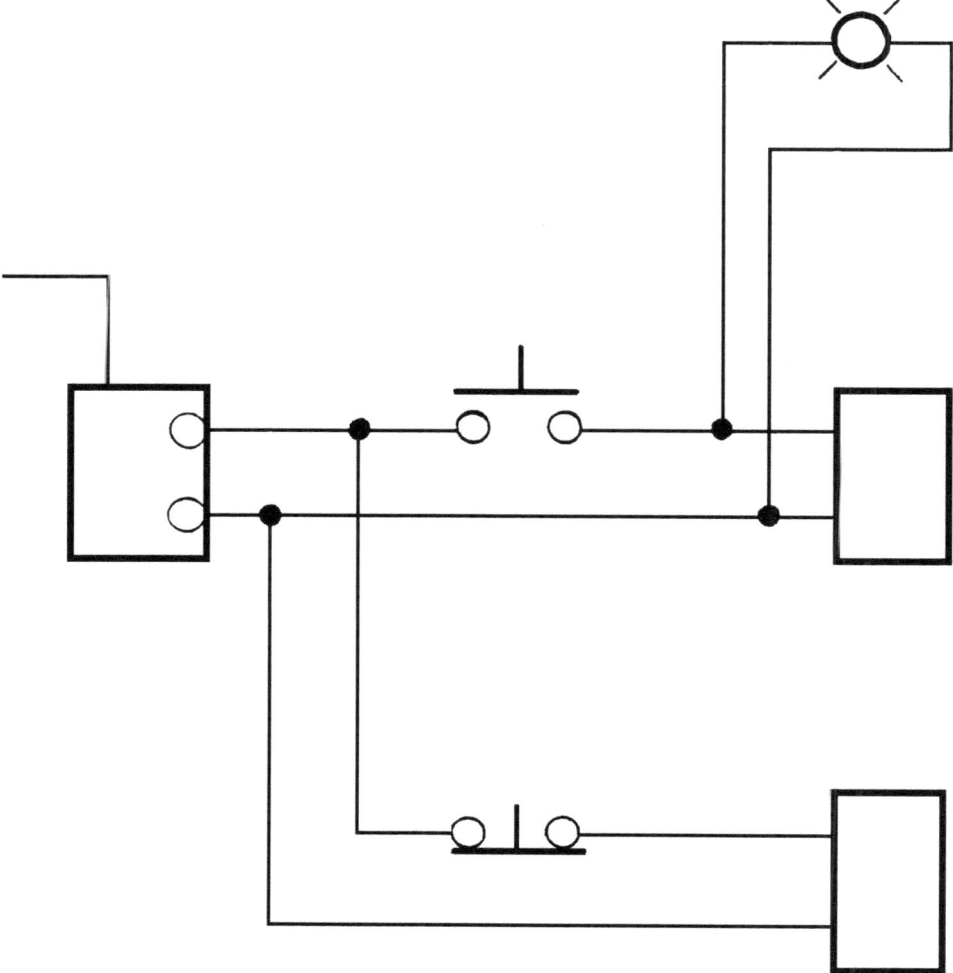

**Figure 4-1**   "Skeleton" of a wiring diagram.

if you are communicating by telephone. Now there is no harm in spelling out names of components, say, power supply instead of PS. Most people experienced at reading wiring diagrams are used to abbreviated identifications, especially if the short forms are standards.

Identifications on the drawing include these:

PS1             Power supply 1. If a second power supply appeared on the
                same drawing, it would be PS2.

SW1, SW2        Switches 1 and 2.

L1              Light (or indicator).

ES1, ES2    ES on this drawing means *electric strike.* Other abbreviations you might see are EM (electromagnet) or EB (electric bolt). These abbreviations are not standard, but are good for quick identifications. Many people will also include the spelled-out word with the abbreviation. Some use the generic abbreviation EL (electric lock) and spell out the type of lock.

The best guidelines for identifications are to

- Check Chapter 2 for standard abbreviations.
- Make one up if it "looks" good.
- Spell out the word if it could be confusing.

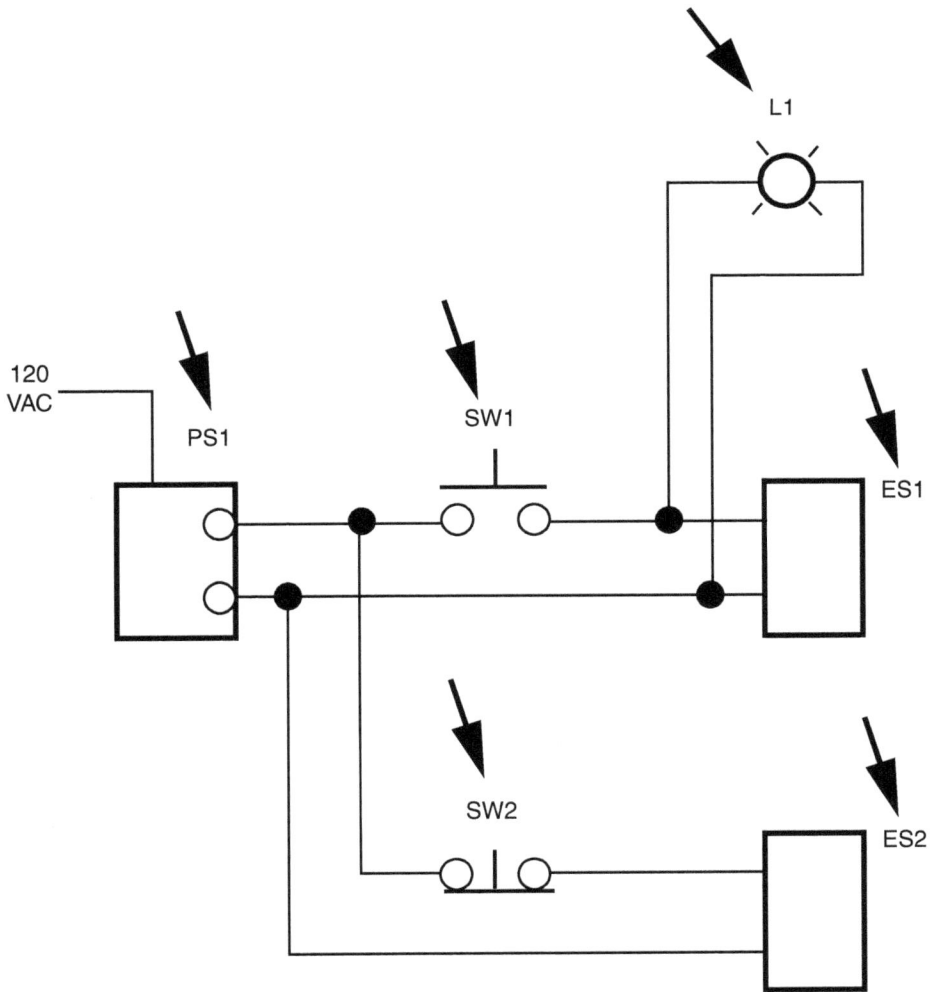

**Figure 4-2**  Typical wiring diagram component identification.

## Functions

Each switch and indicator light should be labeled as to its function in the circuit, as shown in Figure 4-3. This information helps the reader understand how the system is to operate. It also helps the installer when "starting up" a completed installation. Many times the switch and indicator identifications on the diagram will coincide with the actual labels on switch plates, monitoring panels, and consoles. Notice also that the color of the light is indicated by a single letter. This is common, as indicators normally fall into a small range of colors. The most prevalent colors used in the security industry are red (R), green (G), and amber (A).

Switches may also be labeled with their actuation type: MOM (momentary) and AA (alternate action), sometimes labeled MAINT (maintained).

Locking devices may be identified as either fail-secure (use power to unlock) or fail-safe (cut power to unlock). This information is helpful as it is not obvious by just looking at the lock symbol. In Figure 4-3 the door location has been added to clarify which type of lock is to be installed at each door. Physical locations may be added to other components if known; for example, a switch or access control might be labeled, "Door 1—Exterior."

## Terminations and Wire Colors

A very important component of any wiring diagram is the identification of wire terminations. This information is vital to the correct hookup of each component in the system. Every component has some sort of hookup provision, be it wire leads, screw terminals, plugs and sockets, or some other device. This information should be apparent in manufacturers' technical literature.

As shown in Figure 4-4, the power supply has two labeled terminals for output power. On direct-current (DC) power supplies it is also extremely helpful to designate the polarity—(+) and (−)—of the terminals. This information helps not only the circuit designer, but the installer and troubleshooter as well.

Note that the high-voltage (120-Volt AC) terminations are not shown. This is common on this type of drawing as it is usually a different trade, the high-voltage electricians, that is separately responsible for hooking up this wiring. Personally, I like to add this information to the drawing if I know it.

Switches and indicators usually have colored wires provided for hookup. These wires are normally 12 inches or shorter, and of course they will probably never reach the component they are to be wired to. What installers do is to add wire of the necessary length. This added wire may be a different color, and it is good practice to leave room on the drawing for the installer to note these colors.

Locking devices usually have wire leads or screw terminations. If the two power input leads are the same color, it usually means that observing polarity is not necessary—they can be hooked up either way. If the device is polarity-sensitive, the leads normally will be different colors and may be labeled (+) and (−).

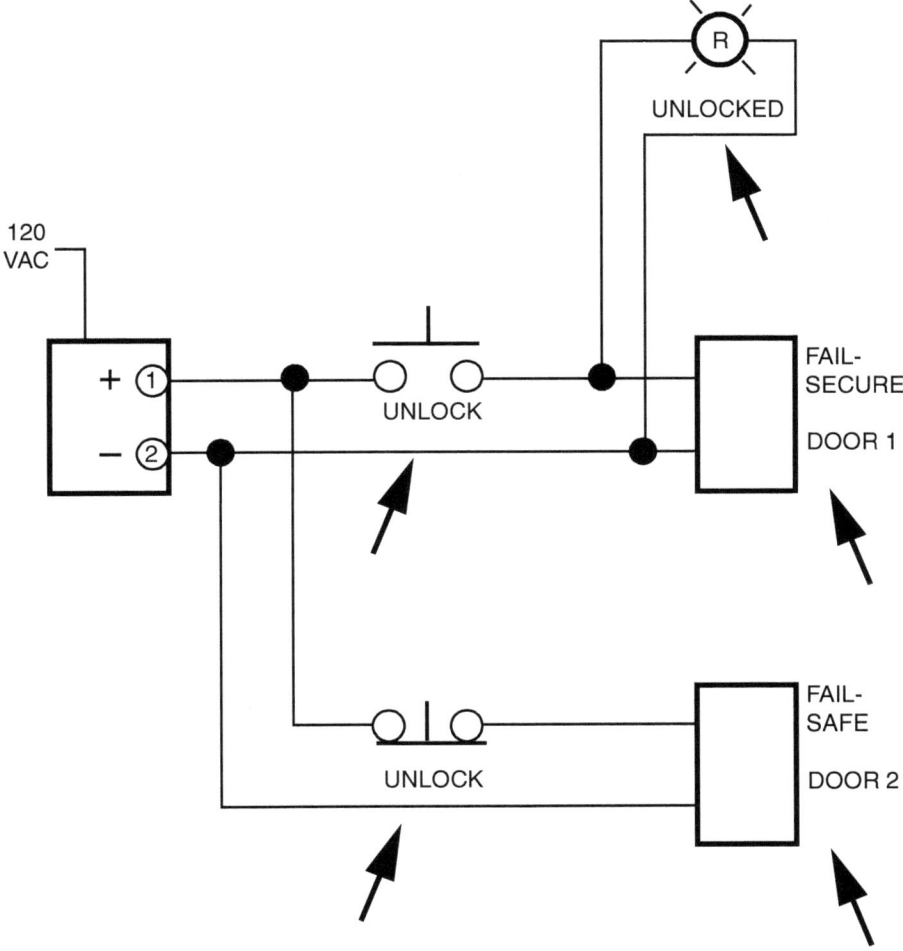

**Figure 4-3** Typical wiring diagram component function identification.

## Power Requirements

Adding the operating voltage and current (ampere) requirements to applicable components is very helpful. Unfortunately, this information is also often left off wiring diagrams.

Any component that uses electricity is called a *load*. In Figure 4-5 these items are the indicator light and the electric locks. As noted, each of these items is labeled 12 volts DC (operating voltages must all be the same if a single power supply is used). Each component is also labeled with how much current it uses. This information is absolutely necessary in order to select the proper power supply. If you have to look it up, you might as well write it on the drawing; it is valuable information, especially during troubleshooting.

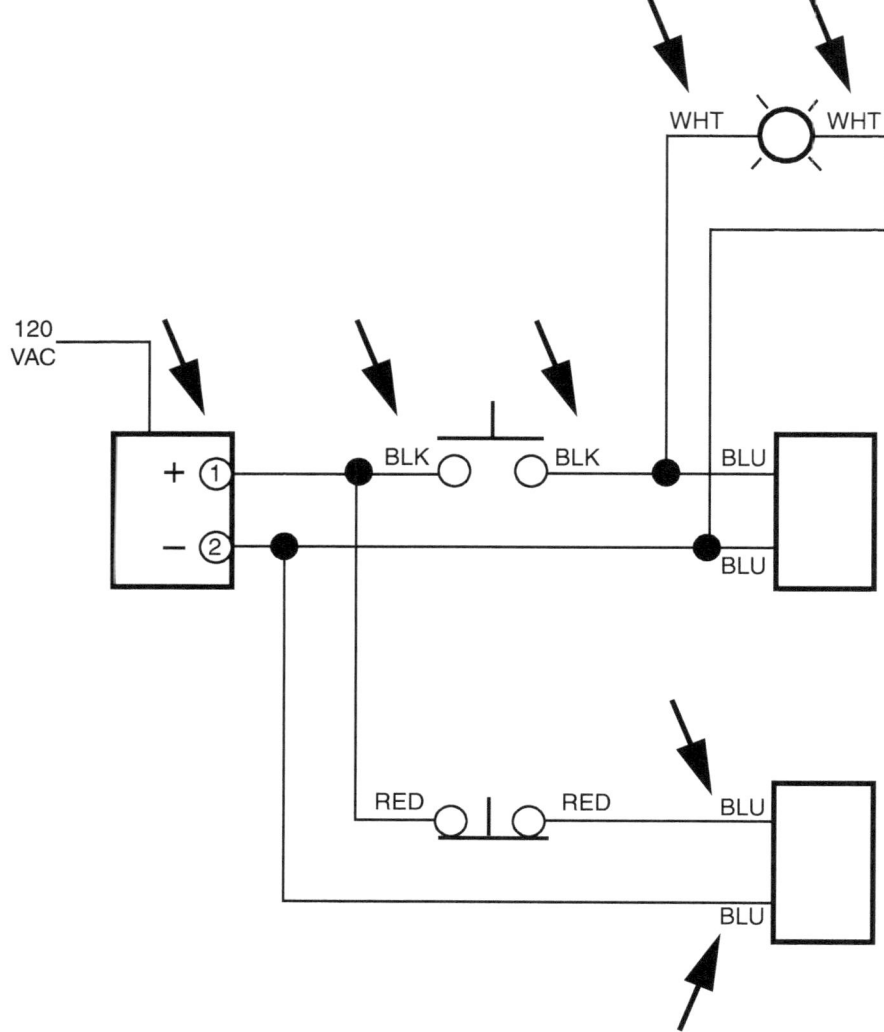

**Figure 4-4** Typical wiring diagram termination identification.

The power supply is also labeled with the voltage and amperage. This indicates how much power is available for all the loads in the circuit. Note that the voltage must be the same as the loads, and the amperage must be equivalent to or greater than the total of all the loads.

You should note here also that basic switches do not *use* power. Their job is to control the flow of power. Switch contacts are rated as to the amount of power they will handle. This information is used during the selection of the switch and is not normally noted on the wiring diagram.

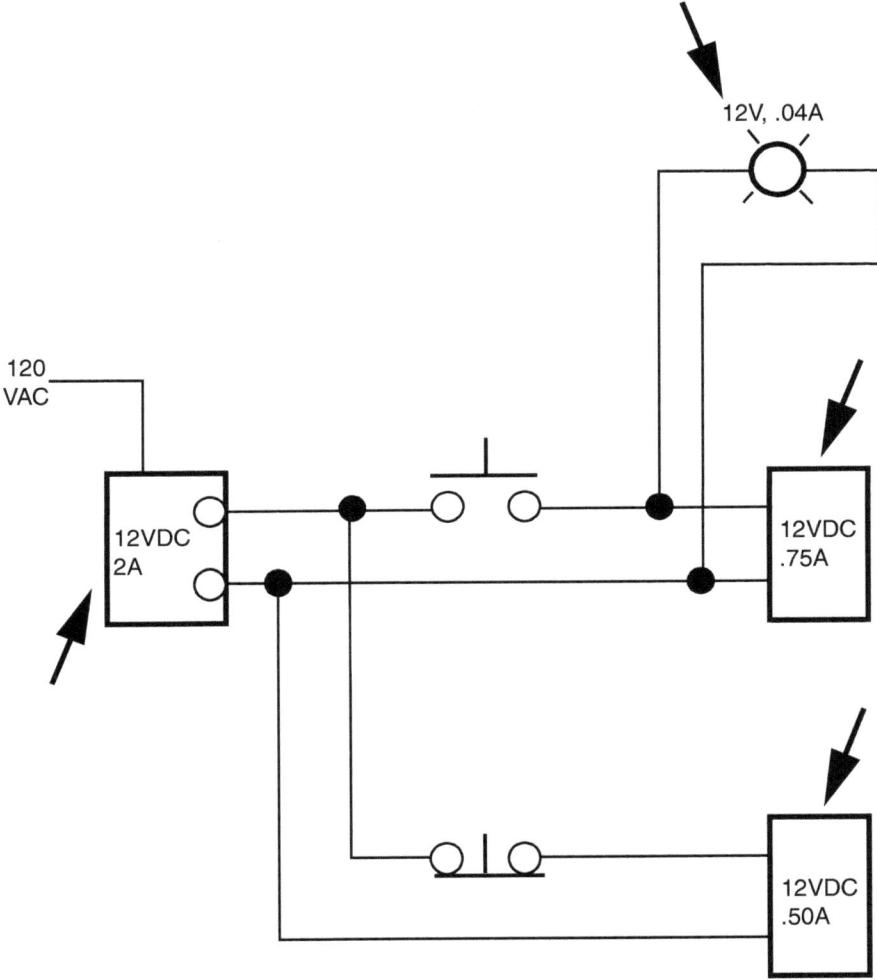

**Figure 4-5**  Typical wiring diagram power requirement identification.

## Power Flow

We next look at how a wiring diagram actually works. Electricity follows paths created by the wires or conductors. The idea is to get electric power from the power supply to the load. Along the way there will be certain devices that control the flow of power.

Earlier I compared electrical drawings to road maps. I sometimes tell students to think of electricity following a path as the same as a car following a road. This analogy may help you understand an electrical drawing.

You most often will be working with direct-current (DC) systems. Like a road map with route numbers, DC power will be moving on paths marked either $(+)$ or $(-)$. Alternating current, or AC power, moves on any "road." Figure 4-6 shows a typical pathway from power to load.

Both $(+)$ and $(-)$ must reach the load in order for it to operate. Notice that $(+)$ can only travel as far as the gap caused by the open switch. I often compare a switch to a drawbridge on a road. You cannot proceed until the drawbridge closes.

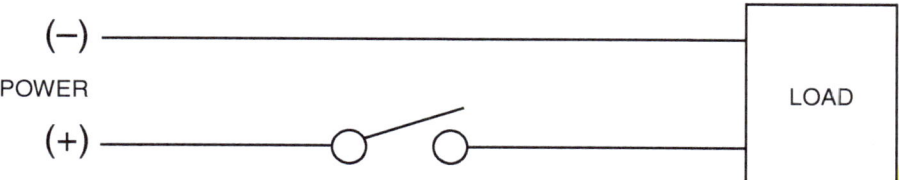

**Figure 4-6**   Typical path for power flow.

In low-voltage work, there are no real national standards as there are with high voltage. So following some of the high-voltage rules is, at least, good practice. One high-voltage "rule" is to always switch the "hot" side of power. So in low voltage, good practice would be to always switch the $(+)$ line.

This being the case, the $(-)$ line is usually a direct unbroken path to the load. The $(+)$ line normally has open or closed controls somewhere along its path.

Let's take another look at our original wiring diagram and see how power "flows" in that system.

## Path of $(-)$

Usually when I am checking a wiring diagram, or even creating a new diagram, I "trace out" the $(-)$ lines first. These are normally direct runs and easy to follow, so I get them out of the way in the beginning.

Figure 4-7 shows the paths that $(-)$ will follow. Note that there are no interruptions or obstacles blocking the flow to each load. The $(-)$ output of the power supply gets to loads L1, ES1, and ES2 simultaneously without hesitation. If you need to send $(-)$ somewhere else in the circuit, you could simply "tap" off a new $(-)$ line anywhere along any $(-)$ path.

## Path of $(+)$

Figure 4-8 shows the paths that $(+)$ will follow. Note that $(+)$ follows the path to SW1 but cannot cross the gap of the open switch. So in this system the normal condition of loads L1 and ES1 is "not energized." When the switch is actuated and closed, $(+)$ will flow to the loads. Indicator L1 will illuminate, and lock ES1 will energize and release. The $(+)$ output also follows the path to SW2, passing through the closed switch, energizing load ES2. So lock ES2 is normally energized in this circuit. If you needed $(+)$ somewhere else in the circuit, you could tap off a new $(+)$ line. The caution here is you would normally

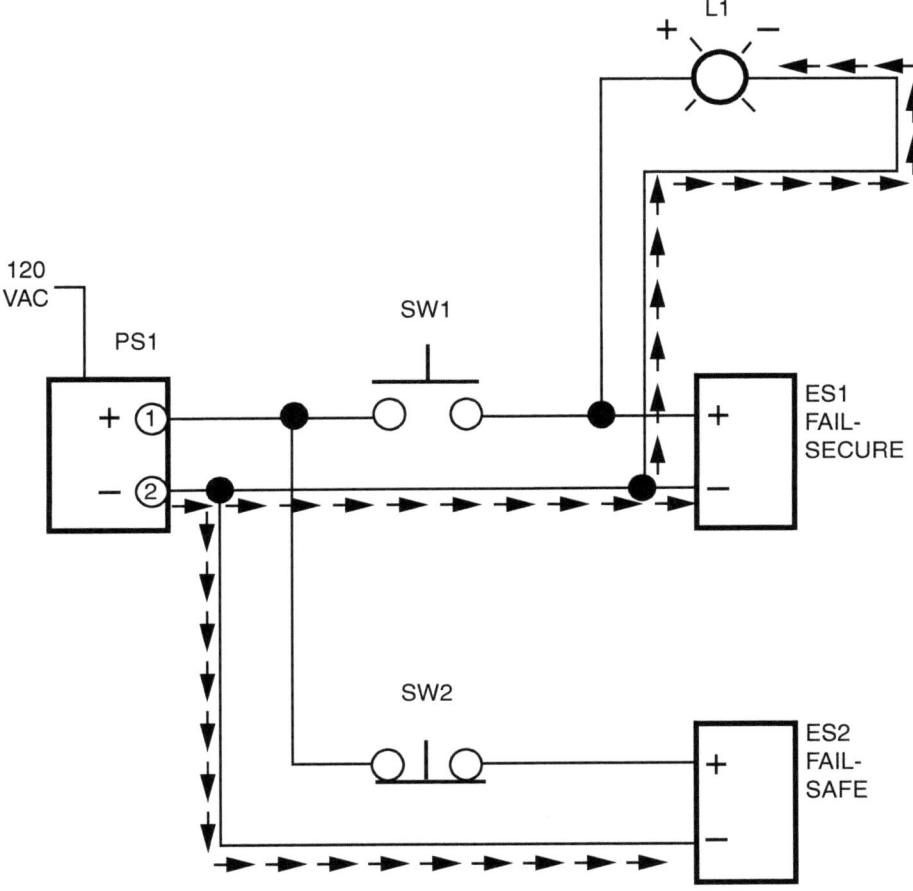

**Figure 4-7**   Paths for (−) in a typical circuit.

tap a new line somewhere between switch SW1 or SW2 and the power supply. This will ensure that you have a (+) line that will not be interrupted by any control switches.

## Summary

Figure 4-9 shows the completed wiring diagram. Keep the following points in mind when making a wiring diagram:

- Lines should be vertical and horizontal.
- Minimize crossover lines.
- Identify components.
- Identify component functions.

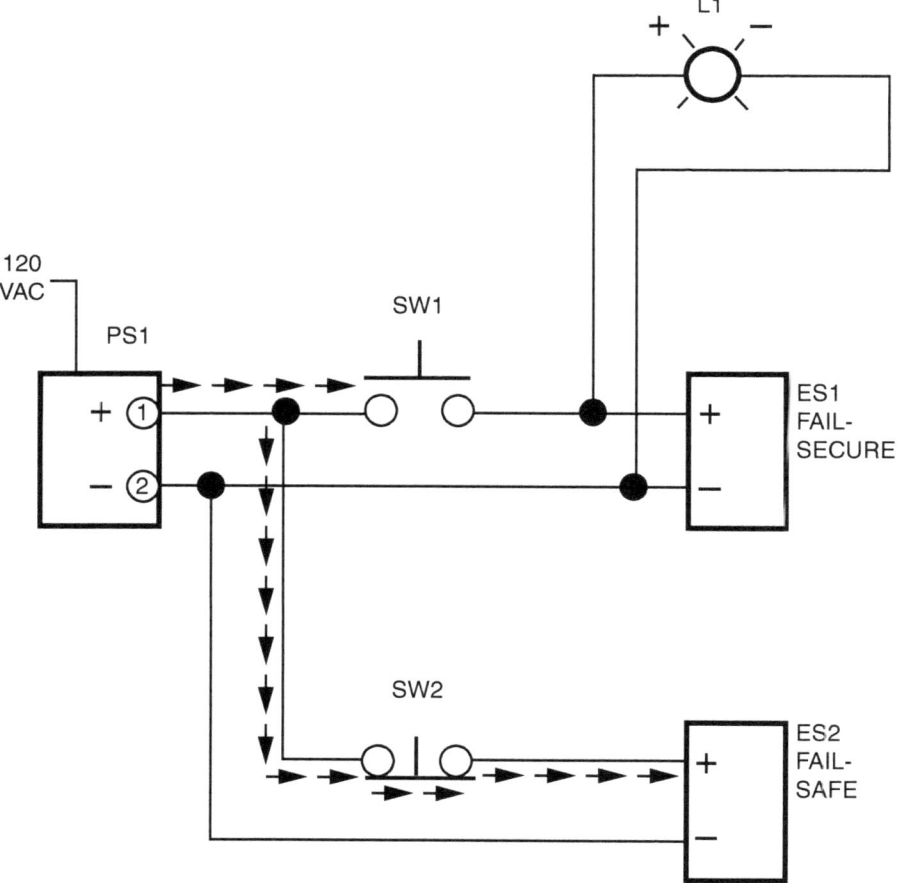

**Figure 4-8**   Paths for (+) in a typical circuit.

- Identify all wiring terminations.
- Label power supplies and loads with the voltage and amperage.
- Put switch contacts in (+) lines only.

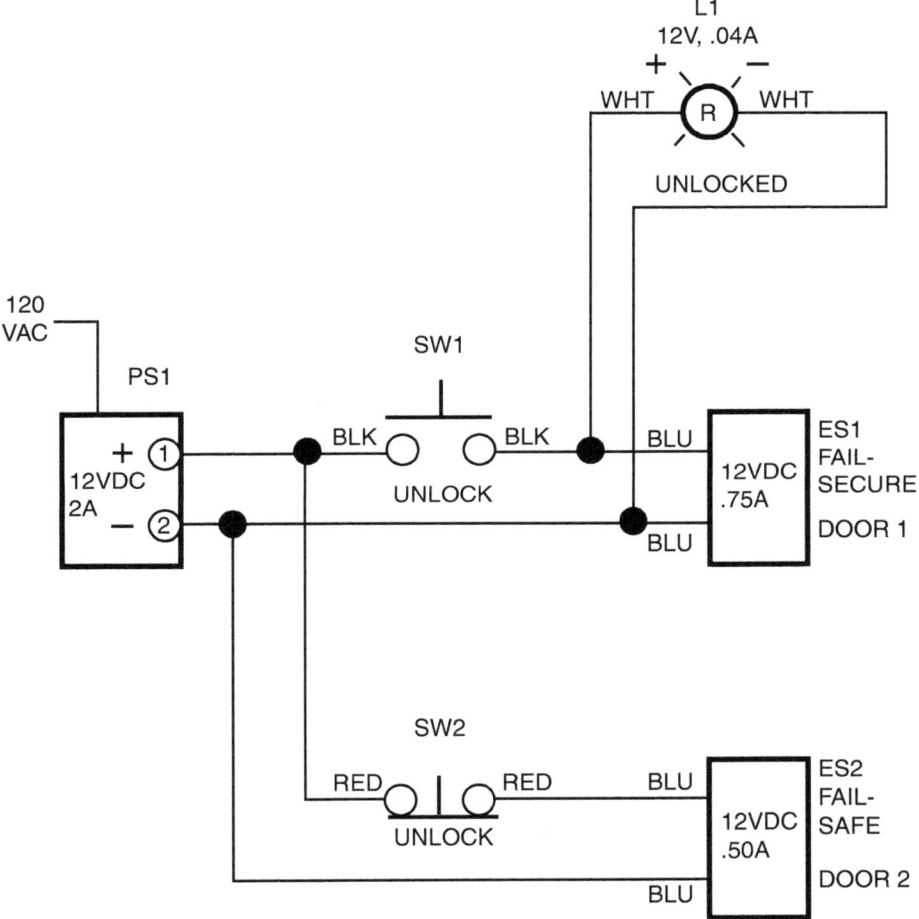

**Figure 4-9**   Complete wiring diagram for a typical system.

# 5

# Making a Wiring Diagram

The most sought after system document is the wiring diagram. Many people can rough out an elevation diagram and even a riser diagram. Just as many people have difficulty producing the wiring diagram. At this point in the book you should have a pretty good idea of how to go about making a basic wiring diagram.

## Major Components

There are a couple of items to always keep in mind. Most wiring diagrams will consist of four major components:

- *The load.* This is the component that uses power, for example, an electric lock or an indicator light.

- *The power supply.* This item supplies power to all the loads in the system. It is the heart and soul of the system. A lot of people leave it to the electricians; but electricians are high-voltage people—this power supply is low-voltage. You select it; it is part of *your* system.

- *Controls.* No matter what you call them—pushbuttons, key switches, card readers, or retina readers—they are all basically switches that control the flow of power. (Some controls contain electronics and are also loads that require power. They will be covered in later chapters.)

- *Conductors (wire).* This ties everything in the system together. The wire is represented by all the interconnecting lines you will be drawing on the wiring diagram.

Are you worried that you can't remember all this? Just look around the room you are in now. The electric light (the load), the switch on the wall (the control), the power supply for the light (120 Volts AC somewhere in the basement or electrical closet) and the wires (conductors) running throughout the walls.

There is a fifth component that is not absolutely necessary, but is absolutely important:

- *Monitoring.* This can be accomplished by audible and visual devices that monitor the condition of components in the system or some event that affects the system. Monitoring will be covered in depth in Chapter 6.

A basic electric locking circuit consists of four components, as shown in Figure 5-1. There may be multiple controls and loads in one circuit, and we'll cover this a little later on.

1. Load (electric locking device)
2. Power source (power supply, transformer, etc.)
3. Control (on/off switch)
4. Conductors (wire run)

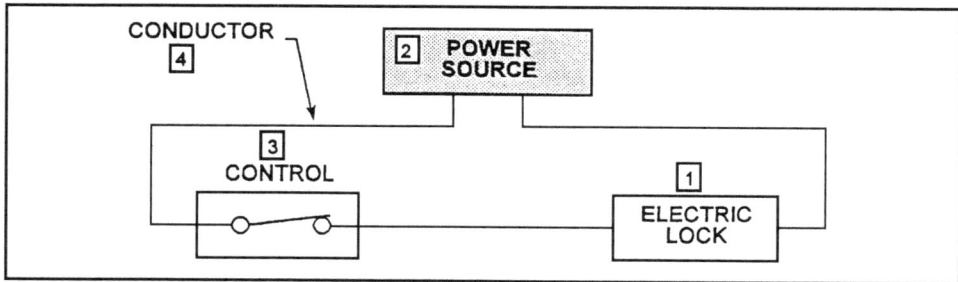

**Figure 5-1**   The four components of a basic electric locking circuit.

Now how do we start a wiring diagram? First we will need some information about the system.

## Gathering Information

The first step in designing a system is to document how it is to function. This can be simply a rough outline developed between you and the customer. Later it can be formalized and can become part of the documentation package. Some common terms describing this information are *description of operation, system operation,* and *theory of operation.*

This information, while necessary to design a system, can also be used to develop a specification. Another use is as an "approval" submittal to ensure agreement between you and the customer. Figure 5-2 lists some basic questions that should be answered to start your description of operation.

Most systems begin with a phone call. I personally like making a rough outline sketch while I am conversing with the caller. A worksheet such as shown in Figure 5-3 can be helpful, although a plain writing pad will do.

Let's run through a typical scenario leading to making a wiring diagram. I'll be referring to Figures 5-4 through 5-14 to illustrate the sequence I use to build a system, up to and including the wiring diagram. Once you see the thought process involved, you'll be well on your way to being able to create a wiring diagram. At the end of this chapter you will find a test exercise to try on your own.

## Initial System Layout

Usually, with a customer call, most of the questions in the Figure 5-2 questionnaire will have to be asked. After you gain some experience, you'll know what to ask from memory. I also always have worksheets (Figure 5-3) or a pad by the phone. I am already sketching while the customer is talking.

In this exercise I have roughed out the system on the worksheet shown in Figure 5-4. I usually start with the sketch, then fill in a brief description of operation.

This particular customer wanted a magnetic lock to secure a private entrance door not used for public egress. The customer wanted a simple key switch for access control. (I later suggested adding an adjustable relock delay to the key switch.) For interior control a wall-mount momentary pushbutton was desired with a red light to indicate when the lock was released. The customer explained that the system did not require a tie-in to a fire alarm system. At my suggestion standby batteries were added to the power supply. Standby power was to have the capacity to keep the lock energized for at least 4 hours during a main line power interruption. (The door was to be dead-bolted at night.) I also have not indicated any type of power supply. The customer has left this to my discretion, although I am already thinking of using 24V (hence the two batteries).

As you can see, the information on the worksheet in Figure 5-4 is short and concise. But as I was filling out this sheet, my mind was already designing the system.

For now this sketch became my guideline. From this information I was able to select products and create a *hardware schedule*—simply a list of equipment necessary to do the job.

Also notice in Figure 5-4 that I ran wire "lines" from each piece of equipment. This is an elevation drawing, and it is not necessary to show wire. I do it out of habit on simple systems. I know I can easily make this drawing a riser diagram, which we will discuss next and is shown in Figure 5-5.

## The Riser

The drawing in Figure 5-5 is identical to the elevation drawing in Figure 5-4; in fact I photocopied Figure 5-4. If your elevation is drawn somewhat neatly, usually it can be easily turned into a riser diagram. Remember, it is the information provided that is important, not the exact physical location of the components. In fact, most likely you will have no idea where the power supply will be located. The important thing is to provide a count of how many wires are needed to tie everything together.

| QUESTIONS TO ASK: | | |
|---|---|---|
| **What is the normal condition of the door?** | ☐ Closed / Locked<br>☐ Closed / Not Locked<br>☐ Held Open | |
| **When is the lock secure?** | ☐ Normally Locked<br>☐ Daytime Hours: _____<br>☐ Nighttime Hours: _____<br>☐ Saturday    ☐ Sunday<br>☐ Other: _____ | |
| **What is the inside release?** | ☐ Pushbutton    ☐ Panic Device w/Switch<br>☐ Presence Detector<br>   Describe: _____<br>☐ Other: _____ | |
| **What is the outside release?** | ☐ Key Switch    ☐ Keypad<br>☐ Card Reader<br>☐ Other: _____ | |
| **Is a relock delay required?** | ☐ No    ☐ Yes: _____ Seconds | |
| **Does the power supply require standby batteries?**<br>**How Many  Hours Backup?** | ☐ No    ☐ Yes<br>☐ 4     ☐ 8<br>☐ 16    ☐ 32 | |
| **Should fire alarm system release the lock?** | ☐ No    ☐ Yes | |
| **Notes:** | | |

**Figure 5-2**  Questionnaire for a basic system.

Now this is not as complicated as it sounds. Each product will only have a certain number of wires or terminals associated with it. If you account for all the wires possible and actually use only some of them, so what! You end up with a few extra wires on the job, which is actually a good practice. Many riser drawings will have a note recommending the pulling of spare wires. It is much better to have a couple of extra wires than to be short of wires.

Now, how did I determine the wire count in Figure 5-5? Without even knowing the exact products I will be specifying, I know, for instance, that the magnetic lock has two wires for input power; any load will have two wires for power.

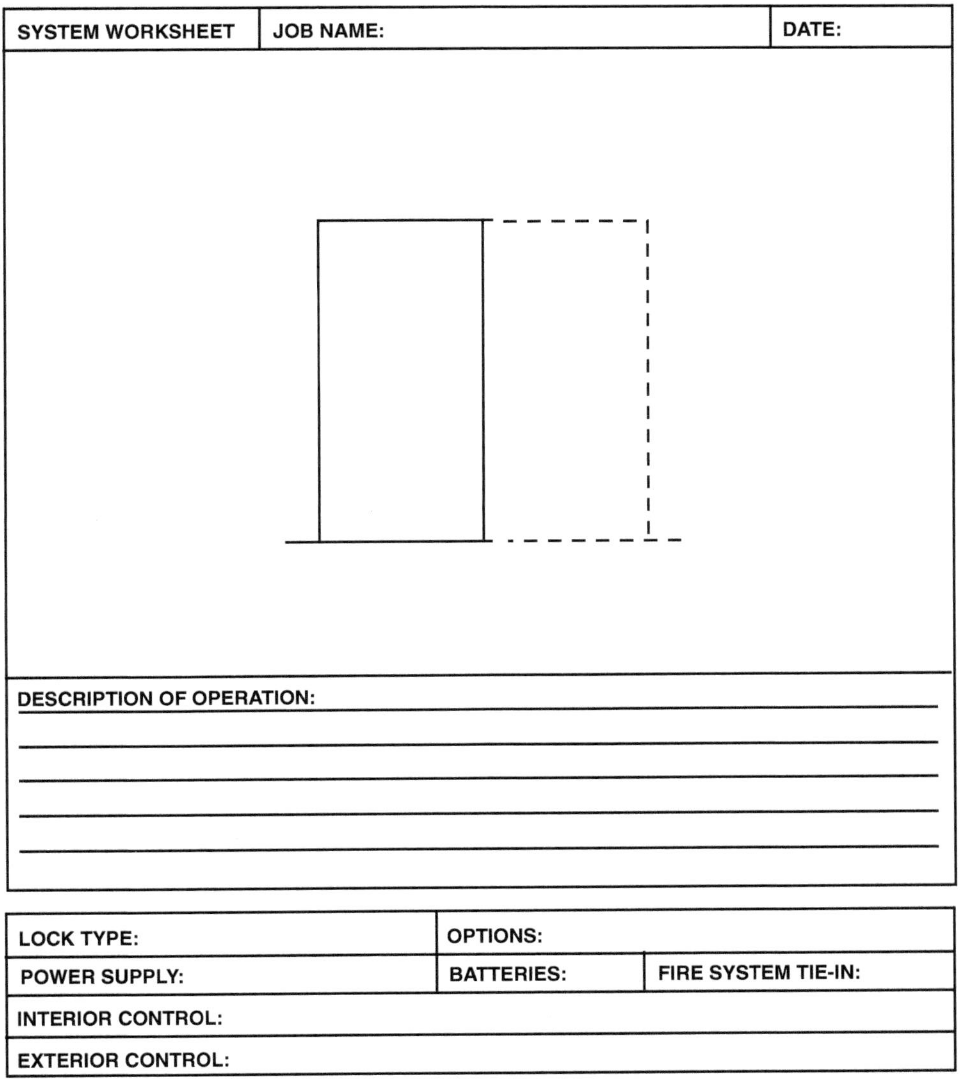

| SYSTEM WORKSHEET | JOB NAME: | | DATE: |
|---|---|---|---|

DESCRIPTION OF OPERATION:

| LOCK TYPE: | OPTIONS: | |
|---|---|---|
| POWER SUPPLY: | BATTERIES: | FIRE SYSTEM TIE-IN: |
| INTERIOR CONTROL: | | |
| EXTERIOR CONTROL: | | |

**Figure 5-3**  System worksheet.

The key switch is simply an on/off switch operated by a key. It also contains an adjustable time delay to allow the user time to get the door open before relock occurs. On this particular unit the time delay is pneumatic and does not require electric power.

A key switch commonly has three wires, as shown in Figure 5-6. Normally I would use only two of the wires to release the lock. In this system I think I need the third wire to energize the indicator light when the lock is turned off. As I mentioned before, count all the wires; it is better to be safe than sorry!

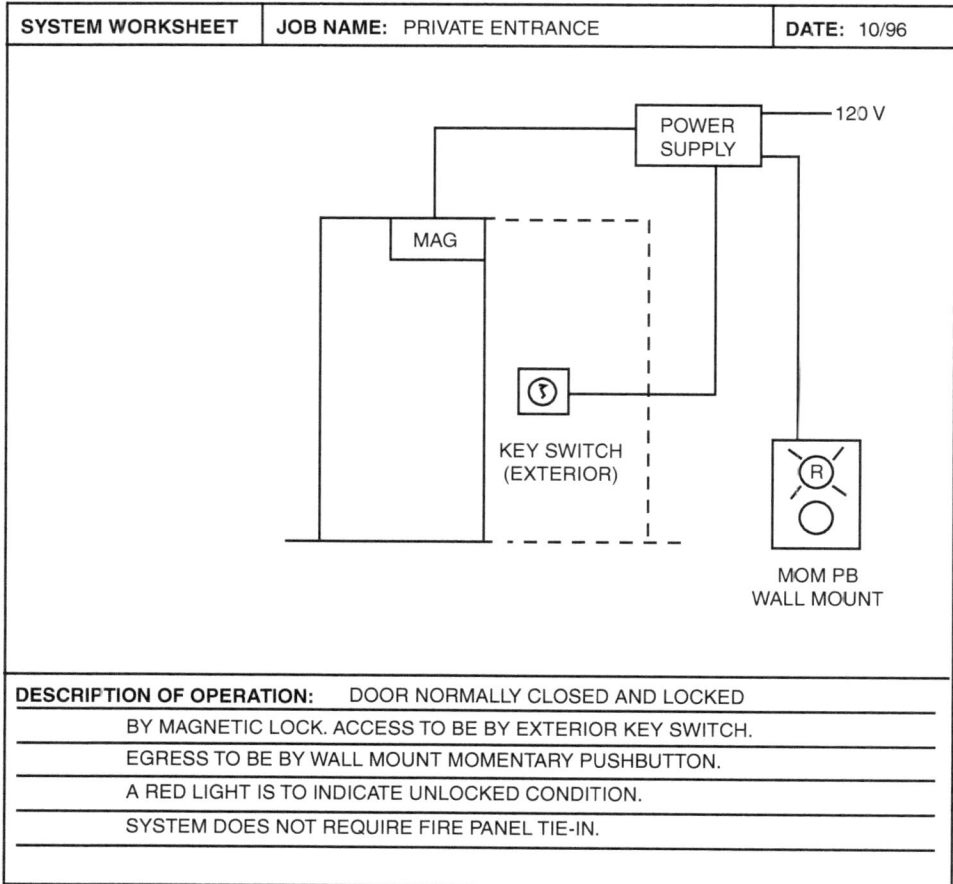

| SYSTEM WORKSHEET | JOB NAME: PRIVATE ENTRANCE | DATE: 10/96 |
|---|---|---|

**DESCRIPTION OF OPERATION:**    DOOR NORMALLY CLOSED AND LOCKED

BY MAGNETIC LOCK. ACCESS TO BE BY EXTERIOR KEY SWITCH.

EGRESS TO BE BY WALL MOUNT MOMENTARY PUSHBUTTON.

A RED LIGHT IS TO INDICATE UNLOCKED CONDITION.

SYSTEM DOES NOT REQUIRE FIRE PANEL TIE-IN.

| LOCK TYPE: MAGNETIC | OPTIONS: | |
|---|---|---|
| POWER SUPPLY: | BATTERIES: 2 | FIRE SYSTEM TIE-IN: NO |
| INTERIOR CONTROL:    MOM PB W/RED INDICATOR | | |
| EXTERIOR CONTROL:    KEY SWITCH W/ATD | | |

**Figure 5-4**  Worksheet layout of a simple locking system.

For the interior control and monitor light I chose a single wall plate that will be mounted near the door. A time delay was not necessary here at the customer's request. As shown in Figure 5-7, the switch will have three wires and the light, two wires.

That completes the riser diagram. It is *not a big deal!* Once you become familiar with different electronic security hardware, you will almost automatically know the wire counts. For unfamiliar components ask for the manufacturer's technical literature for the product. Remember, when in

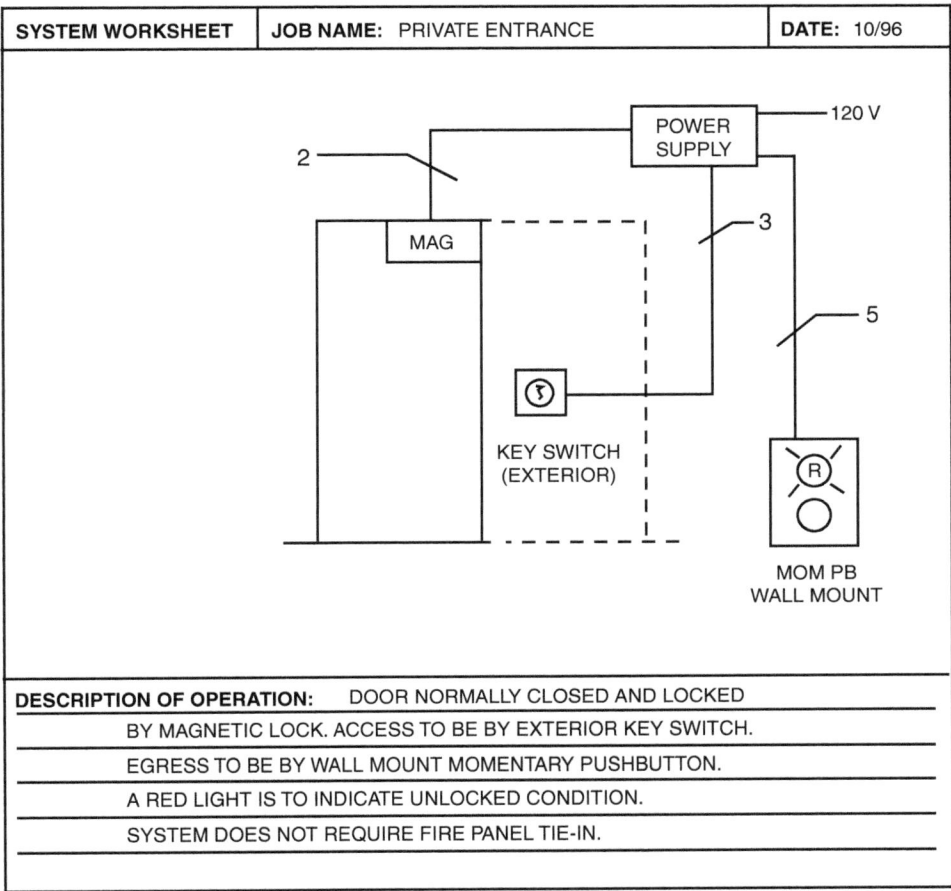

| SYSTEM WORKSHEET | JOB NAME: PRIVATE ENTRANCE | DATE: 10/96 |
|---|---|---|

**DESCRIPTION OF OPERATION:**    DOOR NORMALLY CLOSED AND LOCKED
BY MAGNETIC LOCK. ACCESS TO BE BY EXTERIOR KEY SWITCH.
EGRESS TO BE BY WALL MOUNT MOMENTARY PUSHBUTTON.
A RED LIGHT IS TO INDICATE UNLOCKED CONDITION.
SYSTEM DOES NOT REQUIRE FIRE PANEL TIE-IN.

| LOCK TYPE:    MAGNETIC | OPTIONS: | |
|---|---|---|
| POWER SUPPLY: | BATTERIES:  2 | FIRE SYSTEM TIE-IN:  NO |
| INTERIOR CONTROL:    MOM PB W/RED INDICATOR | | |
| EXTERIOR CONTROL:    KEY SWITCH W/ATD | | |

**Figure 5-5**  The riser diagram.

doubt as to how many wires are actually needed to hook up a component, count them all.

One question that sometimes comes up here relates to why I show all component wires headed to the power supply. One is never sure exactly what route the installer will take when running wire.

It has been my experience that most of the wire gets pulled to a single location. Many times this will be a junction box, where all the wires are interconnected. Often the power supply enclosure doubles as a junction box. This practice also facilitates troubleshooting that may have to be done after installation.

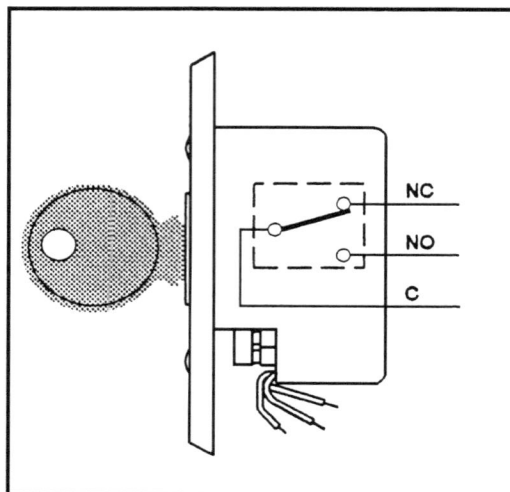

**Figure 5-6**   Key switch wire count.

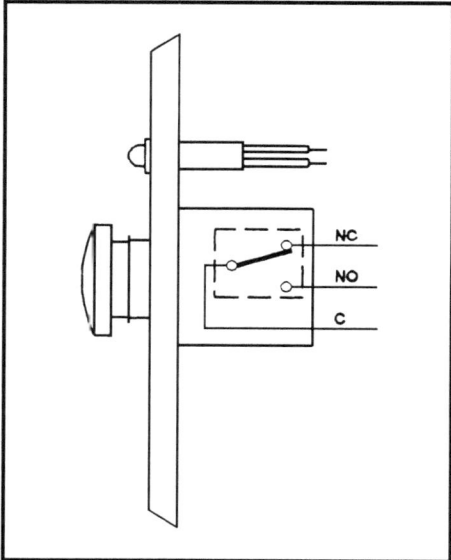

**Figure 5-7**   Control and monitor wire count.

In some cases a system designer will actually do the wiring diagram first and count the wires for the riser from the wiring diagram. So let's look now at how we create a wiring diagram for this system.

## The Wiring Diagram

There are actually two ways in which a wiring diagram could be created. One method I call the *generic* wiring diagram. Here the designer would know how things should be wired; for example, the load gets two wires, a closed or open switch contact is in one of the wire runs—general type information. A drawing of this type could be used by an installer. The trade-off is that the installer would have to identify each manufacturer's termination identifications. In other words, the installer would fill in information such as terminal numbers and wire colors. This type of drawing might be produced by a system designer who did not have access to manufacturers' literature or know which specific manufacturers were to be used.

The second type of wiring diagram would be complete, with all termination information. This is the preferred and most common method for wiring diagrams.

Normally, when I start a wiring diagram, I am thinking "generically." At this point I just want to establish the basic layout of the system. I usually spread out a bunch of symbols, representing each item in my system as shown in Figure 5-8.

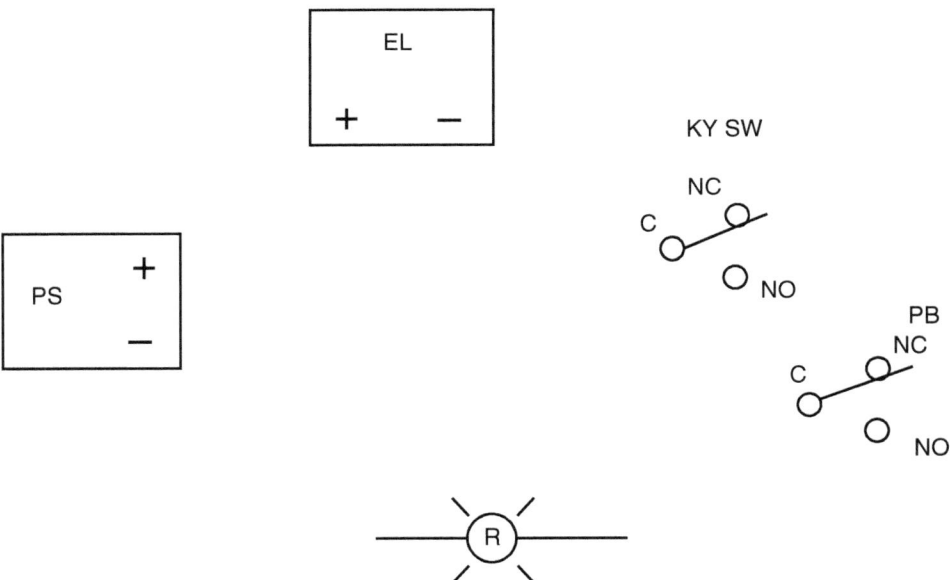

**Figure 5-8**  Initial wiring diagram component layout.

For this layout I don't even have to know the exact products I'll be using. I know the power supply has an output, (+) (−). I know the lock and indicator have inputs, (+) (−). I think I'll need single-pole double-throw switches.

Now I attempt to connect everything together to make the system work. Remember earlier I noted that we could get rid of all the (−) hookups first? In Figure 5-9 I have wired the (−) output of the power supply to all the loads.

Now, in Figure 5-10, I bring (+) from the power supply to the electric lock. Naturally I want to route (+) through all the switches that will control the lock. Since it is a fail-safe lock (power must be on to lock), I use the normally

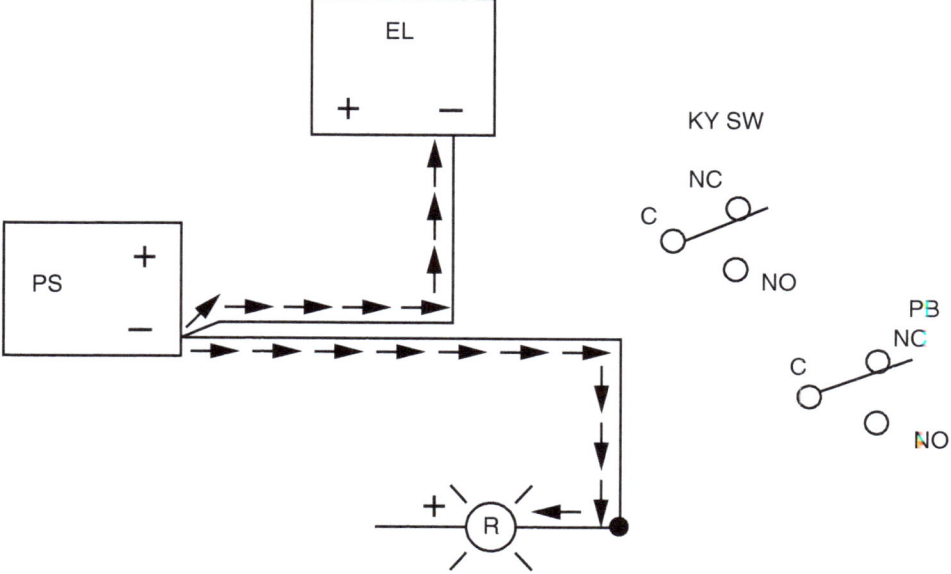

**Figure 5-9**    Wiring the (−) output.

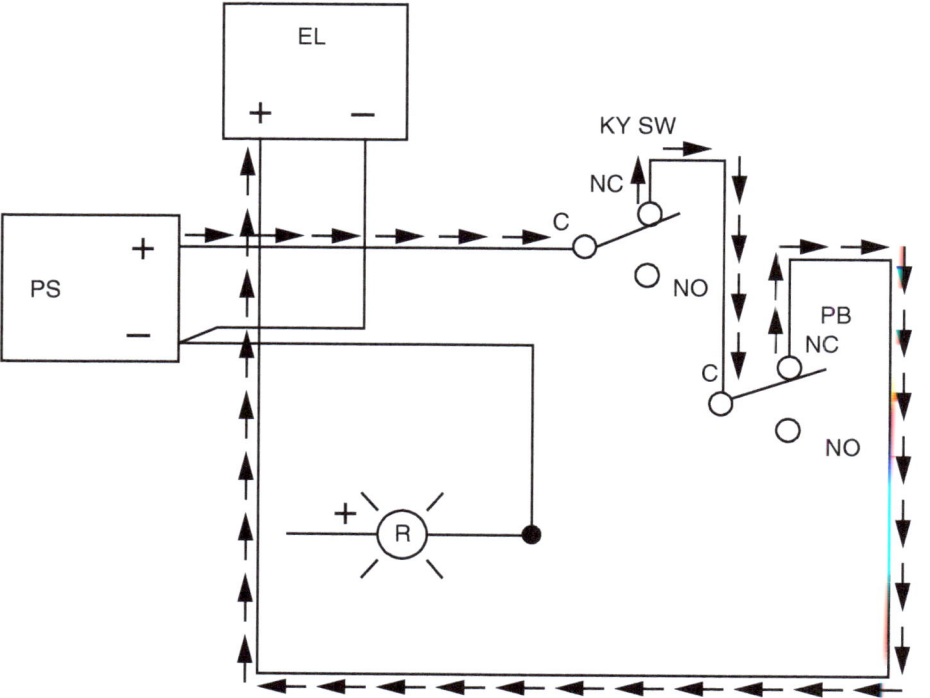

**Figure 5-10**    Wiring the (+) output to the lock.

closed contacts of all the switches and wire them in series. Note that operating any of the switches will break power to the lock.

Next, in Figure 5-11, I find a route for (+) that will light the indicator whenever I turn off the lock. Now I can use the other "half" of the switches—the normally open contacts. Note that operating either switch will close the normally open contact and allow (+) power to flow through to the red indicator.

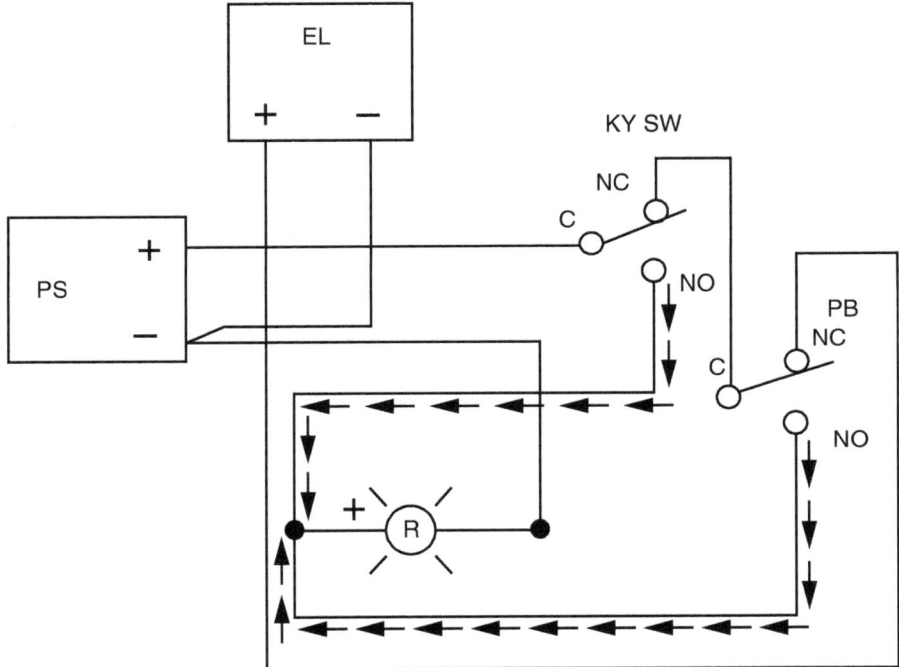

**Figure 5-11**  Wiring the (+) output to the indicator light.

Figure 5-12 shows the finished diagram, as far as wiring goes. It is a little messy, but it works.

Earlier I mentioned two items: neatness and elimination of crossovers. In Figure 5-13 I have rearranged things a little. This took two to three attempts, but in the end I finally eliminated *all* the crossovers! It is the same "electrical" drawing as before, except it is a *lot* neater.

Granted, this was a pretty simple diagram. A more complex system might take many attempts and a whole pad of paper! Once my wiring layout is complete, I select the actual products I will recommend be used. (Product selection could also be done first.)

By looking at the technical literature for each product I can start to fill in wire colors, terminal numbers, voltage and current requirements, and any other pertinent information. The finished wiring diagram is shown in Figure 5-14.

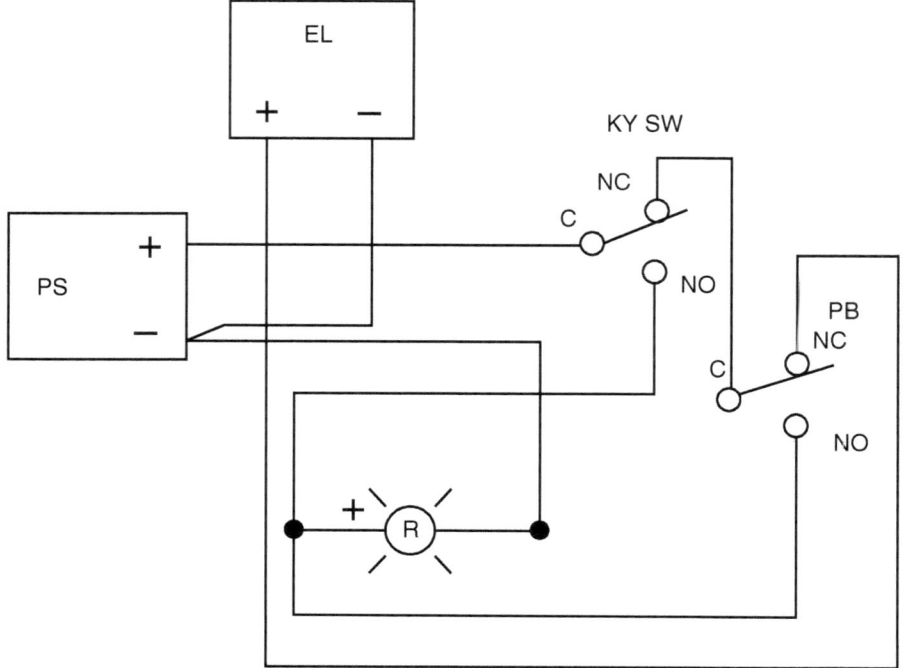

**Figure 5-12**   First draft of the wiring diagram.

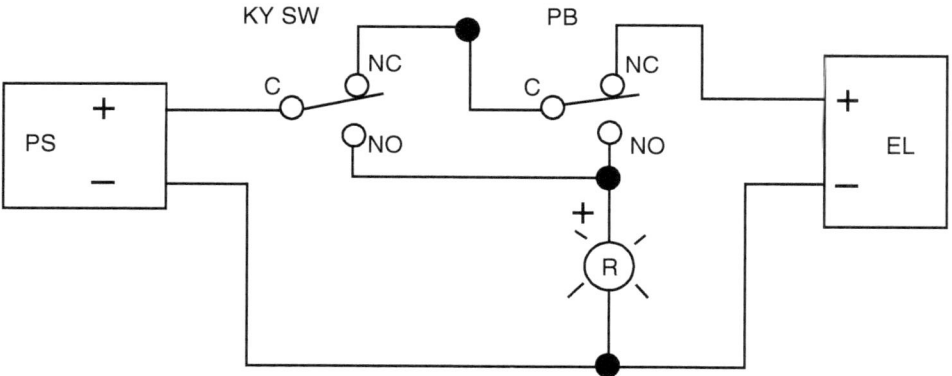

**Figure 5-13**   Final layout of the wiring diagram.

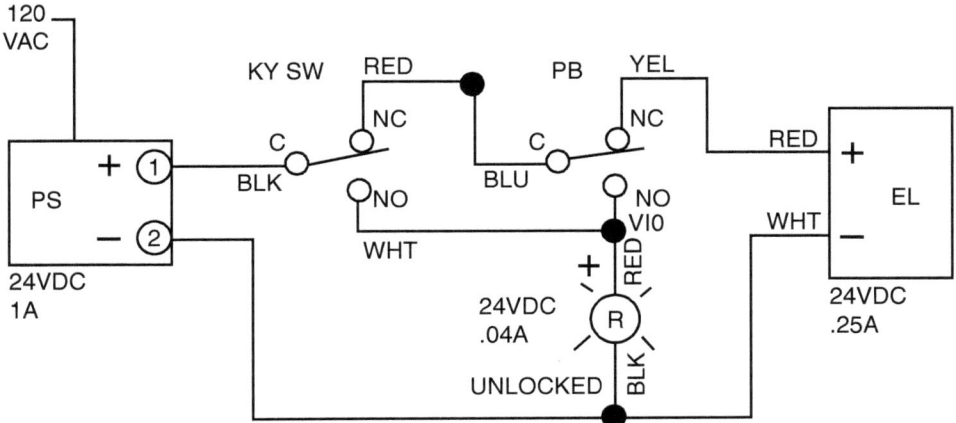

**Figure 5-14**   The finished wiring diagram.

Well, I think it is time for you to try one on your own! Figure 5-15 provides a test exercise for you to try. Your task is to fill in the wiring for this system. I recommend that you make four or five copies of Figure 5-15 before starting; you'll need them! Take your time and be careful; it is a little tricky. One hint: *Fail-secure locks require power to unlock.* A step-by-step solution is provided in Appendix A.

## Test Exercise Problem

A customer has two individual high-security doors. Each door is normally closed and secured by a fail-secure (power is needed to unlock) electric lock. Each door is individually released by an interior single-pole, double-throw (SPDT) momentary key switch. A green indicator outside each door will illuminate when the lock is released. A master double-pole, double-throw (DPDT) maintained key switch is located at the guard station. This key switch will release both locks simultaneously.

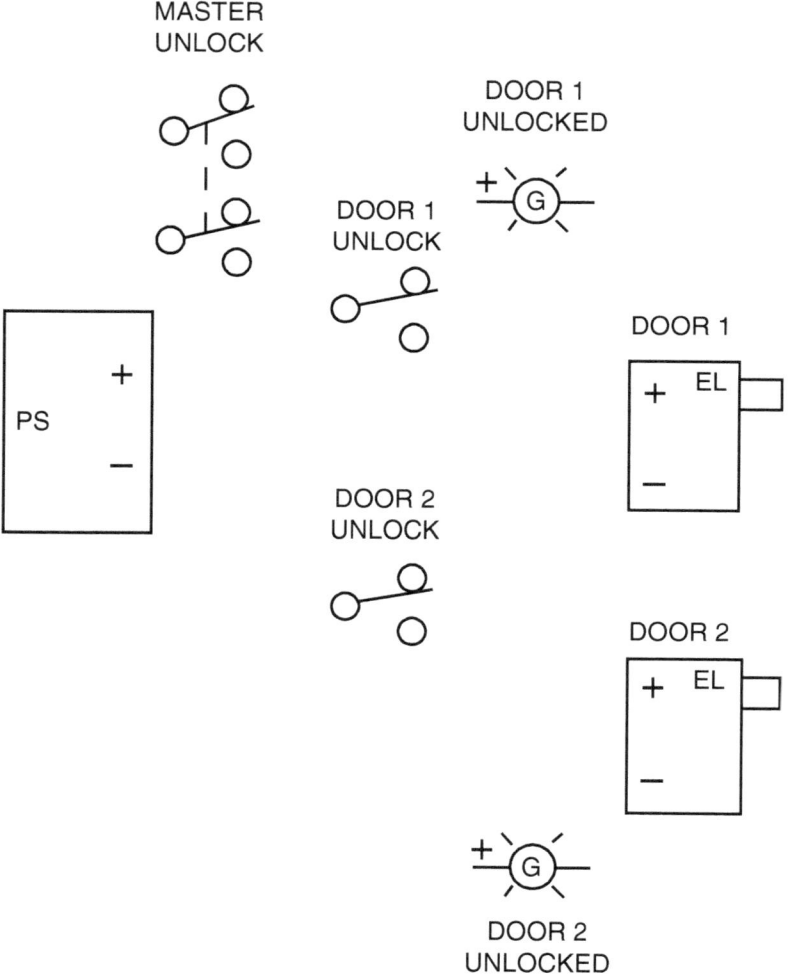

**Figure 5-15**  Test exercise.

# 6

# Other Things

Hopefully your Chapter 5 test exercise went well. It sometimes takes a while to get the hang of it. Keep practicing; it *will* come to you! I debated about adding this chapter, but thought there were some "other things" you should know about. As you become more proficient in reading and designing higher-level wiring diagrams, you will certainly come across these other things.

## The Junction Box

The J/box, as it is commonly called, is normally used to enclose electrical connections. Its primary purpose is to prevent bare wires or "hot" terminals from being touched by personnel or short-circuited by foreign material, especially in high-voltage work. It is also used to contain sparks or heat that may be created by electrical connections. The metal or plastic boxes behind switch and outlet plates in your home or office are typical examples of J/boxes used in high-voltage work.

In low-voltage security work, the J/box offers these same protections, but also provides another function. Many times it serves as a common place to make multiple wiring connections. One of the first places you will come across a J/box is on a *riser diagram,* as shown in Figure 6-1.

Including a J/box is common practice when the designer isn't quite sure how the electricians will actually run the wire. The assumption here is that there will have to be electrical connections made somewhere. Used in Figure 6-1, the J/box is simply a convenient place for the designer to run all the wire runs. It may or may not be how the system is actually wired. In some riser diagrams the wire runs may all be shown connecting to the power supply. It is possible in some cases that the power supply enclosure will be used as the J/box.

When you design *wiring diagrams,* there has to be a purpose to showing a J/box. The designer, in many cases, will not know if J/boxes have been provided

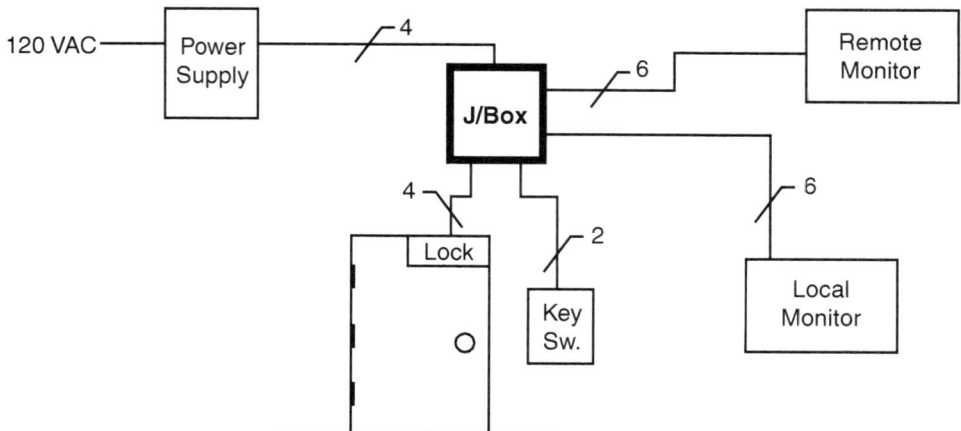

**Figure 6-1**   Use of a junction box on a riser diagram.

or will even be used. On simple wiring diagrams I personally don't believe the designer should show J/boxes unless the designer is knowledgeable about, or has some control over, the actual installation. Rather, the required terminations should simply be shown wherever they fall on the diagram. (The Chapter 5 wiring diagrams use this method.) The installer should know proper procedures for locating and making simple wire connections.

On larger, more complex systems, J/boxes may be purposely shown. This would be due to a large number of electrical connections to be made external to system components. In these cases the system designer would have to be more knowledgeable about the actual installation, as his or her diagram will be leading the installer. The J/box may also be selected or custom-designed to include terminal blocks to make the connections. The example shown in Figure 6-2 is one way this might appear on a wiring diagram.

In this example the designer has run wires and cables from several components into the J/box and coordinated their hookup with terminal numbers. To do this effectively, the designer should also select the cable and wire to be used. In this manner the designated wire colors can be shown on the drawing. If you did not know the wire colors, each wire would have to be shown run separately to its associated device, or each labeled with its other end termination. Figure 6-3 shows this method. The colors of the wires actually used should be filled in at installation.

Large systems with a substantial amount of wiring may even use *wire lists* rather than drawings with wire running all over the place. Figure 6-4 shows a simple example of this method.

The wire list simply gives the installer a wire color and denotes where each end is to be terminated. The use of the wire list requires that each component and its terminations have a simple but clear identification. Of course, this is good practice on wiring diagrams in general.

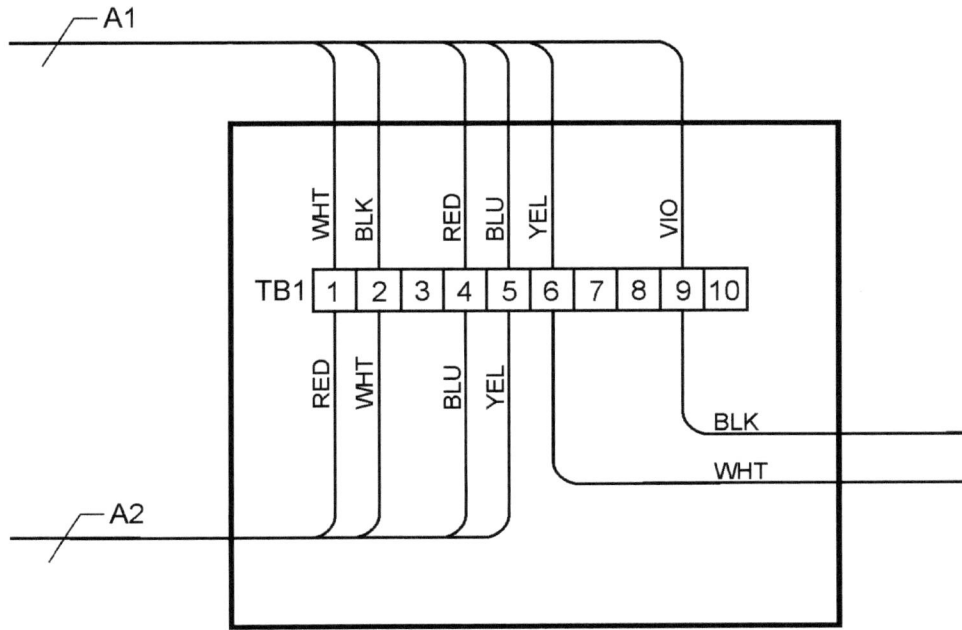

**Figure 6-2**  Use of a junction box with wire colors known.

## Fire Panel Interface

In my opinion this is one of the most discussed, yet least understood subjects in a security locking system. In the following coverage of handling fire panel hookup, I am addressing the security system designer who is not familiar with fire alarm systems. For those who are knowledgeable about fire alarm systems and include them in their design, there is little need to read this section.

It is common in the security industry to believe you must *always* include reference to the fire panel interface on wiring diagrams. It is also a common belief that you must include specific hardware, that is, a control relay, to interface security equipment with the fire panel. But I have found, after making a few mistakes of my own, that these beliefs are in many cases either wrong or at least overkill.

For a long time I always included a reference to fire panel hookup on all my drawings. For awhile I even insisted that low voltage power supplies feeding multiple locking circuits include a relay to interface with the fire panel. At some point I asked myself (or maybe was asked by someone in the industry), How do you know for sure that you need to reference the fire panel on a particular design?

That question prompted me to interview personnel in a couple of companies who actually provide fire alarm systems. At first they were a little suspicious of my motives. They knew I was from a part of the security industry that designed and provided hardware for locking and access systems. I do believe

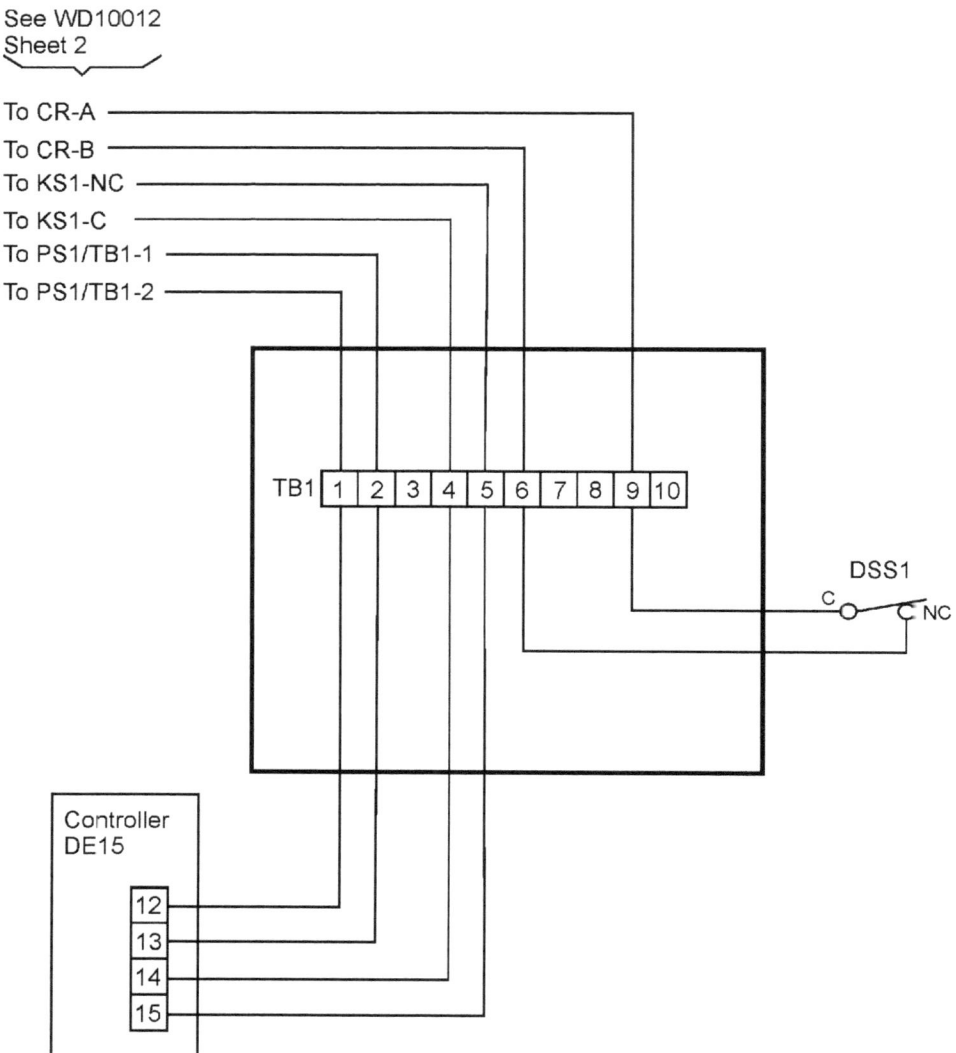

**Figure 6-3** Use of a junction box with wire colors unknown.

they thought I had intentions of infringing on their end of the business. My explanation set them at ease. I told them I simply wanted to better understand their business in order to keep the hardware people *out* of their business.

In brief, they explained that fire alarm systems were a totally separate business, requiring specific expertise and following a whole different set of codes and rules. As a correlation, they mentioned that just as hardware people had to research construction plans to accommodate particular specifications, they had to examine the same plans to accommodate fire protection. The plumbing

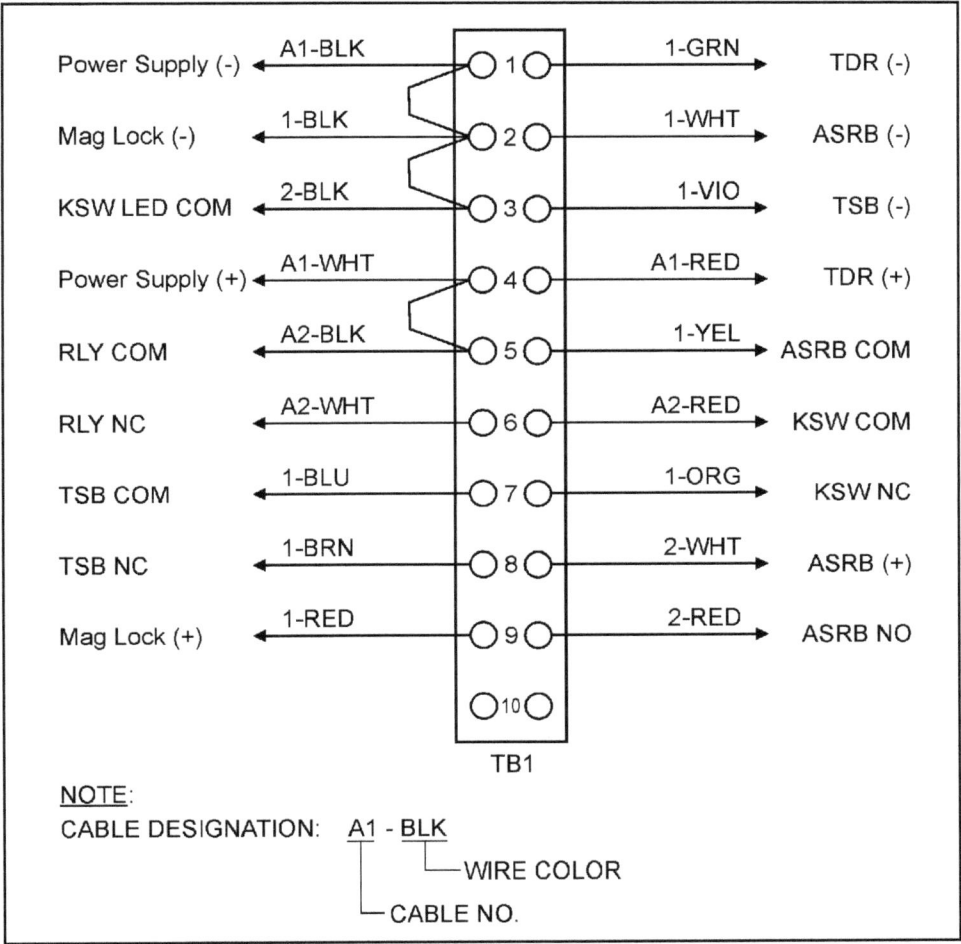

**Figure 6-4** Wire list for junction terminal board.

contractor, the heating, ventilation, and air conditioning (HVAC) contractor, every different trade studies the same construction documents. Each trade is aware of the job to be done, and each has the expertise to provide the proper equipment to interface with relevant equipment of other trades.

The fire system people explained that they had to control certain equipment within a building during a fire emergency. As an example, during a fire alarm the fire panel would shut down the heating and air conditioning systems, return elevators to the ground floor, and ensure that certain doorways were unlocked.

They mentioned that even if a locking system wiring diagram did not reference a fire panel interface, and there was a door that should be electrically unlocked during a fire alarm, they would find it and take care of it!

Now I'm not advocating completely forgetting about showing the fire panel interface on a wiring diagram. But I am saying that it shouldn't be arbitrarily planted on every drawing. It can create some real confusion if hookup to a fire panel is shown when it is not necessary. If in doubt and you want to play it safe, use this notation on the wiring diagram: *When required, interface the security system with the fire alarm control panel.*

Yes, we are passing the buck, but we have also covered our behinds! Now, how does the fire panel actually turn off the electric lock? Do we need to provide a relay, or some other means of doing this? The fire system people told me no.

As a simple example, the fire panel consists of a bunch of relays providing switching contacts as outputs. The panel is programmed to "switch" these relays during specific events. The relay contacts are wired to external equipment, such as the elevator control, the locking system, etc. In the case of an electric lock, I was told that the fire panel relay might switch off the high-voltage line to the lock power supply. If the power supply included battery backup, the low-voltage line to the lock would simply be switched off. The fire system people explained that fire panel relay contacts are pretty heavily rated, say, 5 or 10 amperes, and are quite capable of switching the control circuit loads of external equipment. In some cases the fire panel may be wired directly to the locking device, rather than to the power supply. There also may be the need for auxiliary fire panel relays to be placed near the equipment that the fire panel is controlling.

I also questioned them about the fact that many security hardware manufacturers include a *fire interface relay* in their power supplies, most often in power supplies with multiple power outputs. The theory behind this is that the fire panel only has to turn off one built-in relay, which in turn uses its contacts to internally turn off multiple power outputs.

Their response to this was, "Ok, thanks, this does make our job easier. But do yourself and us a favor: Don't label that built-in relay with any name that includes the word *fire*. You are not in the fire alarm business. Call it something else!"

And they are right! A term such as *fire alarm relay* implies that it is part of a fire alarm system. Most of the time it isn't, *unless* the equipment it is built into has successfully gone through specific Underwriters Laboratory testing for that purpose.

So, in conclusion, I recommend the following:

1. When you are designing a security system, find out if a notation should be included referencing fire panel tie-in. Check with the local authority having jurisdiction, building code officials, or the electrical contractor for the job.

2. If you are including a fire panel interface relay, make sure it is tested and approved for that function.

3. If you show a wire run from an interface relay to the fire panel, *don't ever* put anything else in that wire run. There is a temptation to include a *master cutoff* switch in that line. I did that once, figuring it would be a good

place to include a master key switch used to shut down a series of locks. I was later told to never use any wiring that has anything to do with a fire panel. That wiring falls under special codes for the type of wire and installation and is also supervised by the fire panel. Any break in that wire, whether by accident or by a switch is reported to the fire panel as trouble.

Figure 6-5 shows several methods of noting fire panel tie-in. To the best of my knowledge, they are sufficient to draw the attention of qualified personnel who will handle the fire system requirements.

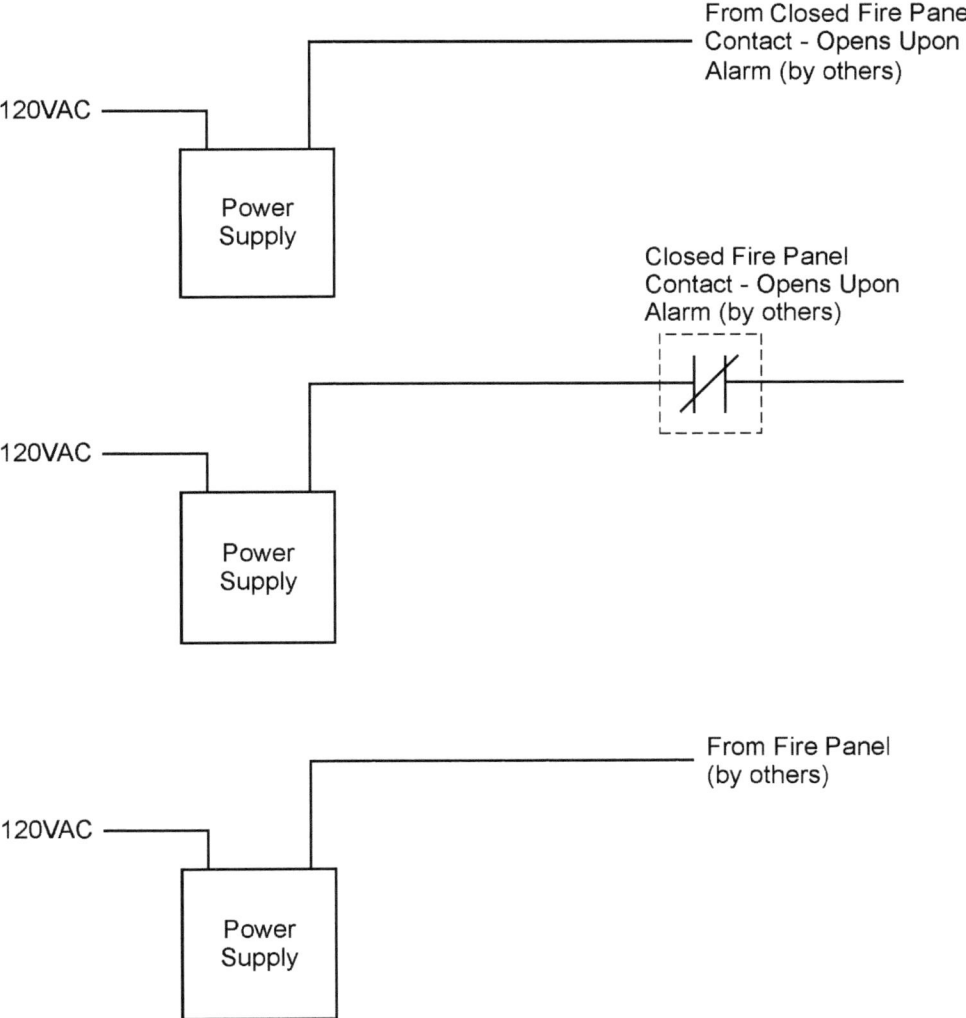

**Figure 6-5**  Typical fire panel interface designations.

There are so many safeguards regarding fire system design, installation, and testing that it is highly unlikely something would be overlooked. What reinforced my view of this subject was a paragraph from a specification I read. The specification addressed electrically locked doors in a hospital redevelopment project. It read in part, "The emergency exit door control system shall be hardwired by the Security Contractor to an addressable relay provided by the Fire Alarm Contractor at each delayed exit door to immediately override the time-delay locking mechanisms through the Fire Alarm Control Panel." I liked this; fire panel interface was covered, but left to the experts!

## Overrides

Occasionally you may be asked to provide functions with names such as *master unlock, master lockdown, lockout, alarm shunt,* or some similar nomenclature. On large systems these functions are done through computer programming or relay logic. On small systems these are functions you may be able to provide through simple switch contacts.

### Unlock override

As an example, let us use a simple locking system made up of a fail-safe electric lock and a momentary NC (normally closed) pushbutton. The pushbutton is used during the day, but needs to be overridden at night by a restricted device. Figure 6-6 shows a method of providing this function with a simple key switch.

The NO (normally open) contact of the key switch is wired in parallel with the NC pushbutton contact. Once the key switch is operated, and the key removed, the closed contact provides an uninterruptible path for (+) to the lock. Depressing the pushbutton will not release the lock.

Providing an override for a fail-secure lock circuit is a little different. As shown in Figure 6-7, we now wire the NC contact of the key switch in series

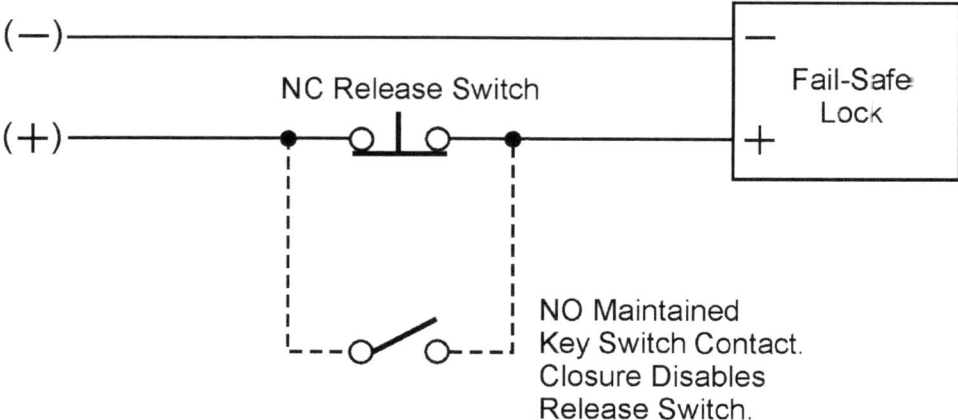

**Figure 6-6**  Unlock override circuit for fail-safe lock.

with the NO pushbutton contact. Now when the key switch is operated, the open contact breaks the (+) path to the lock. Depressing the pushbutton will not release the lock.

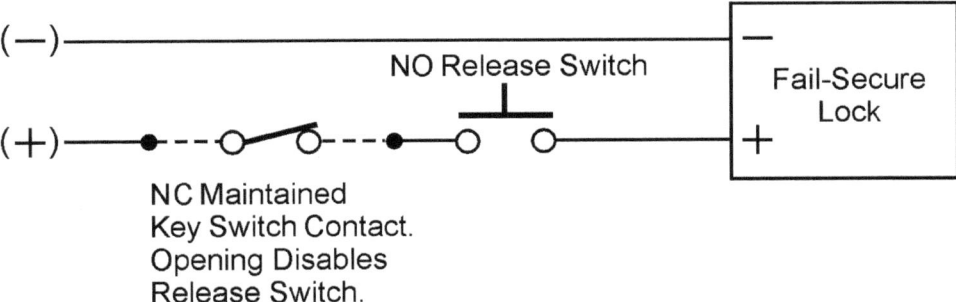

**Figure 6-7** Unlock override circuit for fail-secure lock.

Overriding the releases for multiple fail-safe locks can be a little more difficult. As shown in Figure 6-8, two lock release controls can be overridden by a device with DPDT contacts. If you have more than two locks with individual release switches, you need an override contact set for *each* release switch. (A single override contact for all release switches will not work!) It is suggested that when three or more release switches for fail-safe locks need to be overridden, multipole relays be used.

Overriding multiple release switches for fail-secure locks is much easier. A single override contact, as shown in Figure 6-9, can override any number of normally open contacts that are wired in parallel. All the override contact need do is to break the (+) line somewhere *before* all the release switches.

Although I have mentioned the use of key switches as override switches, many other devices with contact sets can be used. Some digital keypads offer maintained contacts that change state when a specific code is entered. Card readers, automatic timers, and control consoles are among other equipment that can be used.

## Master unlock

Some systems may call for a master control that unlocks all locks simultaneously. This function is pretty much the opposite of the unlock overrides we just studied. For fail-safe lock systems it is a simple matter of a single contact set killing a main powerline. As shown in Figure 6-10, breaking the main powerline somewhere before it starts to split up will kill power to everything downstream.

Releasing multiple fail-secure locks would require multiple contact sets and assurance that power would be available. This situation is not common in commercial security work.

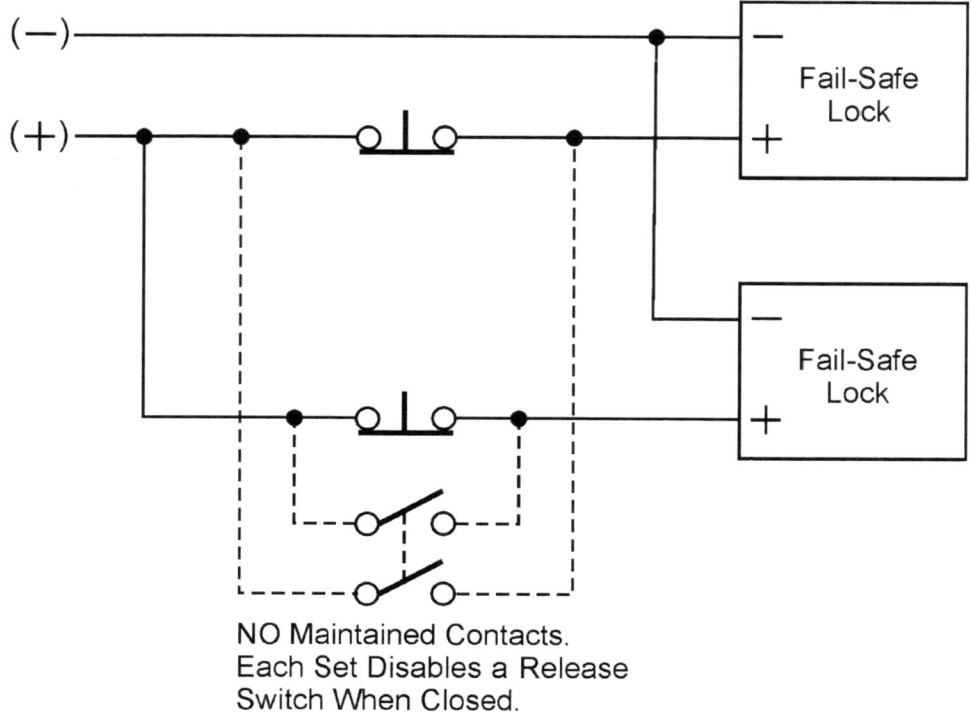

**Figure 6-8**   Unlock override circuit for multiple fail-safe locks.

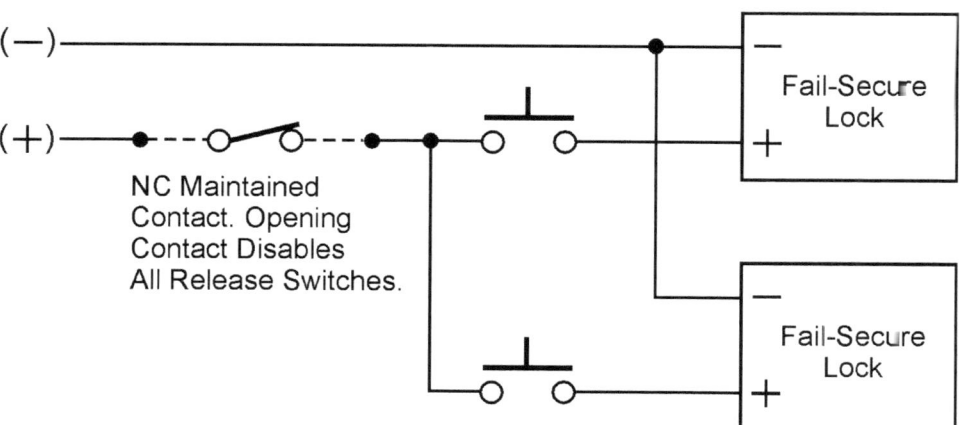

**Figure 6-9**   Unlock override circuit for multiple fail-secure locks.

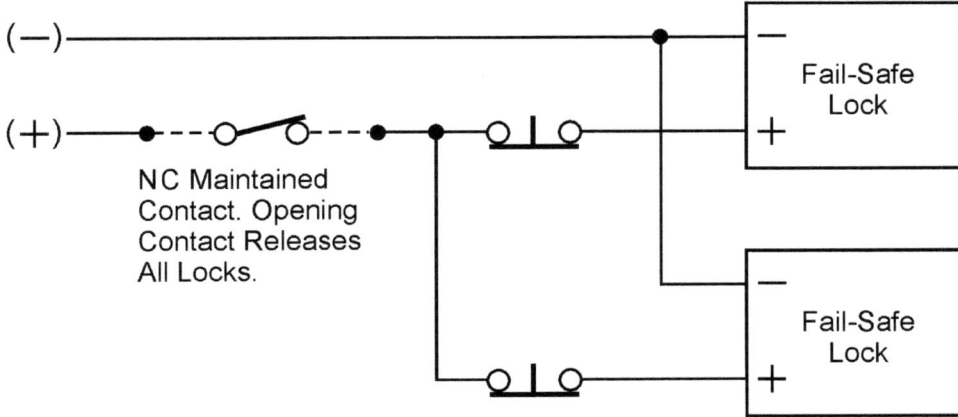

**Figure 6-10** Master release circuit for multiple fail-safe locks.

A simple example is shown in Figure 6-11 just to give you an idea of how this override would work. Multiple NO contacts, in a single device, would simultaneously close, sending (+) power to each individual lock, releasing all locks.

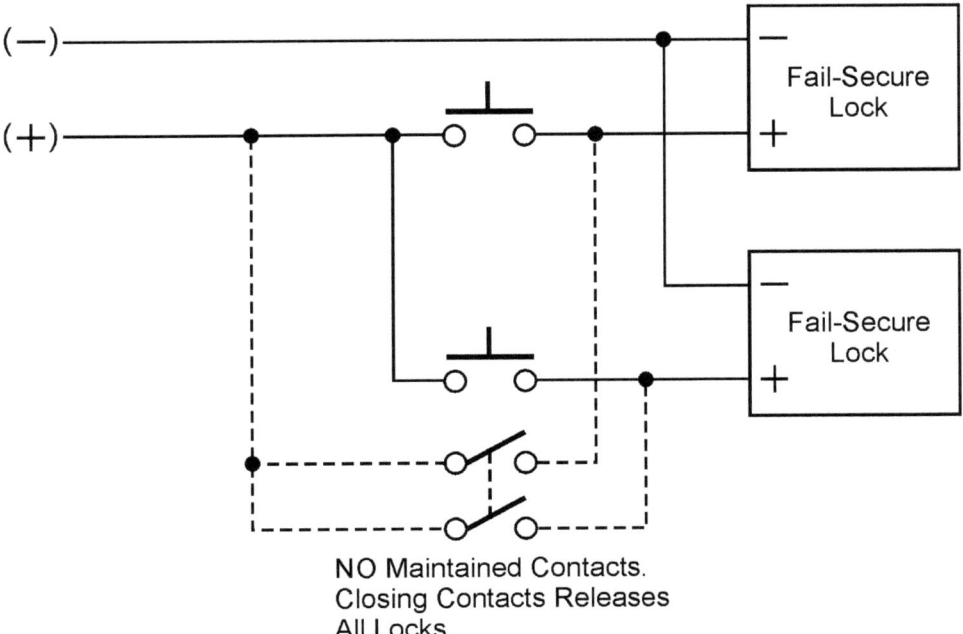

**Figure 6-11** Master release circuit for multiple fail-secure locks.

## Alarm shunt

Another function that fits into the override category is the *alarm shunt*. This function is used to prevent a false alarm signal during an authorized event. A simple example would be the authorized access or egress through an alarmed door.

In Figure 6-12 an exit door is equipped with an exit device with a DPDT switch or relay or two SPDT switches. When the exit device is pushed, the switch contacts open. One set of contacts releases the magnetic lock. The other set of contacts opens the wire run to the alarm, preventing the door open status signal from reaching the alarm panel. If the door were somehow ripped open without using the exit device, the alarm would sound.

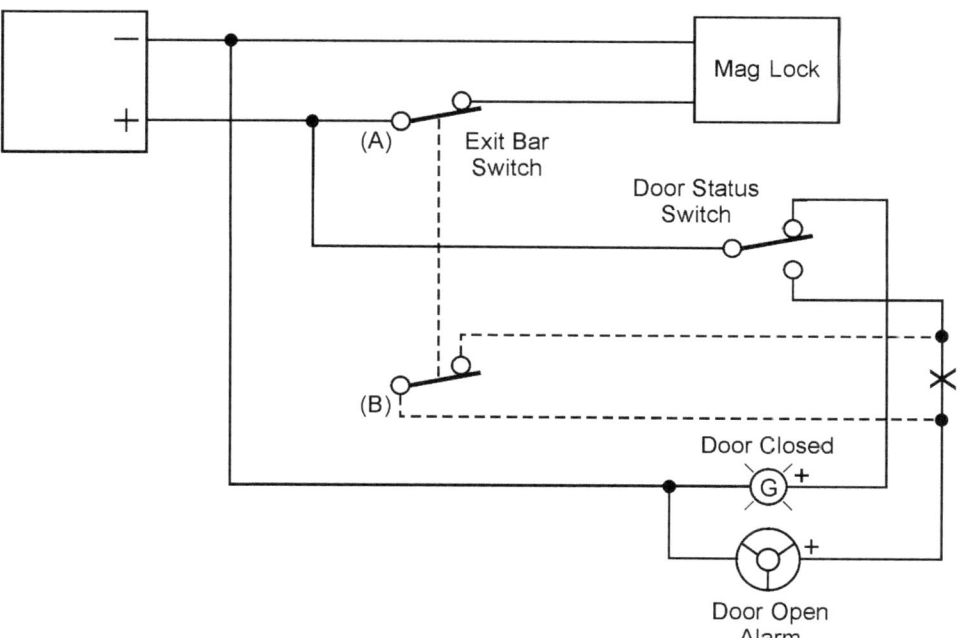

**Figure 6-12**　Perimeter exit door with alarm circuit.

In Figure 6-13 an entrance door is equipped with a card reader for access and an exit device for egress. The lock release switch from the exit device can usually be wired to a request-to-exit feature in the card reader. Using a valid card or operating the exit device causes the card reader release contact and alarm shunt contact to open. In this manner the lock is released, and the alarm circuit shunted, preventing an alarm signal.

## Master lockdown

This function is the opposite of the master unlock function. It is highly unlikely you will run across this function unless you are involved in detention or prison

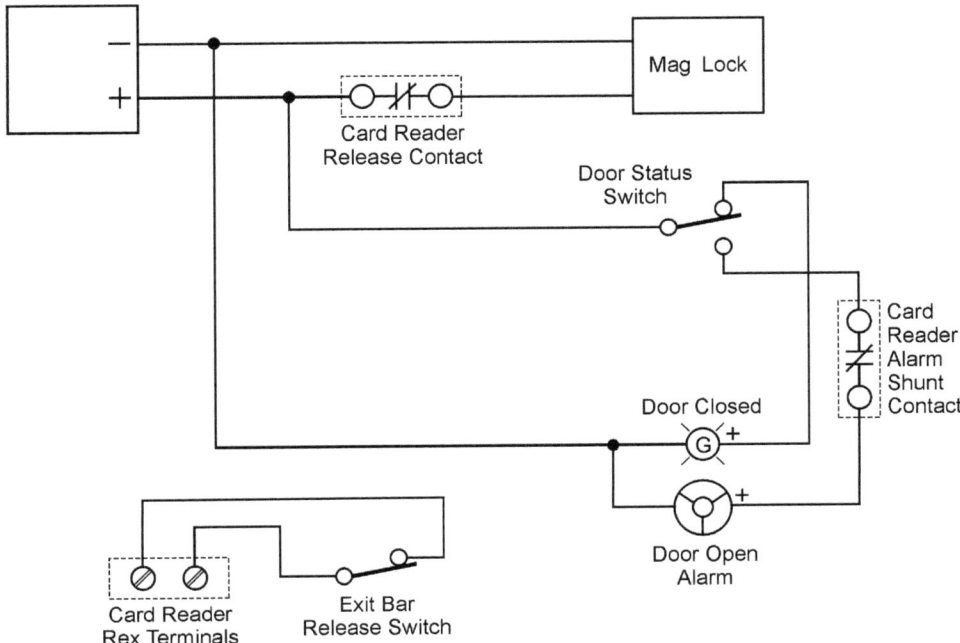

**Figure 6-13** Entrance/exit door with alarm circuit.

work. Simultaneously locking multiple doors would most likely be a factory-built function in a control console.

## Lockout

This is another function that is normally a factory-built feature of a control console. Usually a key switch is provided to manually lock out the use of individual control switches. It is commonly used to prevent the use of the console when left unattended.

All the examples given in this chapter are pretty simplistic. The primary purpose of these exercises is to show the logic of override functions. Although these are legitimate scenarios, many system components are available with built-in or ready-to-wire override features.

# 7

# Monitoring

As mentioned earlier, monitoring is not an absolutely necessary component of an electric locking security system. But monitoring is such an important feature that I consider a locking system incomplete without it. In fact, many systems consist *only* of monitoring, especially in the alarm industry.

The objectives of this chapter are to familiarize you with monitoring functions and devices and to show you how to add monitoring to a wiring diagram.

## Monitoring Functions

The purpose of monitoring in a locking system is to let personnel know, either locally or remotely, the status of a door or locking device. More sophisticated systems can even provide live or recorded viewing of events at a door or monitored area.

One of the most difficult processes in monitoring is deciphering what an end-user really wants to monitor. In classroom study I am constantly emphasizing the difference between door status and lock status monitoring; and they *are* two distinctly different conditions. The following paragraphs provide a description of several monitoring functions.

## Power status

Is power on to a specific piece of equipment, say, an electric lock, or to a complete system? This monitoring function gives the least information as it indicates only the presence of power, not the condition of any specific piece of equipment. One good use of power status monitoring is to indicate when mainline power has failed and standby battery power has been activated.

### Door status

Is a door open or closed? This is all that the door status indicates. It does *not* imply that a door is *locked*.

### Lock status

Is a lock in the secure or not-secure position? This is all that lock status indicates. It does *not* imply that a door is *closed*.

### Door closed *and* secure

This is a combination of two functions and requires two monitoring devices. It better indicates a secure condition as two conditions must be present before a secure status is indicated.

### Door open *or* not secure

This is a combination of two functions and requires two monitoring devices. It better indicates a "trouble" condition as *either* of these undesirable conditions will indicate a problem exists.

### Door closed *and* not secure

This combination of two functions is sometimes used when a door is left unlocked for specific periods of time but should be kept in a closed position when not in use.

These monitoring functions cover the majority of the conditions you will face in basic security work. There are many other conditions that can be monitored. Devices are available that monitor gas presence, water level, people presence, heat, fire, and smoke; and all can be included in security system design. It is up to the system designer to choose the best monitoring functions for a specific application.

Now that we have defined several monitoring functions, we need physical equipment to detect and indicate these conditions. We separate this into two sections:

- The actual signaling device that *monitors* a condition
- Devices that are used to *indicate* a condition

## Signaling Devices

The output of almost all signaling devices used for monitoring is based on switch contacts. There are a few devices that provide voltage signals, but they are not common enough to describe in this manual. They are commonly used to signal microprocessor-based equipment but can be converted to switch contacts through a relay. We already know how to draw switch contacts, so our only task here is to identify the signaling devices!

**Power monitor**

There are several ways that "power on" may be monitored. The presence of power can be signaled by tapping directly off the output power or using output power to trip a relay. Figure 7-1 shows how this might look on a wiring diagram.

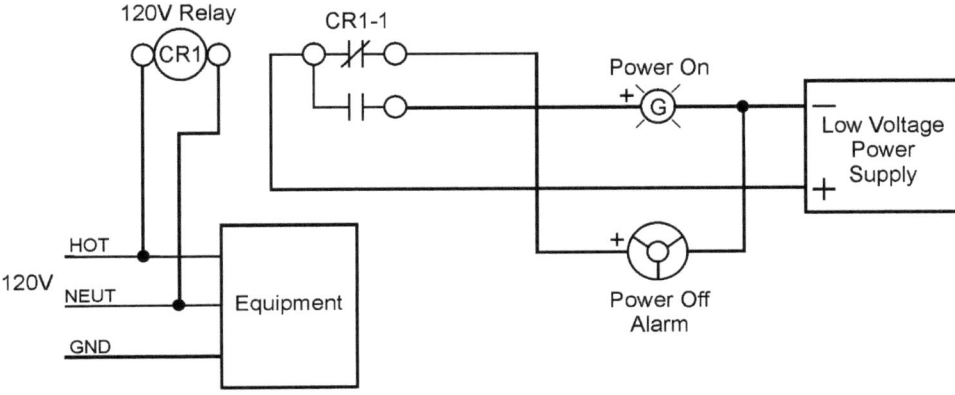

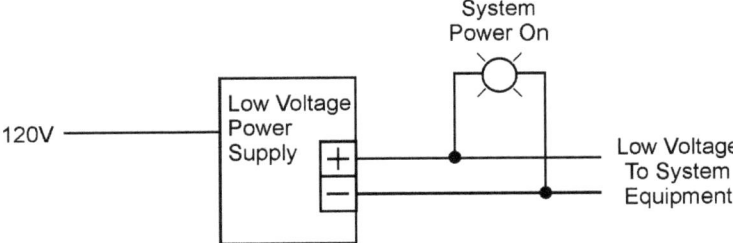

**Figure 7-1**   Typical "power on" monitoring diagrams.

It is a good bet that if a relay dry contact output is desired to monitor power status, it is offered as built in by the manufacturer of the equipment. This saves you the trouble of designing it into the system.

**Door status monitor**

As can be summarized, a device used to monitor a door is called one of the following:

Door status switch (DSS)

Door status monitor (DSM)

Door position switch (DPS)

Regardless of its name, it is a device that physically interacts with the door. Figure 7-2 shows two of the most common of these devices.

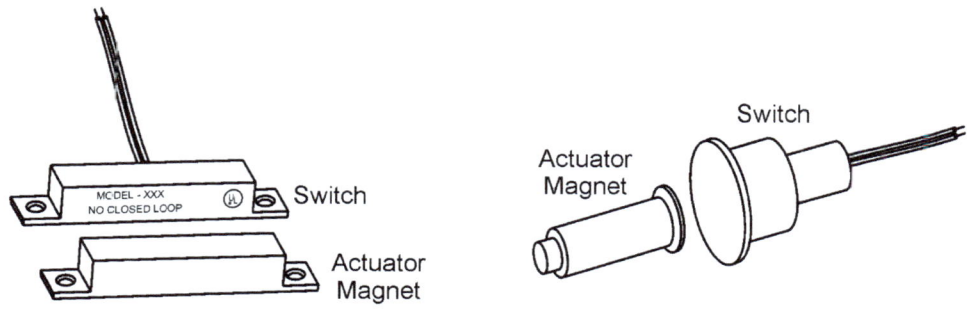

Wiring Diagram Symbol

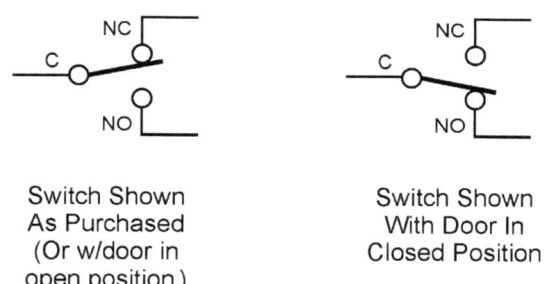

Switch Shown
As Purchased
(Or w/door in
open position)

Switch Shown
With Door In
Closed Position

**Figure 7-2**  Typical door status switches.

The switch itself is mounted to the door frame. It is usually a small reed switch that is triggered by its mating half—a small permanent magnet mounted to the door. The switch contacts are very small and are attached to thin metal blades. When the permanent magnet gets close to the switch, it attracts the blades, opening or closing the contacts. The gap at which this attraction can occur varies with the type of switch selected, commonly available from $\frac{1}{2}$ to 1 inch.

Door switches are available with only open contacts, only closed contacts, or both. Just remember, when the door is closed, the switch contacts are in the opposite state from when you purchased the switch. That is, normally open contacts are held closed, and normally closed contacts are held open.

Also keep in mind that due to their small size, these switch contacts cannot handle very much power running through them. The contacts are commonly rated to handle 24 to 30 volts, $\frac{1}{4}$ to 1 ampere. Usually they are used to switch power directly to small indicators that draw very little power. They could also switch relays that draw little power but whose contacts are rated higher than those of the door status switch. The relay contacts are then used to switch heavier monitoring loads, for example, sirens, horns, or other equipment, as shown in Figure 7-3.

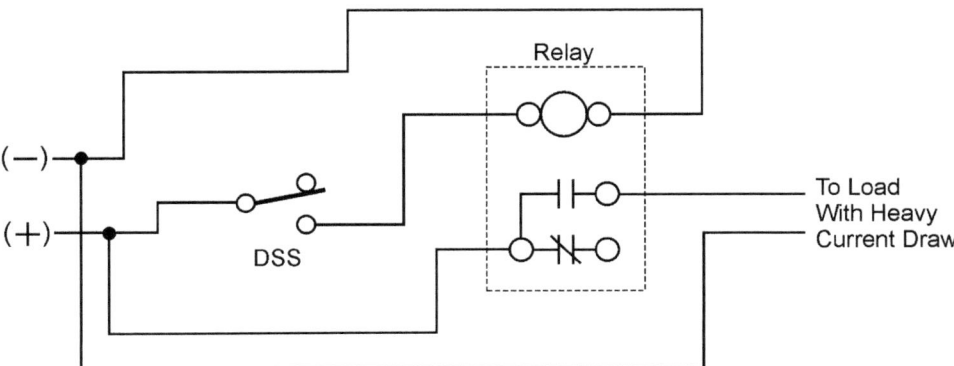

**Figure 7-3** Converting low rated switch contacts to higher rated switch contacts.

Door status switches are also available built into some locking devices. They work the same as door- and frame-mounted switches but are somewhat concealed to help prevent tampering. Usually these devices are also reed-type switches with some locks offering heavier-duty ball-type switches.

Figure 7-4 shows how the use of door status switches might appear on a wiring diagram.

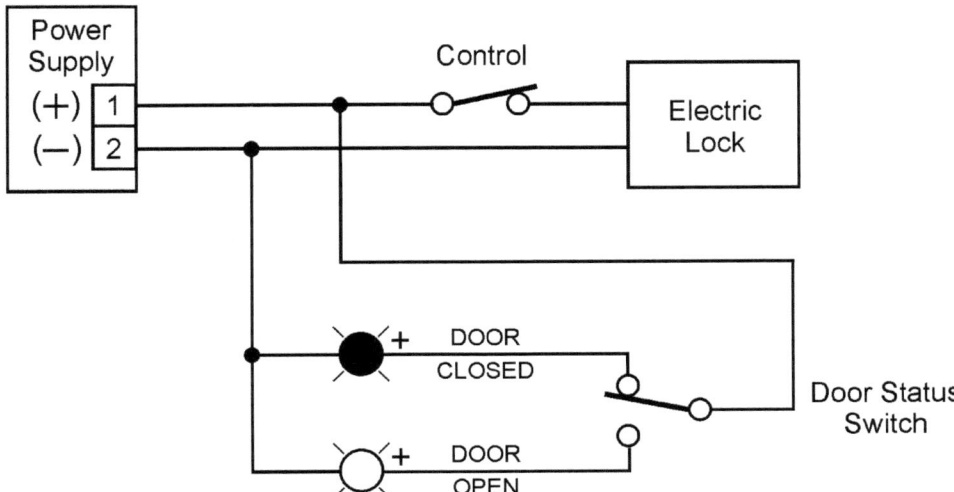

**Figure 7-4** Typical door status monitoring diagram.

You will also be confronted with monitoring pairs of doors. This type of opening is usually monitored as if it were a single door rather than two independent doors. The wiring diagram in Figure 7-5 shows how to wire a pair of door status switches.

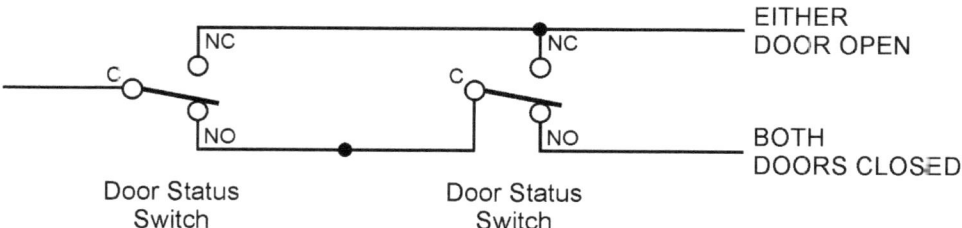

**Figure 7-5**  Typical monitoring diagram for a pair of doors.

Other than monitoring the status of a door, door status switches can be used to control other equipment in the system and to create door interlocks.

## Lock status monitor

This function is commonly overlooked and is a distinctly different monitoring function from door status monitoring. Many people think that once they have selected door status monitoring as a function, they are done. They mistakenly assume the door is *secure* if the door status switch is indicating a "door closed" status. Many are unaware that most every electric lock can be specified with some sort of output that signals the secure/not secure condition of the *lock*. Most of the time the lock status monitoring device is a type of switch providing a dry contact output similar to that of a door status switch.

Occasionally you may find a lock that provides a low-voltage output to indicate lock status. Voltage output to monitor lock status is not as common and is usually associated with magnetic locks. Usually this type of output is low voltage, say, 5 volts, and is interfaced with computer equipment. If needed, voltage output can be converted to dry contact output through the use of a relay.

Electromechanical locks, for example, electric strikes, electric bolts, and electrified hardware, usually incorporate some sort of microswitch within the lock. The switch physically interfaces with a mechanical part of the mechanism that is instrumental in causing a lock/unlock condition. Magnetic locks usually incorporate a reed-type switch to indicate a secure status. The nomenclature of the switch may vary with each manufacturer, but generically it would be called a *lock status switch*.

When using electric strikes you have to be careful to select the right options for true secure status indication. Electric strikes usually offer two types of monitoring switches. For simplicity we will call one the *latch bolt monitor* and the other the *locking cam monitor* (LCM). The latch bolt monitor is a switch that indicates whether the lock latch bolt is extended into the electric strike cavity. This feature doesn't necessarily indicate a secure condition. The strike keeper, sometimes called the *lip* or *gate,* may not be in the locked state. I consider the latch bolt monitor switch more as a door status switch. If the latch bolt is triggering the switch in the strike cavity, you know the door is closed.

The locking cam monitor tracks some mechanical part of the strike that locks the keeper in place. Although this feature indicates a locked keeper, the door could be in the open position!

The problem here is that either monitoring switch alone will not indicate a true locked or secure condition. The answer is to use *both* monitoring switches. This would require two conditions to exist before a secure indication would occur: (1) The door must be closed with the latch bolt extended, and (2) the keeper must be in a locked state.

The latch bolt monitor and locking cam monitor are commonly offered as separate options. This allows for cases where it is only necessary to monitor a single condition. Several manufacturers offer both monitoring features as a single option to monitor a true secure status.

On electric bolts the lock status switch is commonly referred to as the *bolt position switch* (BPS). In many units this is a metal roll pin driven through the back end of the bolt itself. As the bolt travels, the pin hits the lever of an internal microswitch. There may actually be two microswitches, one indicating when the bolt is fully extended (locked condition) and one indicating when the bolt is fully retracted (unlocked condition). This method ensures positive monitoring of either condition.

Magnetic locks are a little different, as there are no moving parts that would mechanically indicate a locked condition. Most of these locks incorporate a small reed-type switch within the magnet body. This switch reacts to the magnetic field created when the door armature plate is fully "bonded" to an energized magnet. The switch is factory-calibrated to indicate a secure/not secure condition.

Builder's hardware, for example, lock sets and panic devices, are sometimes monitored by using a monitoring strike. This device is simply a strike plate with a switch that indicates whether the latch bolt is extended into the strike cavity. Monitoring the actual secure condition of the lock would require a switch internal to the locking device. This requires wiring the door through a door cord or electric hinge or pivot.

By now you get the point that a switch-type lock status monitor is available on electric locks. Figure 7-6 shows how the use of lock status switches might appear on a wiring diagram.

Note that this diagram is identical to Figure 7-4. The only difference is that the door status switch is now called the lock status switch. Now, although monitoring lock status is a step up from knowing the status of the door, this still doesn't tell a complete story. Door status doesn't tell you if the door is locked; lock status doesn't tell you if the door is closed! The ideal method of monitoring, especially in high-security work, is to monitor *both* conditions. The diagram in Figure 7-7 shows how to wire for full monitoring of a secure/not secure condition.

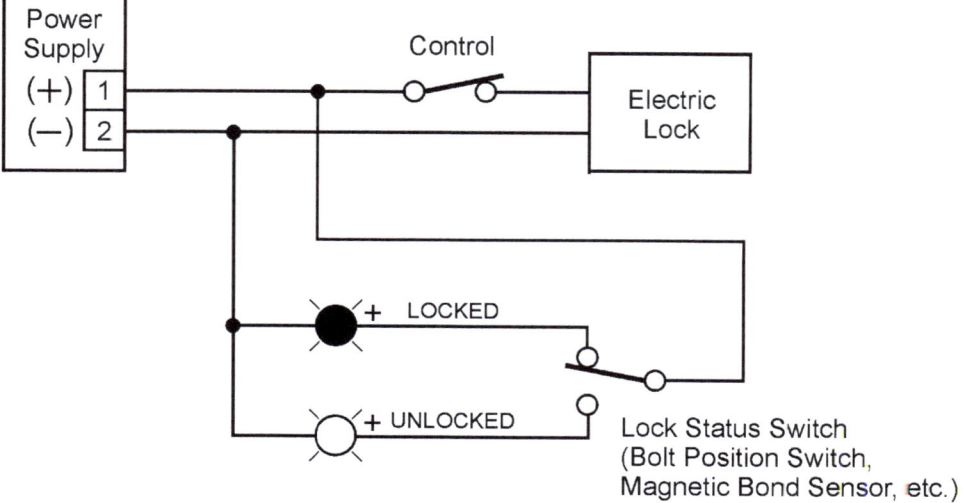

**Figure 7-6**   Typical lock status monitoring diagram.

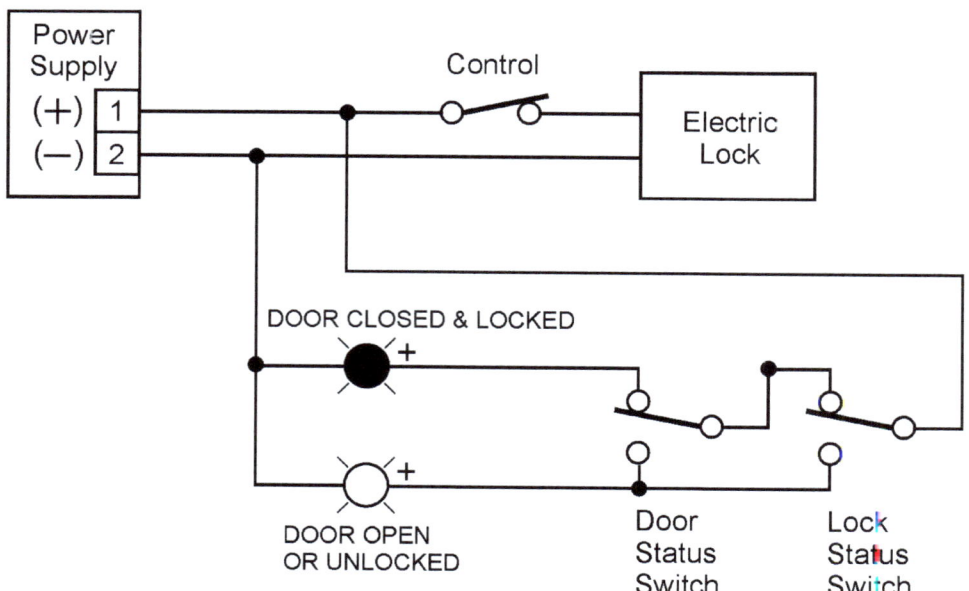

**Figure 7-7**   Typical high security monitoring diagram.

## Indicating Devices

We also need to look at the devices that are used to *indicate* the conditions being monitored. These normally are audible or visual devices, and we already know something about these devices. We have seen the symbols used for them, and we know they are "loads." They need power.

Indicator lights are available in a wide assortment of sizes and colors. The most commonly used indicator is the *light-emitting diode* (LED). It usually ranges from $1/8$ to $1/2$ inch in diameter, is available in a good range of low voltages, and draws very little current.

There are a wide range of colors available, but normally red, green, and amber satisfy most needs in security monitoring. Believe it or not, there can be more discussion over color selection than overdesigning the entire system. Should green mean secure? Or should red mean secure? There is no real standard, and normally it is left to the owner's preference.

Audibles are selected on the basis of the sound level desired. They are available from low level irritants up to levels that would drive a person from a room. Sound level is measured in decibels (abbreviated as dB) and qualified by the distance from the device at which it can be heard. As an example, a remote audible at a guard's desk might be rated at 80 decibels at 3 feet. This type of device is loud enough to alert a person nearby that a problem has occurred and needs a response.

Audibles are available in many voltages but are more apt to be selected to match the system voltage, usually 12 or 24 volts. Some audibles are operated at a specific voltage; others can accept a range of voltage, thereby offering a range of decibel levels.

Audibles and indicators are offered by security product manufacturers. These products range from LED indicators and audibles on single-gang wall plates up to highly sophisticated control and monitoring consoles. The important thing is to determine what you think is required and then check what is available. Many times, with a little compromising, you can find a standard product that will fit your requirements.

## Summary

As stated earlier, monitoring is not always required, but it is important enough to be discussed when designing a security system. Each system should be reviewed as to what conditions are desirable to be monitored. Once those conditions are identified, the monitoring devices can be selected along with the indicating devices, such as lights and audible devices. It is also important to cover how the monitoring system should operate. For example, should alarm conditions be manually reset? Is an alarm condition override needed?

Following this summary is a *monitoring system design worksheet*. It should help you determine what you need for the monitoring part of a system. This worksheet should be used along with the basic system questionnaire (Figure 5-2).

# MONITORING SYSTEM DESIGN WORKSHEET

| Condition To Be Monitored | Indicator Light | | | | | | Audible Device | | | | | |
|---|---|---|---|---|---|---|---|---|---|---|---|---|
| | Color | Voltage | Steady | Flash | Latching | How Reset | dB/ft | Voltage | Steady | Pulse | Latching | How Reset |
| ☐ Power On | | | | | | | | | | | | |
| ☐ Power Off | | | | | | | | | | | | |
| ☐ Door Open | | | | | | | | | | | | |
| ☐ Door Closed | | | | | | | | | | | | |
| ☐ Lock Secure | | | | | | | | | | | | |
| ☐ Lock Not Secure | | | | | | | | | | | | |
| ☐ Door Closed and Secure | | | | | | | | | | | | |
| ☐ Door Closed and Not Secure | | | | | | | | | | | | |
| ☐ Door Open or Not Secure | | | | | | | | | | | | |

Comments/Description of Operation:

**Test Questions for Chapters 6 and 7**

1. A device used to house electrical connections and contain sparks and heat is called a _____.

2. After installation the wiring diagram should be marked with
   *a.* Wire size
   *b.* Wire color
   *c.* Wire length
   *d.* All the above

3. One purpose of fire panel interface is to
   *a.* Control specific building equipment
   *b.* Lock stairwell doors
   *c.* Keep elevators operational

4. Every security system drawing should show an interface to a fire panel, true (T) or false (F)?

5. An override is provided to
   *a.* Override the fire panel
   *b.* Override or change a specific function
   *c.* Override the authority having jurisdiction

6. Two important conditions that should be monitored are_____ and _____.

7. Monitoring devices can provide _____ or _____ indication of a condition.

8. Door status switches are commonly available built into locking devices, true or false?

9. Monitoring devices can also be used to create_____between multiple doors.

10. Spell out the following abbreviations.

    DSS   _____    _____    _____
    BPS   _____    _____    _____
    LCM   _____    _____    _____
    DPS   _____    _____    _____
    LED   _____    _____    _____
    dB    _____

# 8

# Relays

I have mentioned relays and relay contacts several times throughout earlier chapters. Many manufacturers offer built-in or optional plug-in relays with their security hardware. I believe that most of the time these features will satisfy your need for a relay within your system. However, knowing how a relay works will help you to make wiring diagrams of a more complex nature.

The main purpose of this chapter is to enable you to understand the use of relays in security systems design. At the very least you may have to draw relay contacts and relay coils on your wiring diagram. The relay contacts will turn electricity on and off to certain loads. They will do this only when some device in your circuit turns electricity on or off to the relay coil. The best way to start is to learn a little about the mechanics of a relay. There are entire books written on relays. In this chapter I will provide a brief review of relays and dwell only on what we'll call the *general-purpose* relay. This is the type of relay that you are most likely to use in security work. By the time you finish this chapter, you will be able to incorporate relay "logic" into your circuit design and add the proper information to your drawing.

## Introduction to Relays

The relay is one of the simplest, yet most useful devices used in electronic security work. Small size and low cost belie its reliability and versatility in the great variety of jobs it performs.

I like to define a relay simply as an electrically controlled switch with one or more contact sets. In more technical terms, it can be described as an electromagnetic switch that opens and closes one or more contact sets to effect the operation of other electrical devices.

The relay can be traced back to 1837, when Samuel Morse made his first working Morse code machine. These early devices were used to extend the range of telegraph signals.

It was also early in the nineteenth century that relays were used in alarm control panels. They provided some of these functions:

- Multiple switching functions
- Separation of control and power loads
- Signal amplification
- Separation of direct-current (DC) and alternating-current (AC) circuits
- Delay of applied signals

In the 1970s solid-state electronics began to take over some of these functions in alarm panel design. These electronic components provided miniaturization and a low-cost method of assembling mass-produced electronic products.

But relay technology also made great advancements over these years and still provides many of the same functions. Modern relays are used extensively for input and output signals between alarm and monitoring panels and external system components. Today there are literally billions of relays in use in electronic equipment.

Without further belaboring the history of relays, we note that relays of all types are available for equipment of all types. Very large relays with high-power consumption are used in high-power switching equipment. Smaller, lower-power consuming relays are found in a multitude of equipment—automotive, electrical appliances, and electronic security equipment. These smaller general-purpose relays offer a great variety of functions, good signal and power switching capabilities, moderate power consumption, and economy of cost. It is this type of relay that we cover in this chapter.

Several other types of relays that may show up in security work will be briefly mentioned. One of these is the solid-state printed-circuit (PC) board mounted relay. Although its functions are pretty much the same as those of the relays we will study, it is unlikely you will design or service solid-state circuits in ordinary security work. Rather, you will be drawing system wiring to and from these relay inputs and outputs.

## How a Relay Works

One of the hardest teaching tasks has been to explain what a relay is, especially to students who never have used or even seen a relay. It is my opinion that these students start off by imagining something much more complex than the real simplicity of a relay. I will further qualify that by noting that a student should not be attempting to study relays without being familiar with switch contacts.

Not knowing where each reader of this chapter stands as far as knowledge of relays is concerned, I will start as simply as possible. If the reader needs to, she or he should first review Chapters 2 and 4 for an understanding of switches and contacts and how they work.

A relay is really nothing much more than a switch that is operated by electricity, rather than by someone pushing a button, flipping a toggle, or turning a knob.

Figure 8-1 shows a simple illustration of how an electromagnetic relay works. The relay coil is simply a small coil of wire with a fixed metal core. When the control switch is closed, power flows through the relay coil, creating a magnetic field. The metal core becomes magnetized, attracting one end of the metal armature. As the armature end is pulled toward the core, the other end pivots upward, sort of like a seesaw. That end is mechanically attached to the swinger of the relay contact set, called the *common* of the contact set. As the common is pulled upward, the *normally closed* contact opens and the *normally open* contact closes. Just as in a switch that has been actuated, a set of contacts has been influenced to change state.

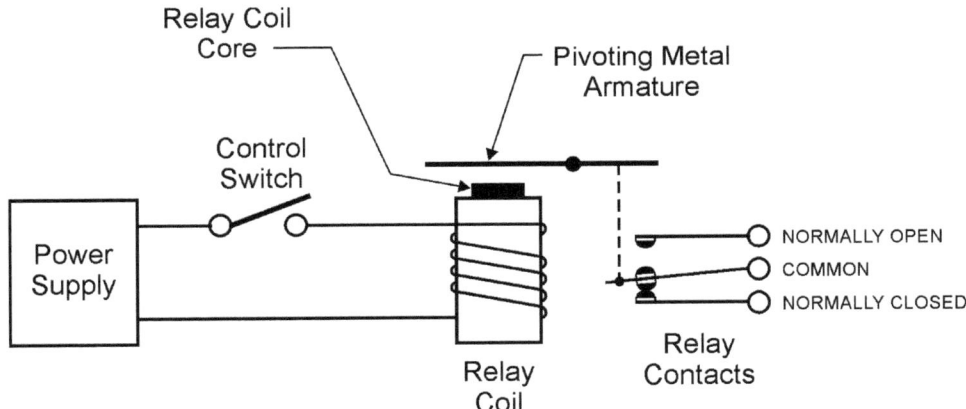

**Figure 8-1** Outline of an electromagnetic relay operation.

Now it may seem silly to use a switch contact just to activate another switch contact. But one good reason might be a control circuit using different power from the power that the relay contacts will be controlling. For instance, the control circuit might be 12 VDC, and the power that the relay contacts will be controlling might be 120 VAC! A simplified example is shown in Figure 8-2.

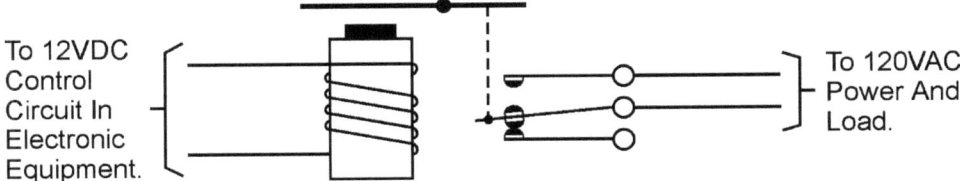

**Figure 8-2** Example of relay contacts controlling power different than the relay control circuit.

But the real benefit of using a relay is more often the use of multiple contacts. In Figure 8-3 we have added two more sets of contacts. The blades of each common contact are all mechanically connected to one another. Now, by using a single control switch, we have the ability to switch three other circuits on or off. Each set of relay contacts is like a single switch, except all three switches are being operated simultaneously. This is one of the prime reasons why relays are used—to provide multiple, simultaneous switching of individual circuits.

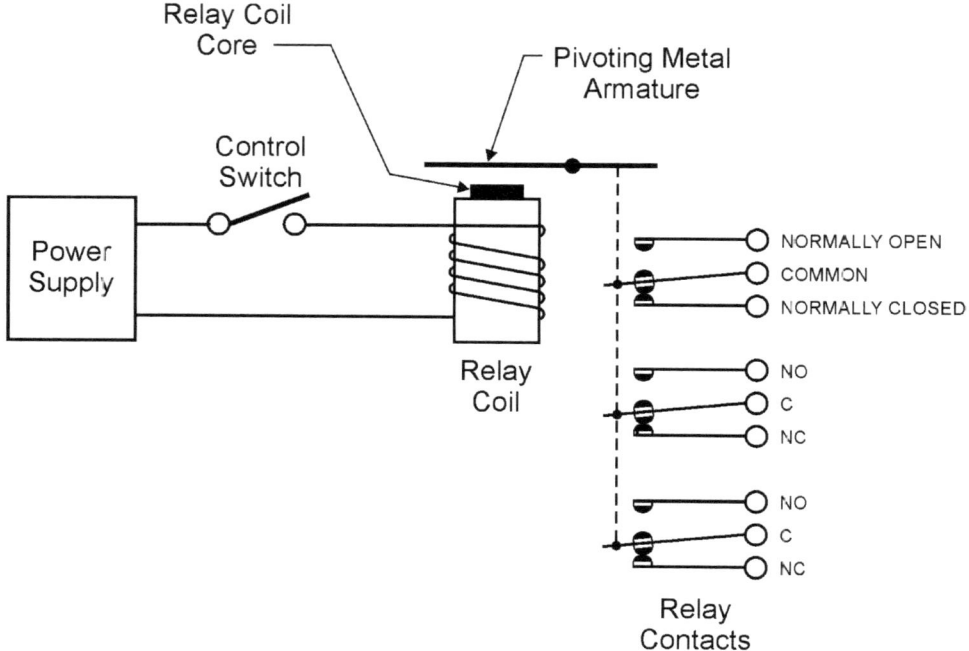

**Figure 8-3**  Outline of multiple contact set relay operation.

This explanation is pretty simple in order to give you a basic understanding of how a relay works. The relay itself is a little more complex, and we should take a brief look at how one is put together and operates.

## Anatomy of a Relay

As mentioned earlier, a relay commonly used in security and alarm work is often referred to as a general-purpose relay. This type of relay is available in many styles of mounting and contact arrangements. A typical relay used in fieldwork is available with contact arrangements from single-pole, double-throw (SPDT) to four-pole, double-throw (4PDT). Contact ratings run from 3 to 10 amperes. A common relay usually is about $1\frac{1}{2}$ inches $\times$ 1 inch $\times$ $\frac{3}{4}$ inch with a clear plastic dust cover. They are commonly flange or socket mount with solder terminals. One popular model plugs into a plastic base with screw terminals for hookup wiring. A typical general-purpose relay is shown in Figure 8-4.

**Figure 8-4**  General-purpose relay with plastic dust cover.

The type of relay we are studying falls into the classification of an electromagnetic relay. This type of relay is distinguished by having a moving armature that is usually hinged or pivoted on the relay frame. Figure 8-5 shows a detailed description of the components of an electromagnetic relay.

The heart of the relay is similar to a solenoid with a fixed core. A coil of wire (1) surrounds a metal core (2). Electric power causes the wire coil to create magnetic flux, magnetizing the metal core. The magnetized core attracts the metal armature (4), putting tension on the armature return spring (5).

The contact swinger (common contact) (6) is attached to the relay armature. As the armature is attracted to the core, the common contact swinger moves, closing the open contact (7) and opening the closed contact (8).

When input power is shut off, the core becomes demagnetized, releasing the armature. The armature return spring pulls the armature back to its normal position. The contacts return to their normal position.

It is important that you understand how a relay works before you add relay contacts to a drawing. As I hope you can see, a relay is simply a multiple-contact switch. You may eventually use or come across other types of relays, say, time-delayed relays and solid-state relays mounted on PC boards. Whatever the type of relay, the functional principle is the same, and the drawing symbols are the same.

## Relay Symbols

We briefly covered relay symbols in Chapter 2 but will expand upon that in this section. As we will see, relay contact symbols are similar to switch contact symbols.

On occasion you may need to add a relay to a circuit, as opposed to using a relay built into the equipment. In this case you will need a symbol for the relay

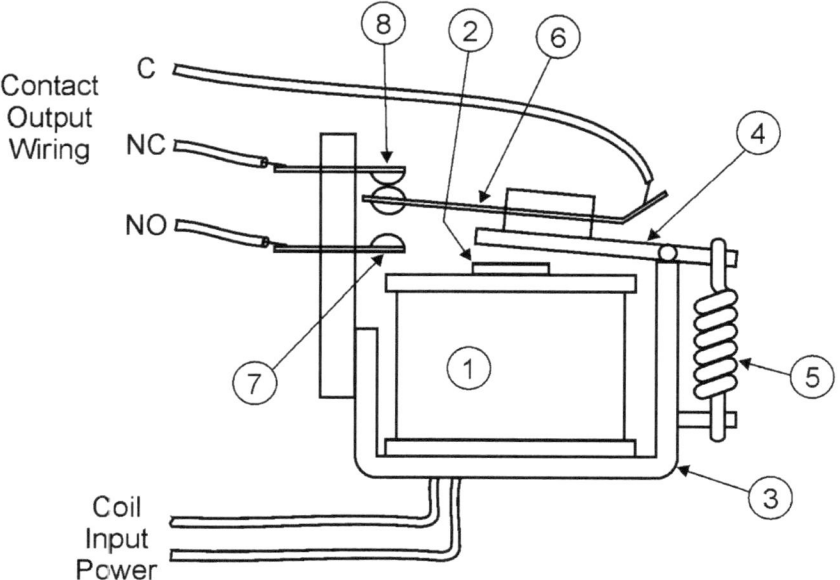

Contact
Output
Wiring

Coil
Input
Power

(1) Coil
(2) Core
(3) Iron Frame
(4) Armature

(5) Return Spring
(6) Contact Swinger (Common Contact)
(7) NO Contact
(8) NC Contact

**Figure 8-5**  Components and operation of a typical relay.

coil. The relay coil needs to be wired to power that is turned on and off to trigger the relay contacts. Figure 8-6 shows the symbol for a relay coil.

Relays are usually identified as CR1, CR2, CR3, etc. Most relays have their hookup terminals identified by numbers embossed or silk-screened on the base of the relay. A typical relay contact set symbol is shown in Figure 8-7.

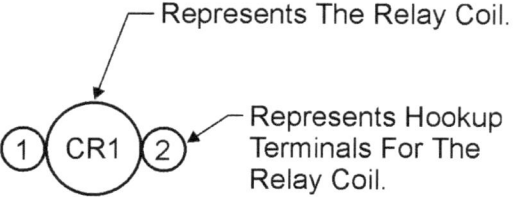

Represents The Relay Coil.

Represents Hookup
Terminals For The
Relay Coil.

**Figure 8-6**  Relay coil symbol.

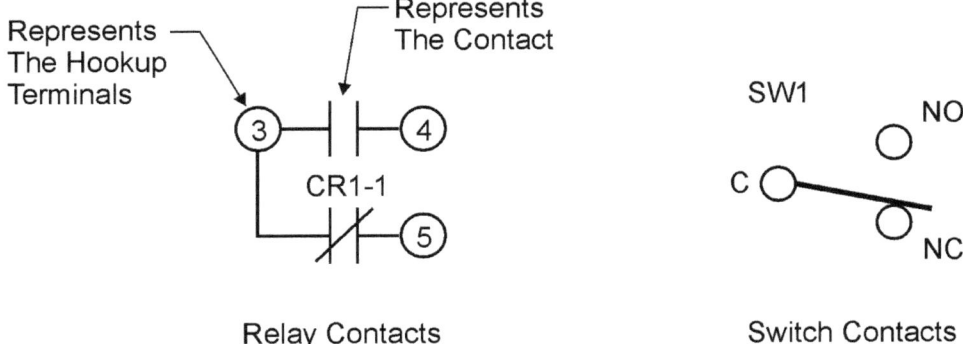

Represents The Hookup Terminals

Represents The Contact

CR1-1

SW1

Relay Contacts

Switch Contacts

**Figure 8-7** Comparison of relay and switch contact symbols.

The relay contact symbol is a little different from a switch contact symbol but is imparting the same information. The symbol shows an open contact and a closed contact with hookup terminal identifications. Compared to the switch contact, terminal 3 would be the common (C), terminal 4 the normally open (NO) contact, and terminal 5 the normally closed (NC) contact. There may be multiple relays in a system. Relay contact sets are identified as to what relay they belong to, for example, CR1-1, CR1-2, CR2-1, and CR2-2. These symbols are normally used in designing a wiring diagram for a particular piece of electronic equipment.

Figure 8-8 shows several relay contact symbols and comparable switch contact symbols.

Often a system will include relays that are factory built in to various system equipment. In these cases the relay coil is commonly energized and deenergized by internal electronic components. The internal components are triggered by some external action, for example, entering a valid card in a reader. In this case the manufacturer will identify only the contacts of the relay. Your drawing will show how these contacts are wired into the system to provide specific functions. Manufacturers usually identify the relay output contacts as shown in Figure 8-9.

## Simple Relay Circuits

Let's look at a basic use of relay symbols. Figure 8-10 shows how a relay can interface a power source from one system to control a power source from a different system.

A call system puts out constant 16 VAC power, energizing the relay coil. The NO relay contact closes, allowing 24 VDC power to flow to the electric lock. When the call system is operated, for example, an entry pushbutton, the 16 VAC is killed, the relay deenergizes, the NO contact opens, and the lock releases. Figure 8-10 is more of a pictorial drawing showing this system. The way it would appear on a true wiring diagram is shown in Figure 8-11.

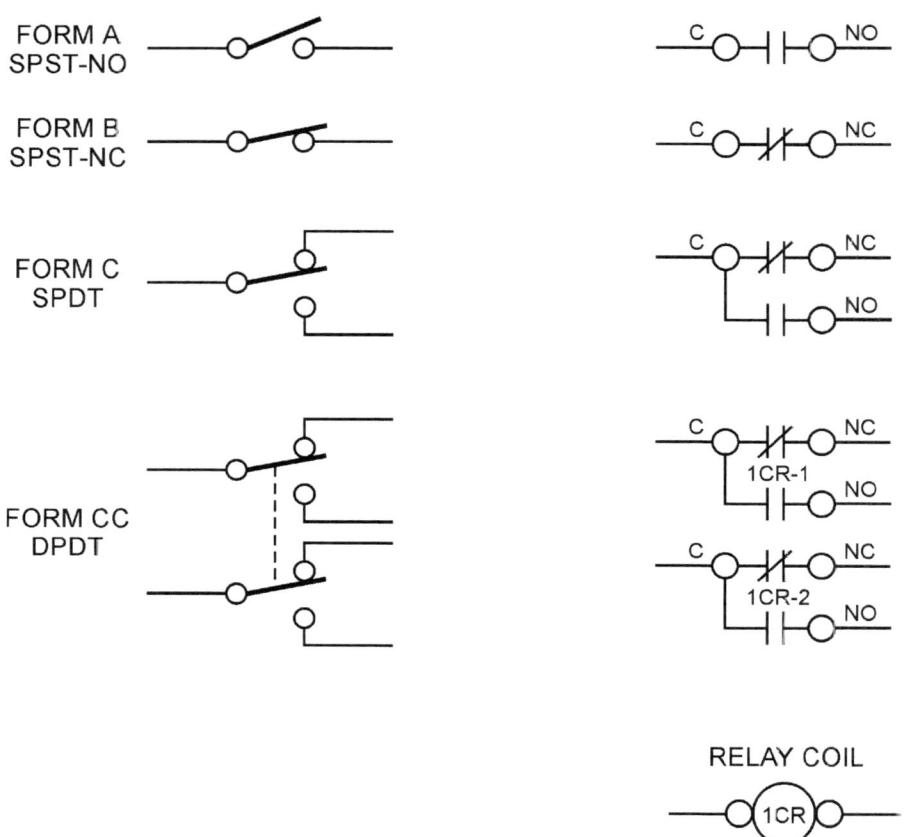

Figure 8-8   Contact forms and symbols.

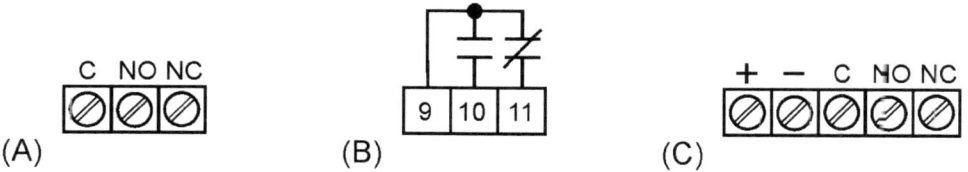

**Figure 8-9**   Symbols for built-in relays. A and B show methods of identifying relay output contacts. C shows identification for a built-in relay that requires external wiring to provide power to the relay coil; $(+)(-)$ terminals.

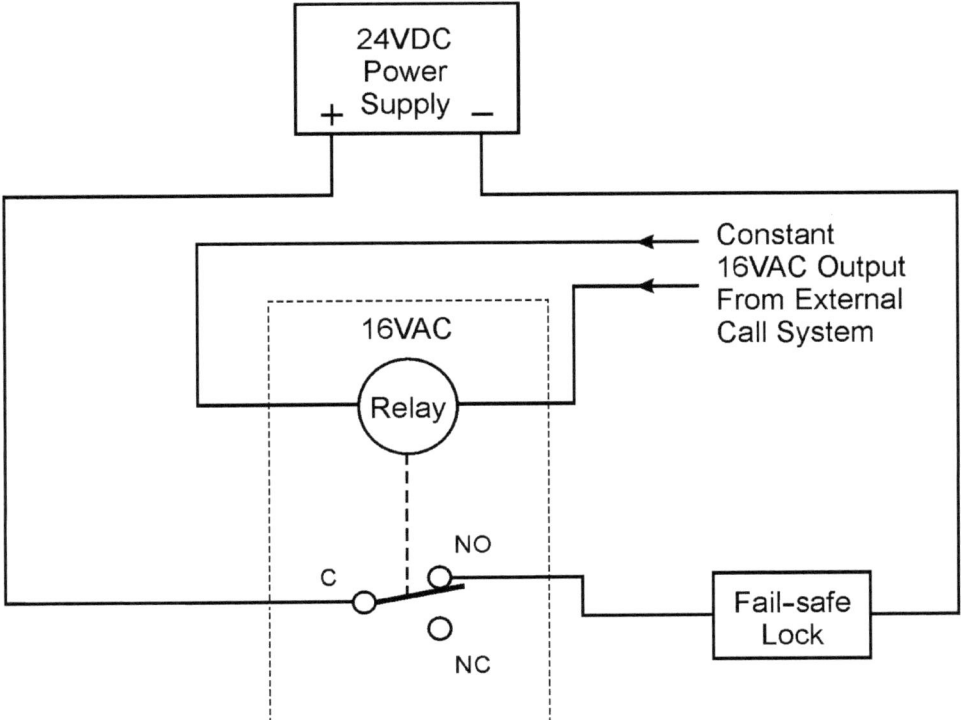

**Figure 8-10** Use of a relay to control an electric lock.

Either of the methods shown in Figure 8-11 would be acceptable for a wiring diagram. Figure 8-11A is used more often on a schematic-type drawing. Figure 8-11B is more common when using a purchased relay module. You should get used to the method shown in Figure 8-11B. Relay output contacts are usually shown this way when the relay is built into electronic equipment.

By now you should have a pretty good understanding of how a relay is used in a system. Most of the time you will find that relays are built into electronic security equipment. In some cases you will only need to show wiring to the relay output contacts. In other cases you may also have to show power wired to the relay coil. We will cover this in greater depth in Chapter 9.

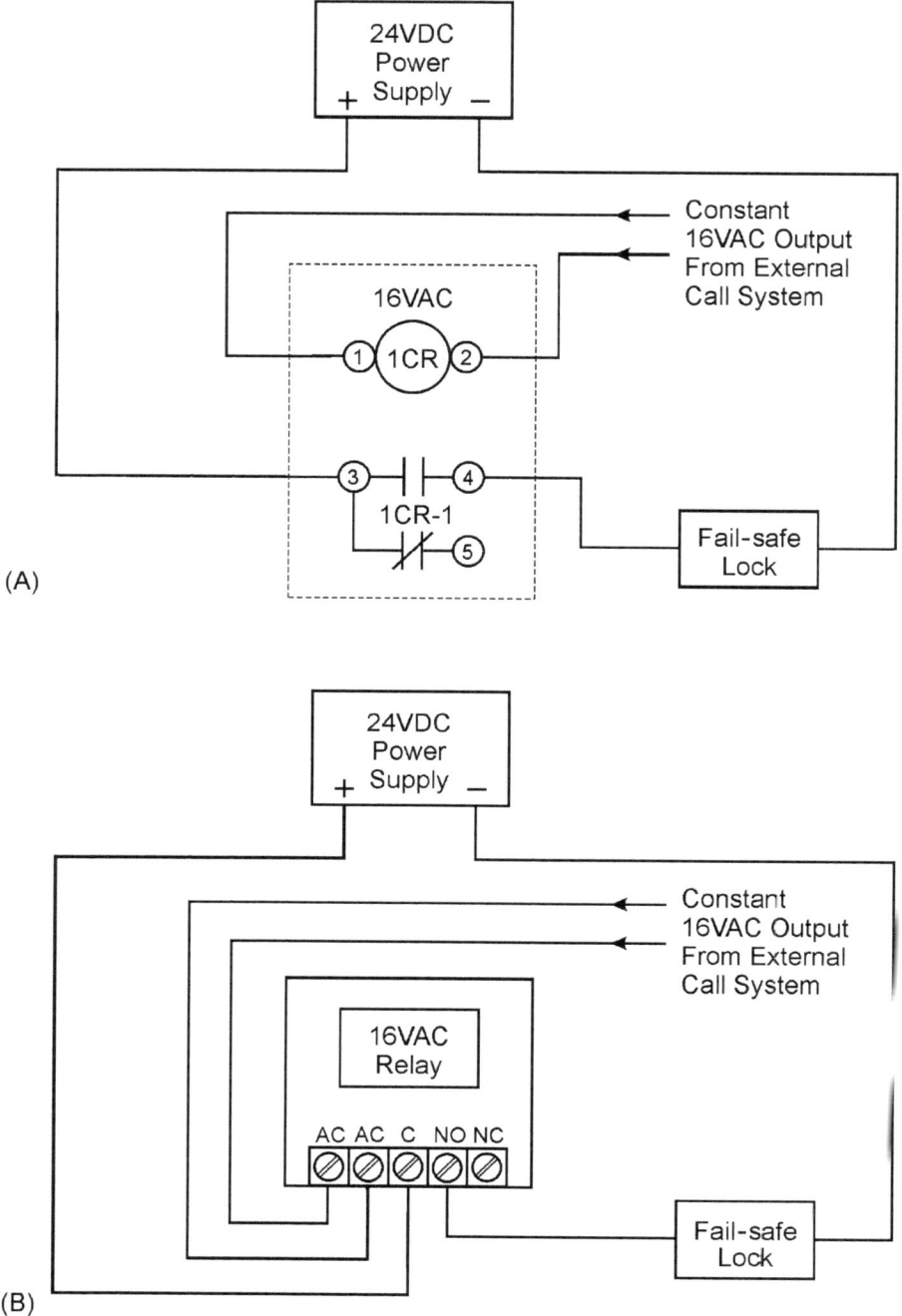

**Figure 8-11**  Use of a relay to control an electric lock.

# Advanced Wiring Diagrams

In the last few chapters we have learned to run power to electric locks and indicator lights. We have also seen how to control power with contacts from switches and relays. In this chapter we will take a look at other things that need power. A high percentage of other things you will be confronted with will be access and egress devices.

Up to now, most of the access/egress controls we covered were simple switch contacts in devices operated by a simple manual turning or pushing operation. But more often you will be working with devices such as card readers, keypads, motion detectors, and other more sophisticated control devices.

I would say that next to mastering simple wiring diagrams, adding these types of devices to the diagram gives the student the most trouble. The following section should make these devices more understandable. This gain in knowledge will enable you to design more sophisticated systems.

## Things That Need Power

Other than the most basic of systems, a good locking system will include monitoring devices and control equipment a little more sophisticated than the switches we used in earlier examples. All this type of equipment is an additional load in the circuit, and the devices will require their own input power to be functional.

Included in this category of equipment are

- Egress sensor devices, for example, PIR, motion detector
- Keypad controllers
- Card and proximity readers
- Biometric access devices
- Monitor/control panels

So, rule number 1 in placing one of these items in your wiring diagram: *Directly wire it for input power.* This can be a simple generic representation, as shown in Figure 9-1.

## Access/Egress
## Equipment

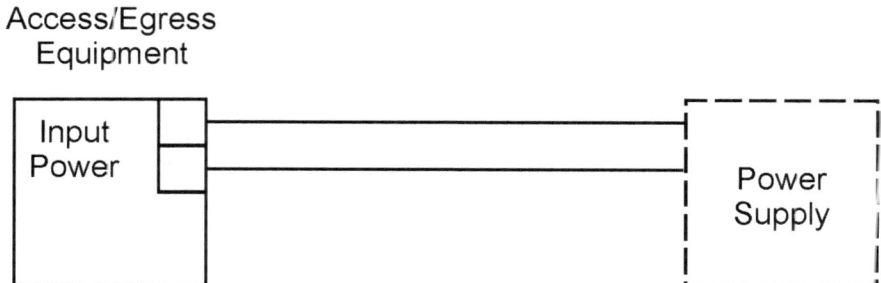

**Figure 9-1**   Input power hook-up to access/egress equipment.

Most of the time, input power will be two wires carrying low voltage, alternating current or direct current. Occasionally it will be three wires carrying 120 VAC. At any rate, start your drawing as shown in Figure 9-1. Normally there will be no break in the input power line, no on/off switch. This is common for low-voltage devices. Devices operated by 120 VAC may include an on/off switch for use during maintenance.

Also, by using manufacturers' technical literature, you can increase the information shown on your diagram. Figure 9-2 shows several typical scenarios for designating input power hookup.

In Figure 9-2A the equipment may be a card reader or large monitoring and control panel operated on high voltage. In all probability this type of equipment has a built-in transformer that internally reduces the 120 VAC to a lower voltage. Not to worry! All you show on the drawing is three wires, or even a single line, and indicate 120 VAC (by others). Usually the high-voltage electricians take care of this.

Figure 9-2B and C shows devices powered by low voltage. The manufacturers' literature should indicate the input voltage and current required and the terminal identifications for hooking up the input power lines. In Figure 9-2B note that DC input power terminations may be marked with polarity symbols (+, −). In some cases, where polarity need not be observed, the manufacturer may not mark the input terminations with the (+) or (−) symbol.

When using low-voltage input power, you may need to use a separate power supply dedicated to just this piece of equipment. If the input voltage matches that of other equipment, for example, the locking device, you may tap off of a single power supply to feed other devices in the system. Be sure the power supply has the current capacity to handle the total of every load you run off of it.

The point I am trying to make here is that you will always find some kind of "input power terminations" on any of these devices that need their own power.

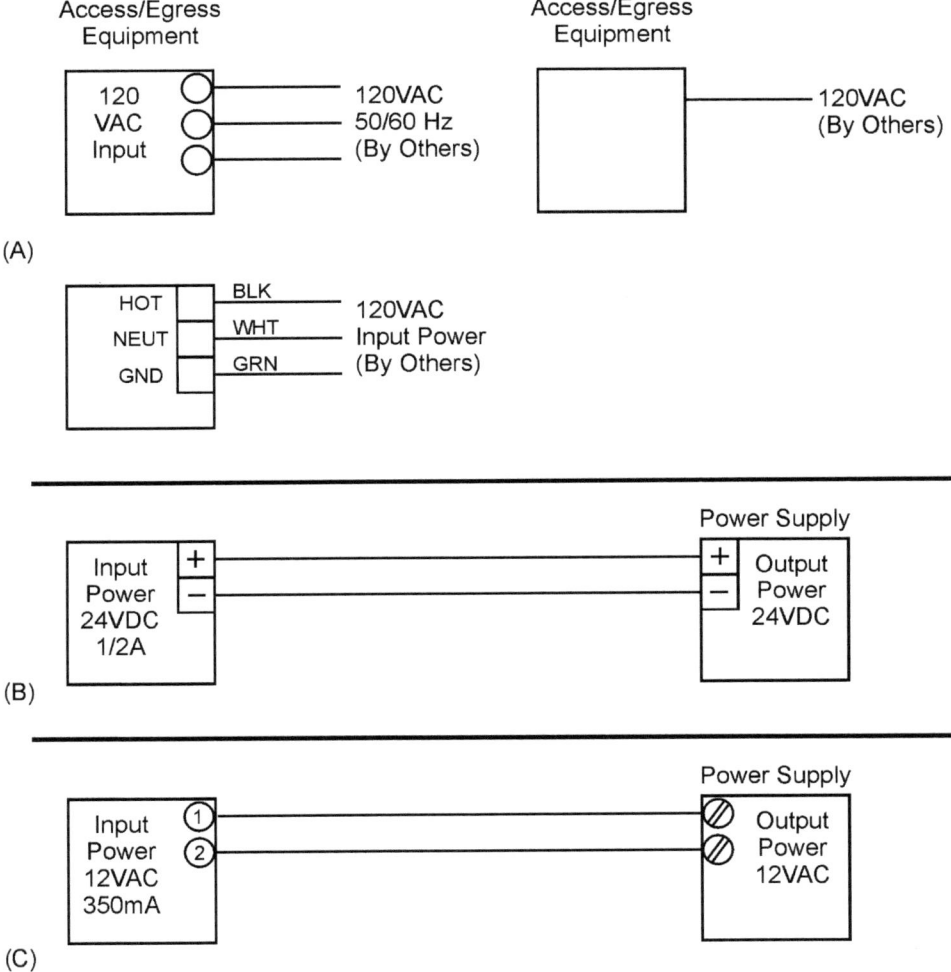

**Figure 9-2**   Common input power designations.

Identify them on your drawing, and run two wires from your power supply *output* terminations.

Now, we are not through yet. All these devices may have *other* input terminals, and most likely *output* terminals. Your job is to identify the inputs and outputs that you will be using for your system and to draw the wiring on your diagram. This is not as hard as it seems; it just requires a little logical thinking. Normally the inputs and outputs are labeled, indicating their function. You may also see areas on printed circuit boards, or separate boards, marked I/O. This is simply the designation for inputs/outputs. The hardest part is usually trying to read the manufacturers' literature! Let's look at some typical examples.

## Inputs and Outputs

Suppose your system requires a card reader as an access device. Up to now I have noted that a load in a system requires power, but a control switch does *not* need power. Well, a card reader can be considered a load and a switch. Inside a card reader there will be some sort of "switch" contacts, most likely relay contacts. These are what you use to control power to your system load, for example, the electric lock. But the other components in a card reader, for example, relay coils, memory circuits, and timer circuits, all need their own power for correct card reader operation.

For that reason, as we just covered, a card reader will always have at least two wires entering it for input power.

Now comes the tricky part. Some card readers use built-in "switches," usually relay contacts, to internally turn *output power* on and off. Figure 9-3 shows output terminals of this type.

**Figure 9-3**  Power output terminals.

This is typical of card reader/electric strike systems. The input power runs the "internals" of the card reader, and some of this power is fed back out to run the electric strike.

The output power is commonly 12 or 24 volts, AC or DC, and you will have to choose a reader with the proper output power for your system. There are also two choices of operating *modes* when using a power output. The output power can be normally on, typically powering a fail-safe electric strike. The voltage is normally DC and is run directly to the strike. When a valid card is used, the card reader output power is internally turned off, releasing the lock.

In the other mode, output power is normally off, typically powering a fail-secure electric strike. The voltage is commonly AC and is run directly to the strike. When a valid card is used, the card reader output power is internally turned on, releasing the lock. In either case a typical wiring diagram for this basic system might be as simple as that shown in Figure 9-4.

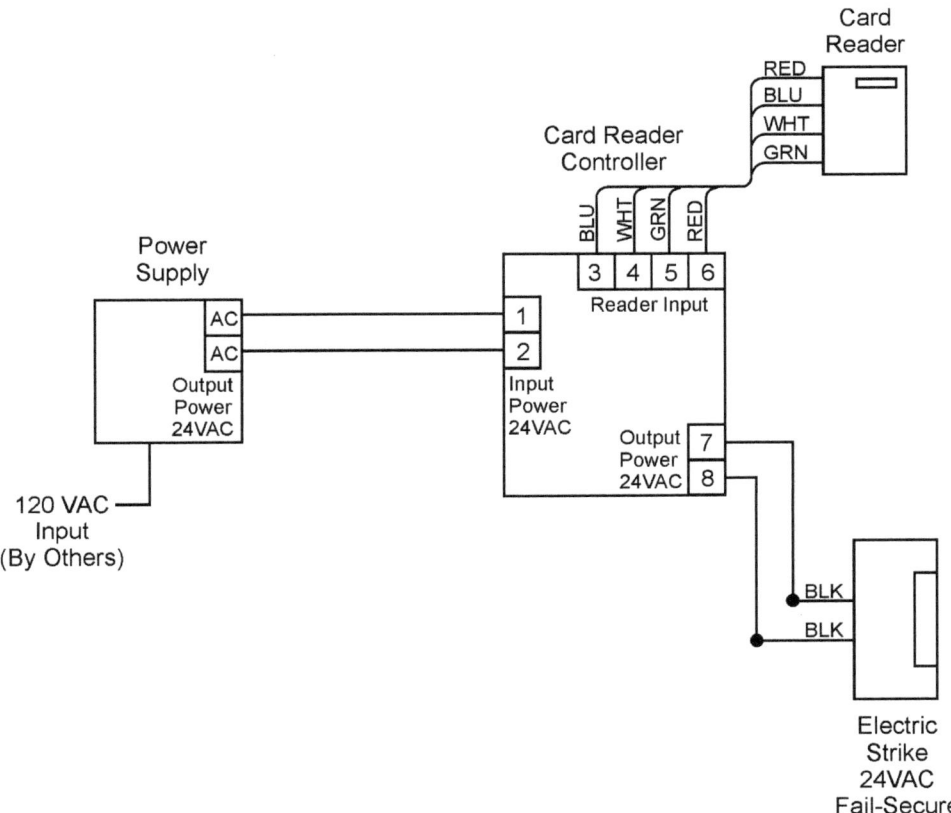

**Figure 9-4**  Basic system wiring diagram.

In this diagram the power supply sends low-voltage power directly to the card reader controller. Inserting a valid card in the card reader trips an internal relay in the controller. The relay allows low voltage to pass through terminals 7 and 8 of the controller, out to the electric strike, releasing the strike keeper.

Another common output for card readers is called *dry contacts*. These are just the relay contacts by themselves; *dry* means that the contacts by themselves are not putting out power. The contacts are used to control whatever power you run through them. In this case you will need a power source for the card reader *and* for the electric lock you are operating. Figure 9-5 shows a wiring diagram for this type of setup.

In this diagram the power for the electric lock is being controlled by the card reader output contacts. One "leg" of the power is run into the common (C) of the relay contacts and comes out of the normally closed (NC) side of the contacts. If you could look inside the card reader controller, it would look something like the diagram in Figure 9-6.

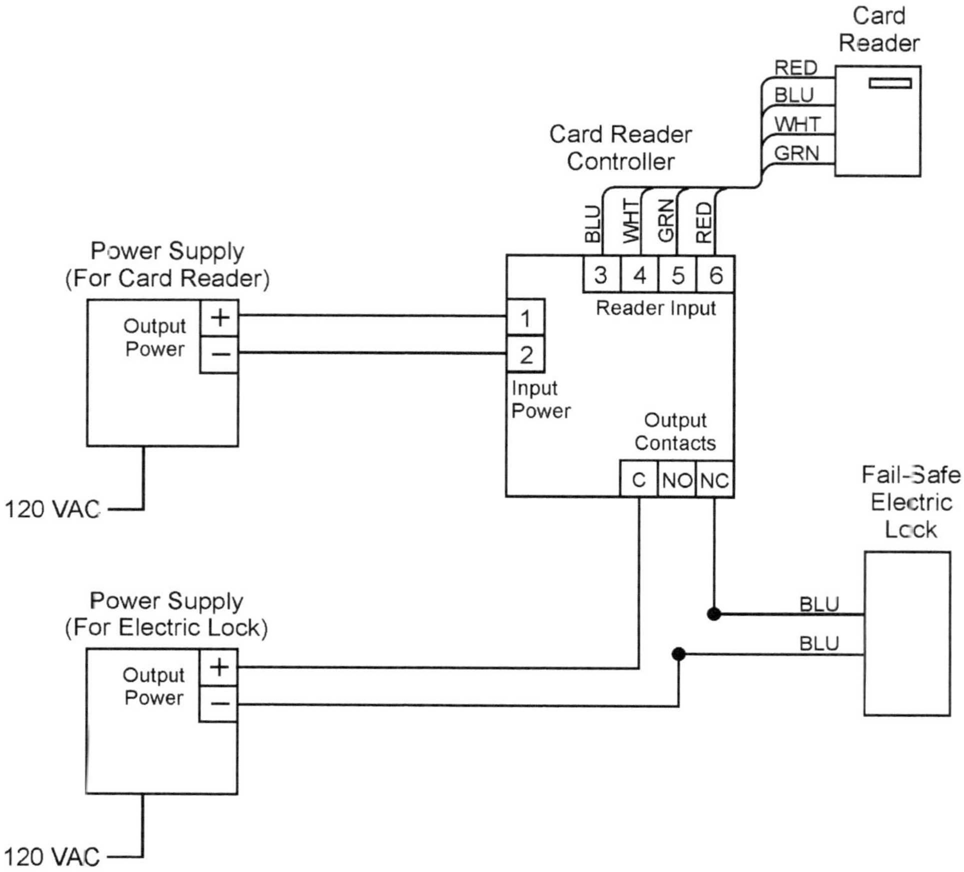

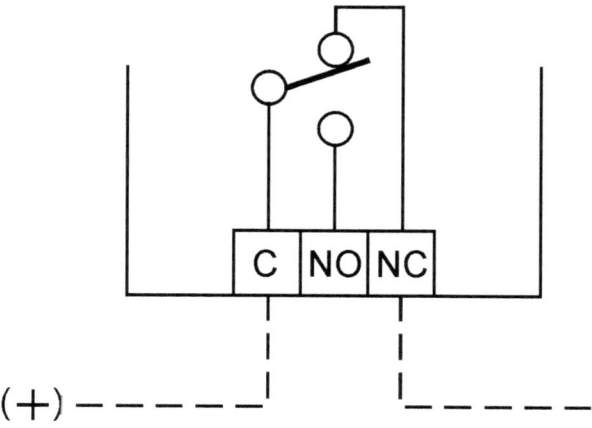

**Figure 9-5** Typical wiring diagram. Dry contact output controlling the lock. (Closed circuit)

**Figure 9-6** Dry contact "switch."

You can see that power is running through the internal relay contacts and out to the electric lock. When a valid card is inserted, the relay is triggered and the closed contacts open. This cuts off power to the lock, releasing the lock.

The point we are making here is that when you see outputs on devices such as card readers labeled C, NO, and NC, think of a switch. Some devices will only have labeling as shown in Figure 9-7A. Other devices may actually show a little diagram right on the device circuit board, as seen in Figure 9-7B and C.

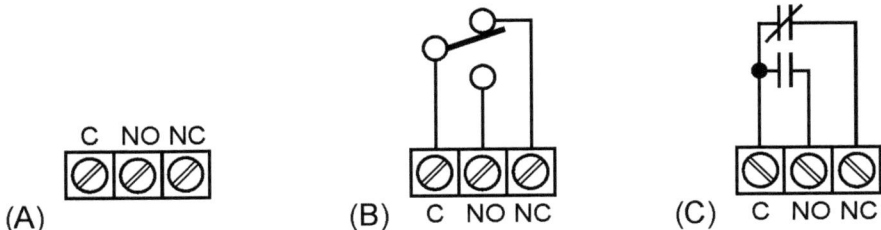

**Figure 9-7**   Common dry contact output designations.

Figure 9-5 showed the use of closed dry contacts allowing power to flow to a fail-safe lock until the contacts open. Figure 9-8 shows how to wire the normally open contacts to a fail-secure locking device.

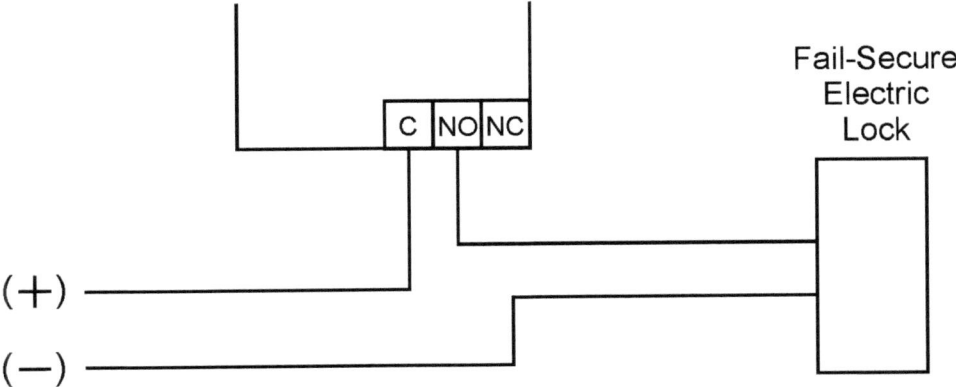

**Figure 9-8**   Typical wiring diagram. Dry contact output controlling the lock (open circuit).

In this diagram the power for the electric lock is being controlled by the open contact set of the card reader.

The lock is normally secure without power applied. Insertion of a valid card triggers the relay, closing the open contacts and allowing power to flow to the lock to release it.

In Figure 9-5 we also showed two separate power supplies. This design is used when the loads (card reader and electric lock) require different types of power. For example, the card reader may operate on 24 volts DC and the electric lock on 12 volts AC.

Usually the system designer will try to select products that all operate on the same type of voltage. This arrangement will allow a single power supply to be used to power both loads, as shown in Figure 9-9.

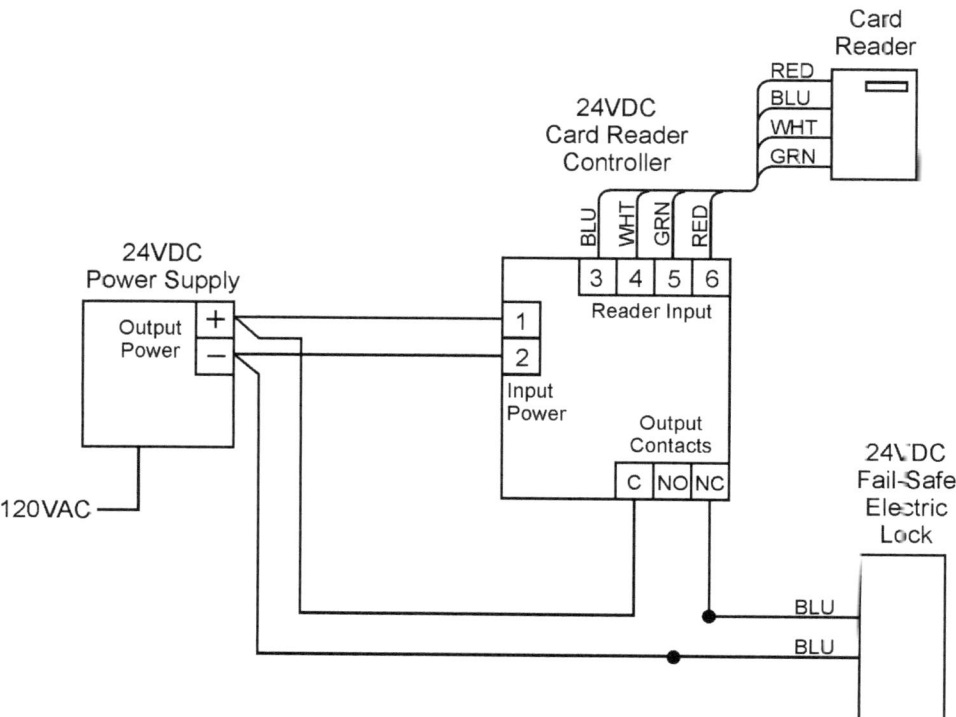

**Figure 9-9**  System with multiple loads and single power supply.

One thing to remember in this design is to specify a power supply that has enough current capacity to handle the total current draw of all the loads it is powering.

The wiring diagram shown in Figure 9-9 is typical of a small single door system. An installer could easily wire up this system by following the information on the drawing. When I am designing a new system, I sometimes begin by laying out the components in a different manner than usual. It is a method used mostly by electrical equipment designers, but I find it useful in trying to "see" how a system works.

This type of drawing could be referred to as a *logic* diagram. It helps lay out the logic for creating all the proper switching sequences to make a system work.

## Designing a System

Let's suppose you are asked to provide a wiring diagram for a system described as follows:

> A single outswinging door is to be normally closed and secured by a fail-safe electromagnetic lock. Entry is to be by a digital keypad programmable for up to 40 individual users. Egress is to be allowed at all times by an interior *passive infrared* (PIR) detector. A red light located in an adjacent room is to illuminate any time the door is open, returning to off when the door is closed.

With this information we can start to design the system. As I noted earlier, I sometimes use a method of layout you will not normally use for field drawings. But this method sometimes helps in understanding how the system works. You can always convert it later to a traditional-type wiring diagram.

I simply start with two vertical lines, as shown in Figure 9-10. These represent the power supply output (there is no need to draw the whole power supply).

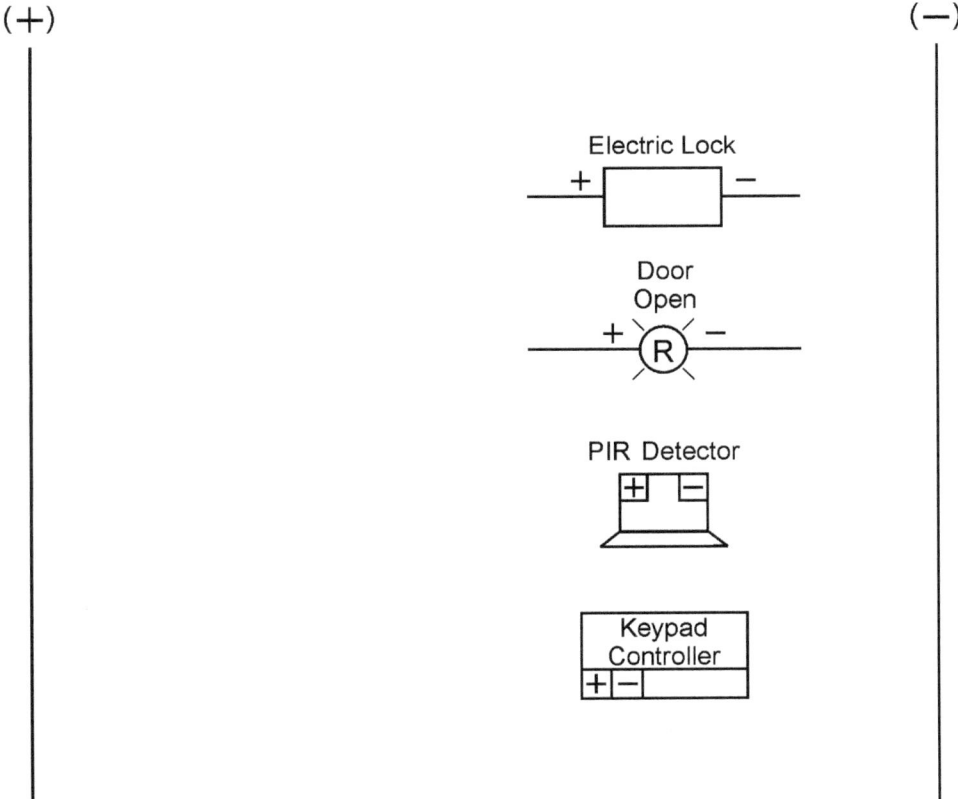

**Figure 9-10**  Component layout for a logic diagram.

I next draw in any component that I know needs power. I line these components up in a vertical column, usually closer to the (−) line. Note that I just
arbitrarily label the component input power terminations (+) and (−). At this
time I do not know what manufacturers' products I will be using or what their
terminal identifications will be. I am simply using generic symbols to design
the system.

My next step, shown in Figure 9-11, is to hook up (−) to all the loads. If
something has to be switched on and off, the convention is to switch the (+)
side. So I get all my (−) connections out of the way first.

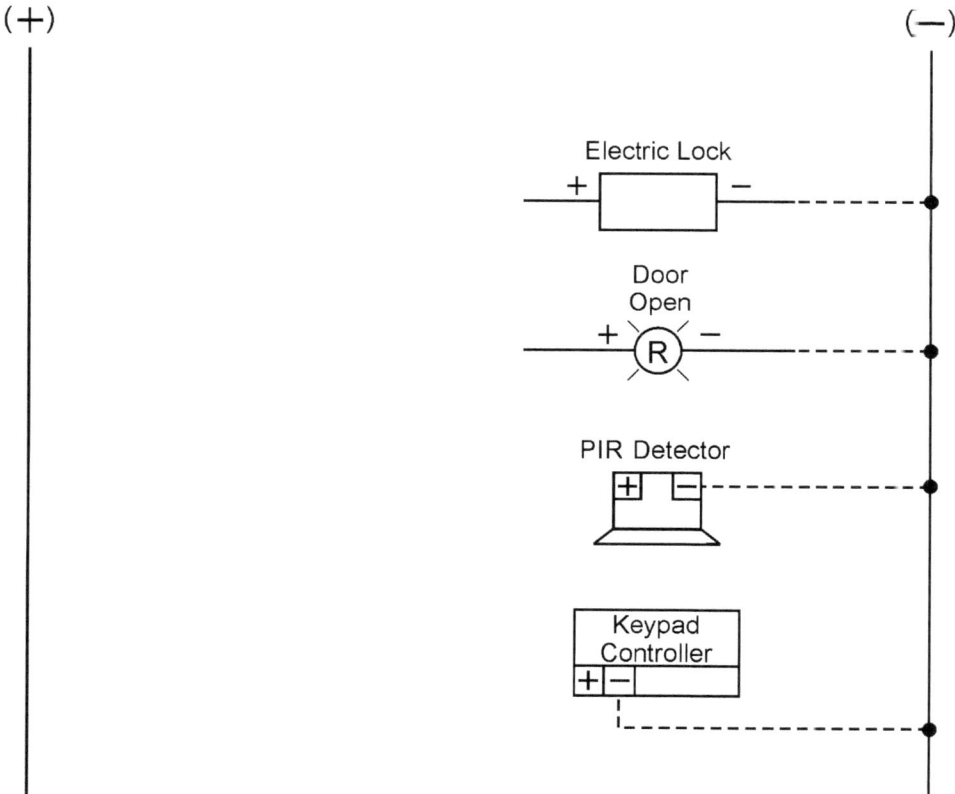

**Figure 9-11**   Hooking-up (−) connections to all loads.

I can also get any unswitched (+) lines out of the way, as shown in Figure
9-12. I know that the keypad controller and PIR units require constant power
to support their internal circuitry and output relays. Since there is no reason to
switch them on and off, I simply wire (+) directly to them.

With a few simple lines I have my wiring diagram half done! Now I have to do
some logical thinking. Here is where I decide how many contacts and switches

(+)                                                                              (−)

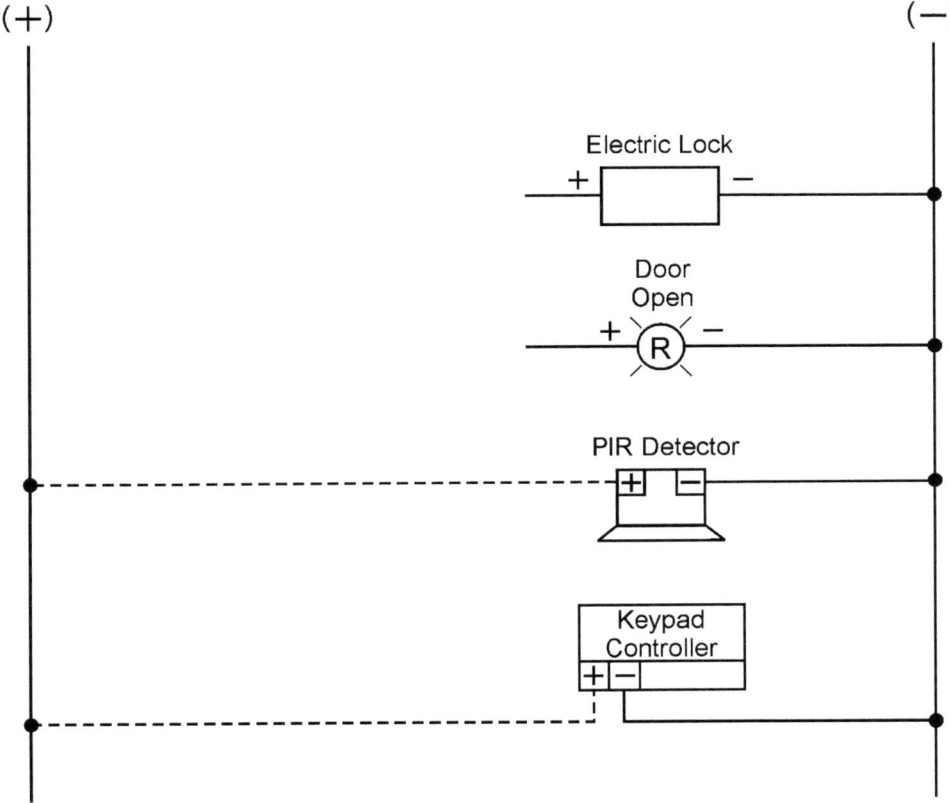

**Figure 9-12**  Hooking-up all unswitched (+) connections to loads.

are needed to do the job. I know that the magnetic lock has to be turned on and off by the keypad system and the PIR detector. (I don't draw in the actual keypad. The keypad controller is what usually provides the control contacts for the lock. The keypad is commonly a multiconductor cable running to terminals on the controller, and I'll show that on the finished field wiring diagram.)

The lock is a fail-safe device requiring power to be in a secure condition. This tells me that I need contacts that are closed when the system is powered and operational. Contacts in control devices such as PIR units and keypad systems are usually relay contacts. So in Figure 9-13 I add two closed contacts in series, using relay contact symbols. I then run the (+) line through these contacts to the electric lock.

Notice that even though the contacts are actually *in* the PIR detector and keypad controller, I draw them anywhere I want on the field of the drawing. This is the value of this type of drawing: You can draw contacts anywhere you need them. As long as you label them as to what piece of equipment they belong to, it doesn't matter where they are placed on the drawing.

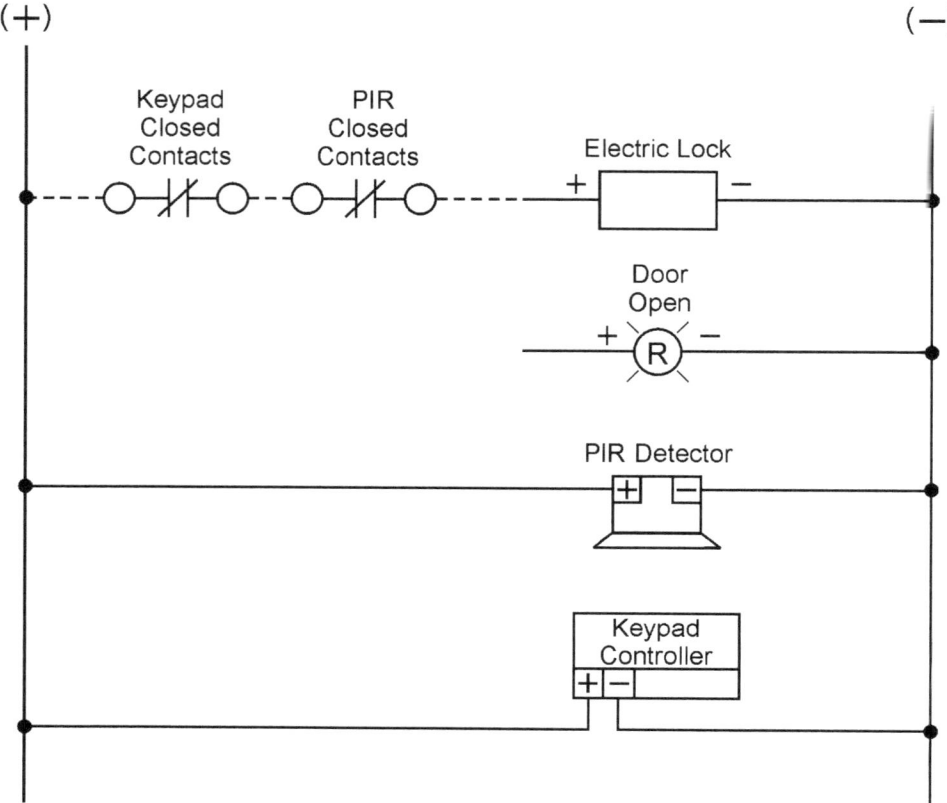

**Figure 9-13** Hooking-up switched (+) to the first load to be controlled.

So in Figure 9-13 we have provided a way to turn the lock off whenever the PIR device or keypad is activated. Opening either set of contacts releases the lock. Now we have to turn the indicator light on whenever the door opens. The logical control to do this would be a door status switch. In Figure 9-14 we add a door status switch to control the (+) line to the indicator. Whenever the door opens, this switch closes, sending (+) to the indicator.

We are in effect finished with designing the function of our system. We know we need a PIR device and keypad system, each with at least one set of normally closed output contacts. We know we will need a door status switch with a set of contacts that close when the door opens. With this information we can make our product selections and make a traditional wiring diagram.

## Making a Wiring Diagram

Using the system we just covered, start by simply making a list of the system components required to do this job.

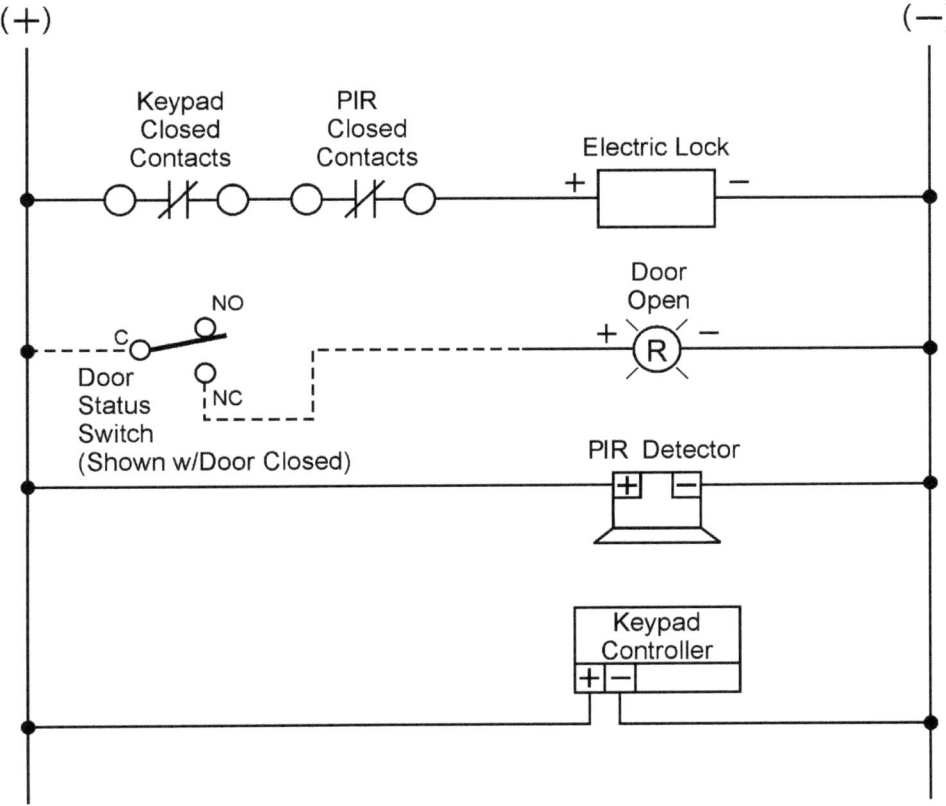

**Figure 9-14** Hooking-up switched (+) to the second load to be controlled.

System components list:

    Magnetic lock

    Keypad system

    Passive infrared detector

    Red indicator light

    Door status switch

    Power supply

At this point you will need to select the exact products you will use to make the true point-to-point wiring diagram. You will need the manufacturers' technical literature for each product. The information you are looking for is typical of that noted on each product shown in Figure 9-15.

All this information is necessary to create a complete wiring diagram that an installer and troubleshooter can use. This information *should* be easily picked off of manufacturers' installation literature. The following list outlines the required information and the reasons for it:

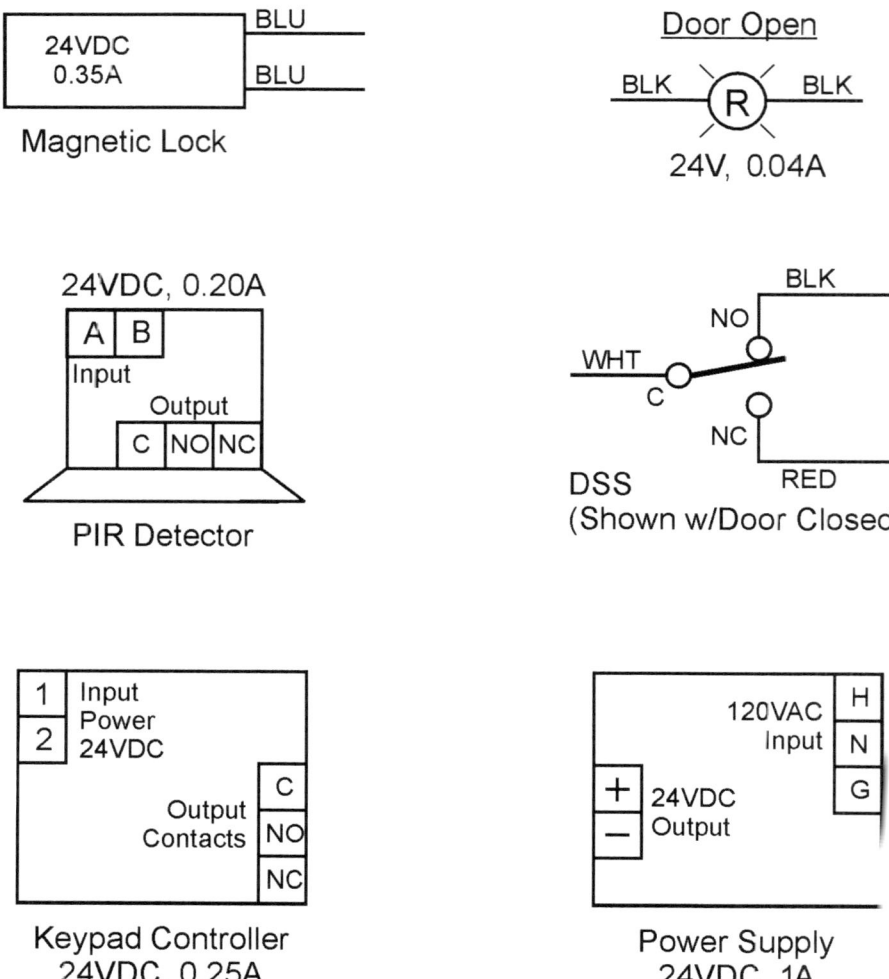

**Figure 9-15** Product information for the wiring diagram.

- Identify the power requirements for all the loads in the system. Any load *uses* power, for example, the magnetic lock, the PIR detector, the keypad controller, and the indicator light. All this equipment must be of the same *voltage* if you are using a single power supply, for example, 24 volts DC (some equipment will work on AC or DC and will be so noted by the manufacturer). You should also note how much *current* each device uses, for example, 0.35 ampere. You will need to know the total circuit current draw to select the proper size of power supply.
- Identify the output power capacity of the power supply. The power supply powers all the loads. Note its *output* power rating. In this system you would select a 24-volt DC unit. If every load in the system were "on," the current

draw would be 0.84 ampere. It is advisable to select a power supply with a little more amperage than you need. A 1-ampere power supply would do fine for this system.

- Identify all input and output terminations necessary for your system. Some will be screw terminals with letters or numbers. Others will be color-coded wire. Identifying terminations that you are *not* using for your particular system may prevent confusion when observing the actual equipment.

In the case of unused screw terminals, it is sometimes more beneficial to show them on the drawing, but not wired. If they are not shown, the installer may wonder if they were accidentally forgotten, or if he or she has the correct piece of equipment. On the other hand, these items may not fit on the drawing, or they may just add to the confusion. In this case a judgment call may be to leave them off the drawing. When the "extra" terminations are physical wires, it is best to show them, along with a field note, as seen in Figure 9-16.

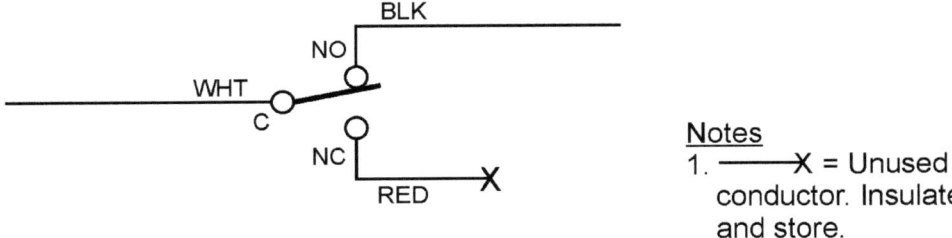

**Figure 9-16** Designation for unused wire leads.

With product information completed, we can proceed to the actual wiring diagram.

## Wiring Diagram Exercise

By the time I have reached this point in classroom work, I figure it's about time for a "test." Rather than walk you through a wiring diagram, I thought it best for you to give one a try on your own. When you have finished, a step-by-step walk-through is provided in Appendix C.

I would normally start this diagram by drawing all the system components, spread out on a sheet of paper, as shown in Figure 9-15. I have provided Figure 9-17 to get you started. The components have been thoughtfully laid out to give you a good start. If you start the layout "cold," you will find the drawing gets messy and has to be laid out several times. This is typical, and I have started a diagram five or six times before finding a reasonable layout. I noted a brief system description of operation separately rather than clutter up the drawing in Figure 9-17. You may want to make several copies of Figure 9-17 to use as worksheets.

## System description for Figure 9-17

Door is normally closed and locked. Access is by exterior keypad. Free egress is by PIR detector. Remote red light indicates any time door is in the open position.

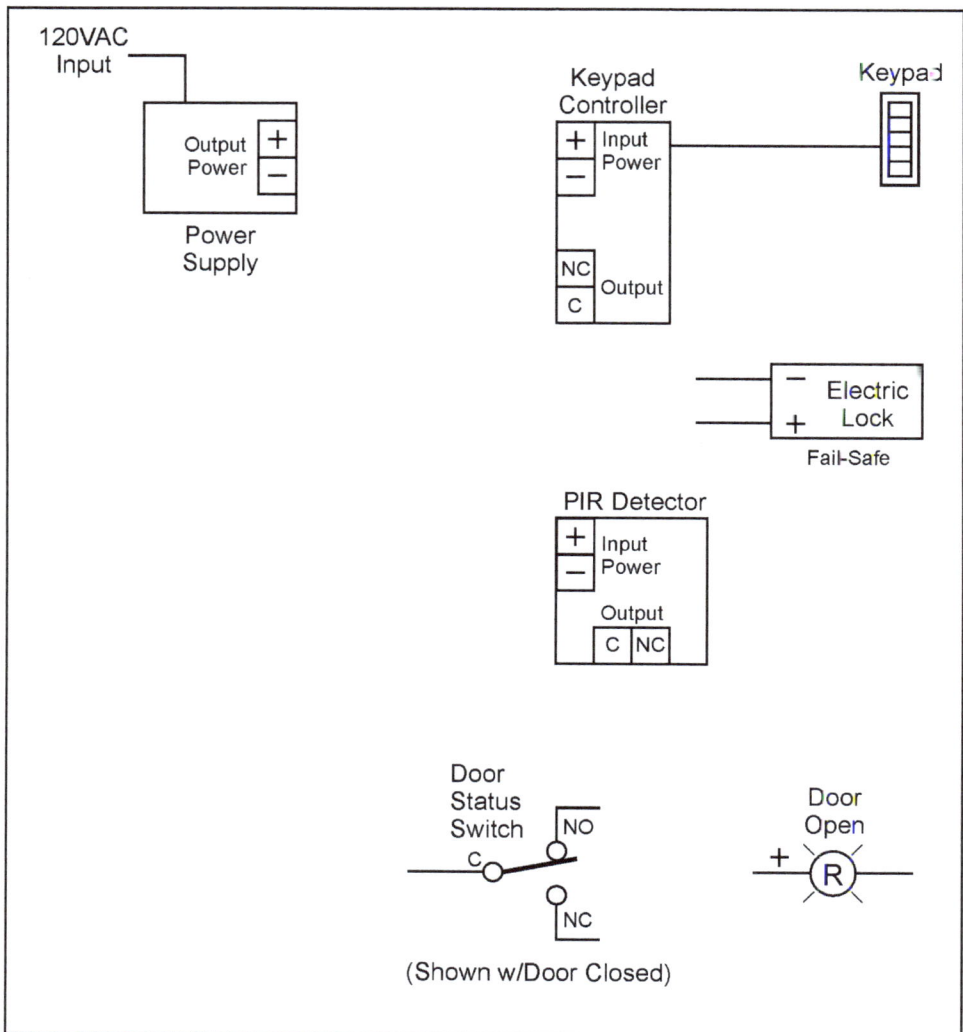

**Figure 9-17**   Component layout for system wiring diagram.

# 10

# Skill Exercises

By now you should have a good understanding of control switches, power flow and control, basic relay logic, and the relationship of system components to one another. In this chapter I thought I would throw some fun exercises at you, as a sort of a test of your new skills. Immediately following each exercise is the solution; but don't give up too soon. The exercises are meant to evaluate what you have learned and maybe to challenge your problem-solving abilities!

## Tune-up

This exercise takes us back to where we first started (Figure 10.1). It is very basic; you should have no problem completing the wiring. The solution is shown in Figures 10-2 through 10-4.

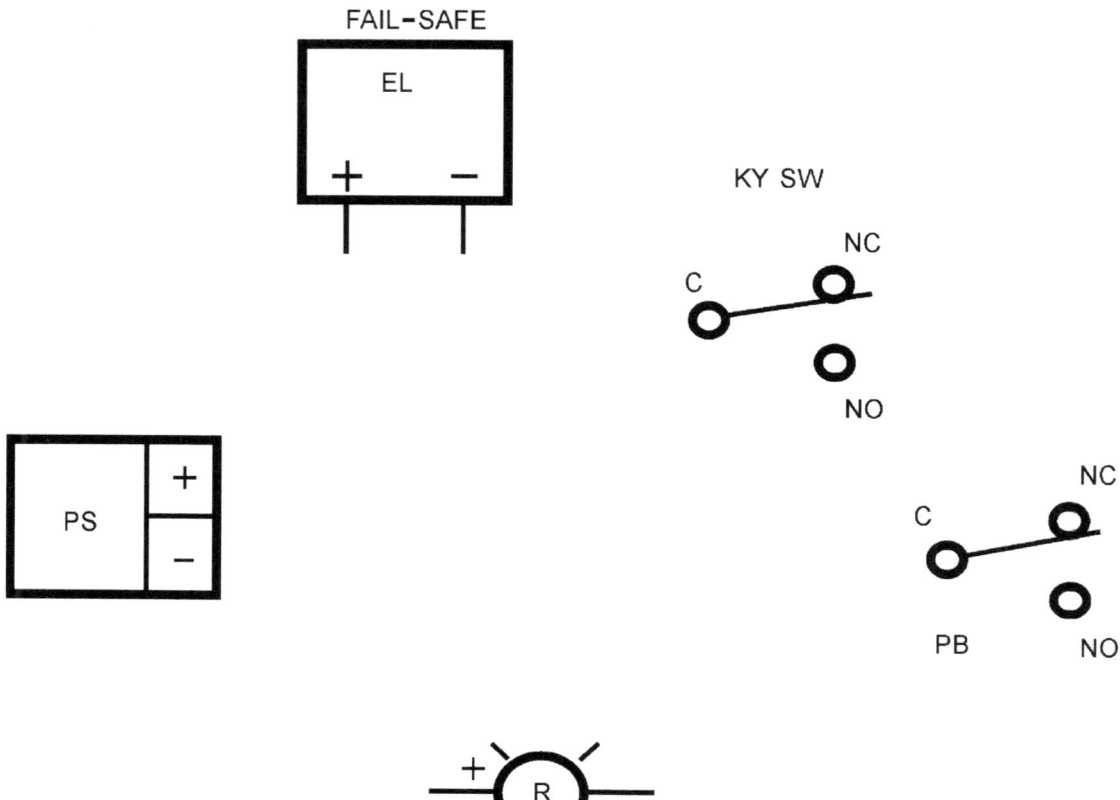

**Figure 10-1**   Basic system exercise.

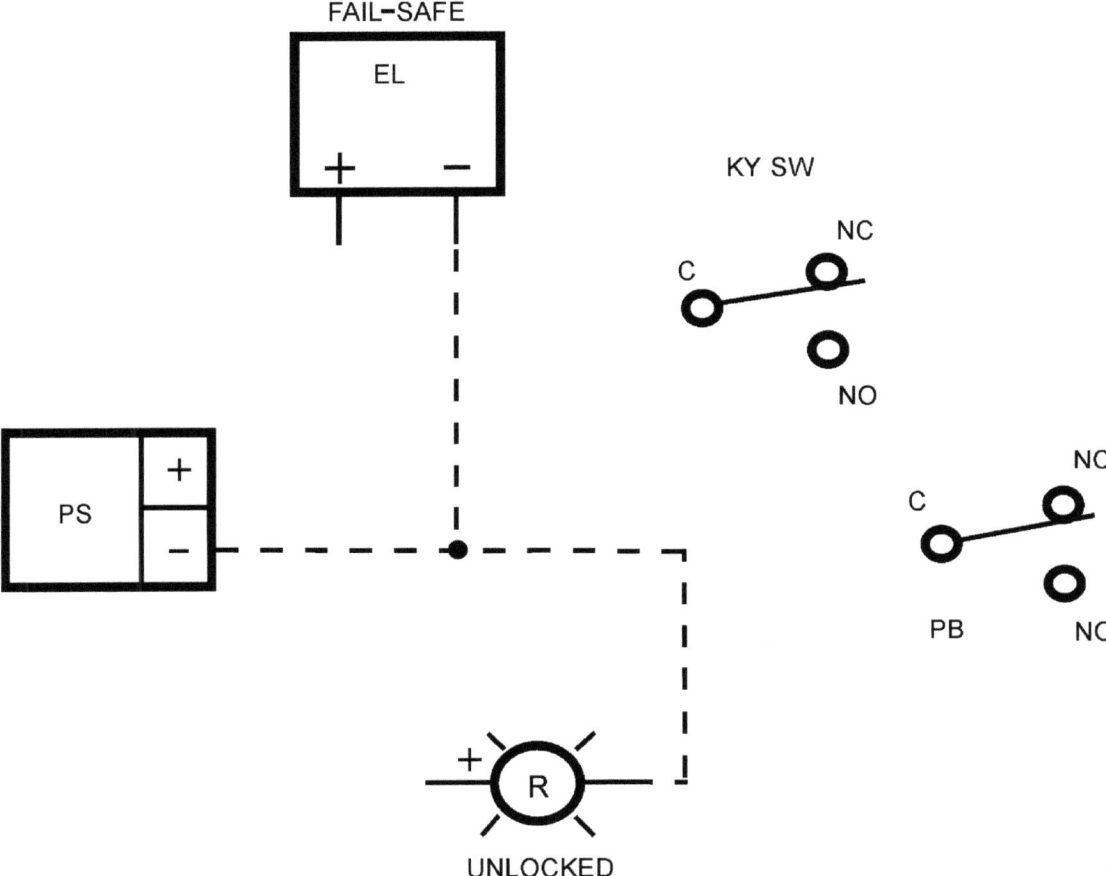

**Figure 10-2**   Hook-up $(-)$ to all loads first.

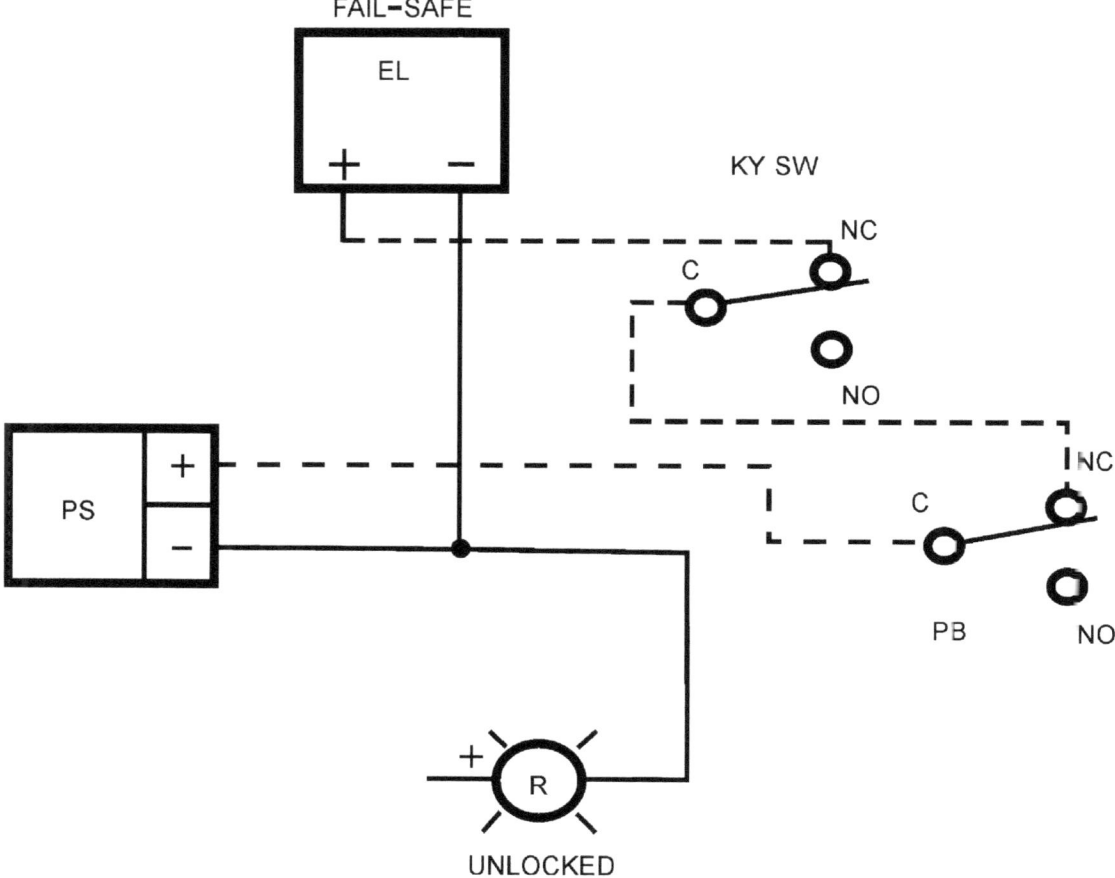

**Figure 10-3**  Route (+) from the power supply to the electric lock through all the control switches.

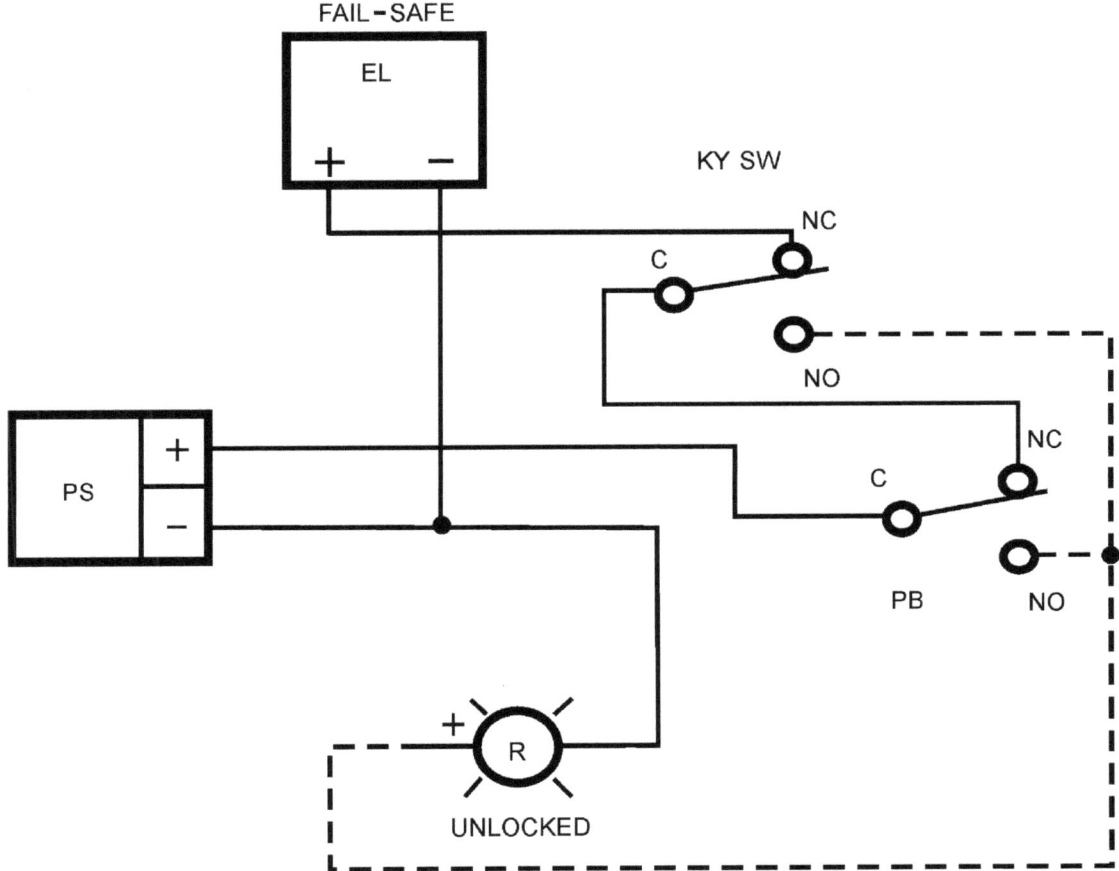

**Figure 10-4** Create a path for (+) to travel so the red indicator lights when the lock is turned off.
Score: ☺ OK—Go to next exercise.
    ☹ Need to solve; if not, reread Chapters 1–9.

## Three-Way Switch Wiring

Did you ever wonder how you could turn the kitchen lights off from one wall switch and back on from another wall switch? This example is similar and is sometimes used in low-voltage work (Figure 10-5). Solution in Figure 10-6.

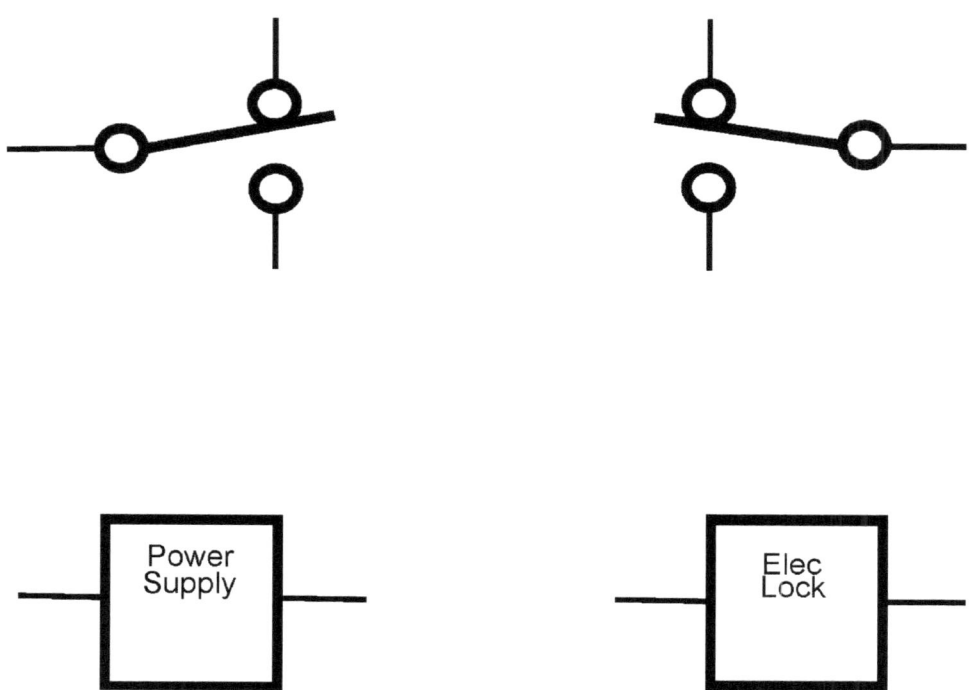

**Figure 10-5** "Three-way" switch wiring exercise. When one switch turns the lock off the other switch can be used to turn it back on. (There are two ways you can "wire" this exercise; either will work.)

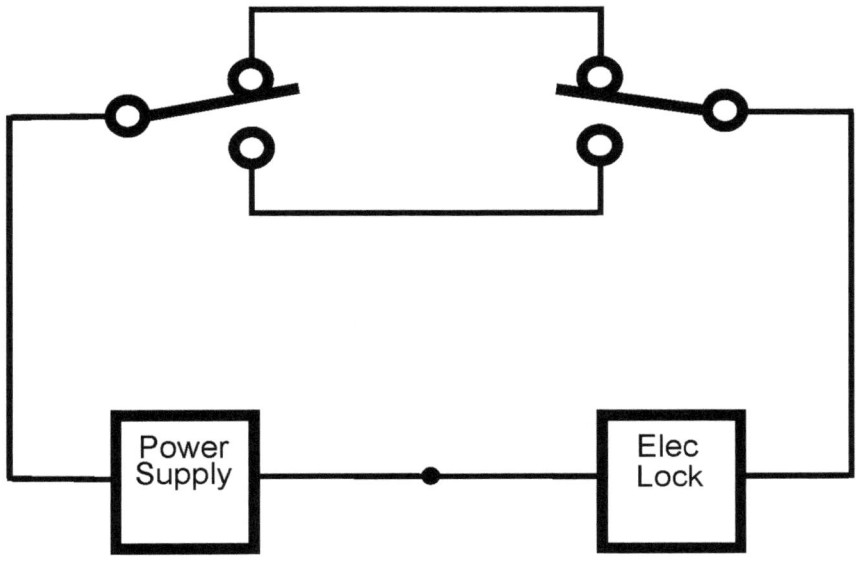

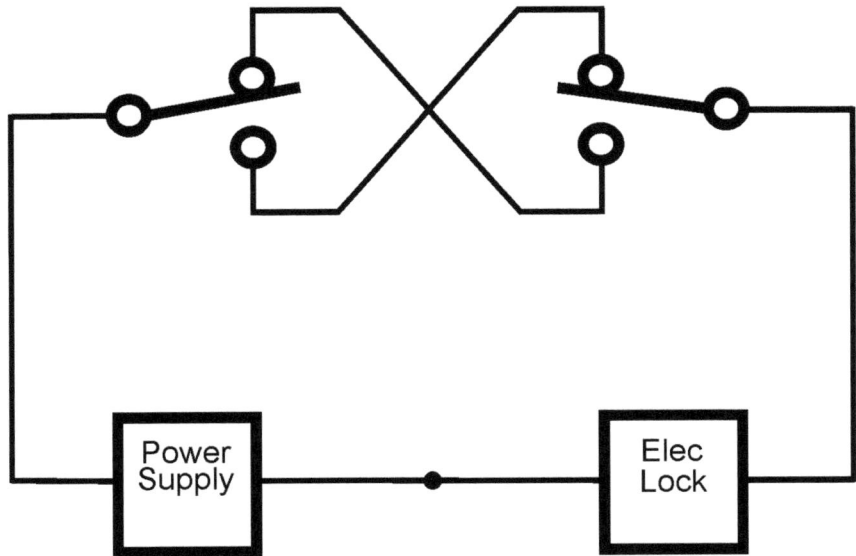

**Figure 10-6**   "Three-way" switch wiring solution.
Score: ☺ This one doesn't count—it is just to get your brain going! Go to next exercise.

## Monitoring Everything!

This exercise (Figure 10-7) tests your ability to control monitoring indicators with the proper switch contacts. It is not common to monitor all these functions on a single panel. Following the solution (Figures 10-8–10-11) to this exercise is a modified solution that is more reasonable for real-world conditions (Figure 10-12).

*Hint:* Run (−) down the right-hand side of the monitor and control panel. It makes wiring much easier.

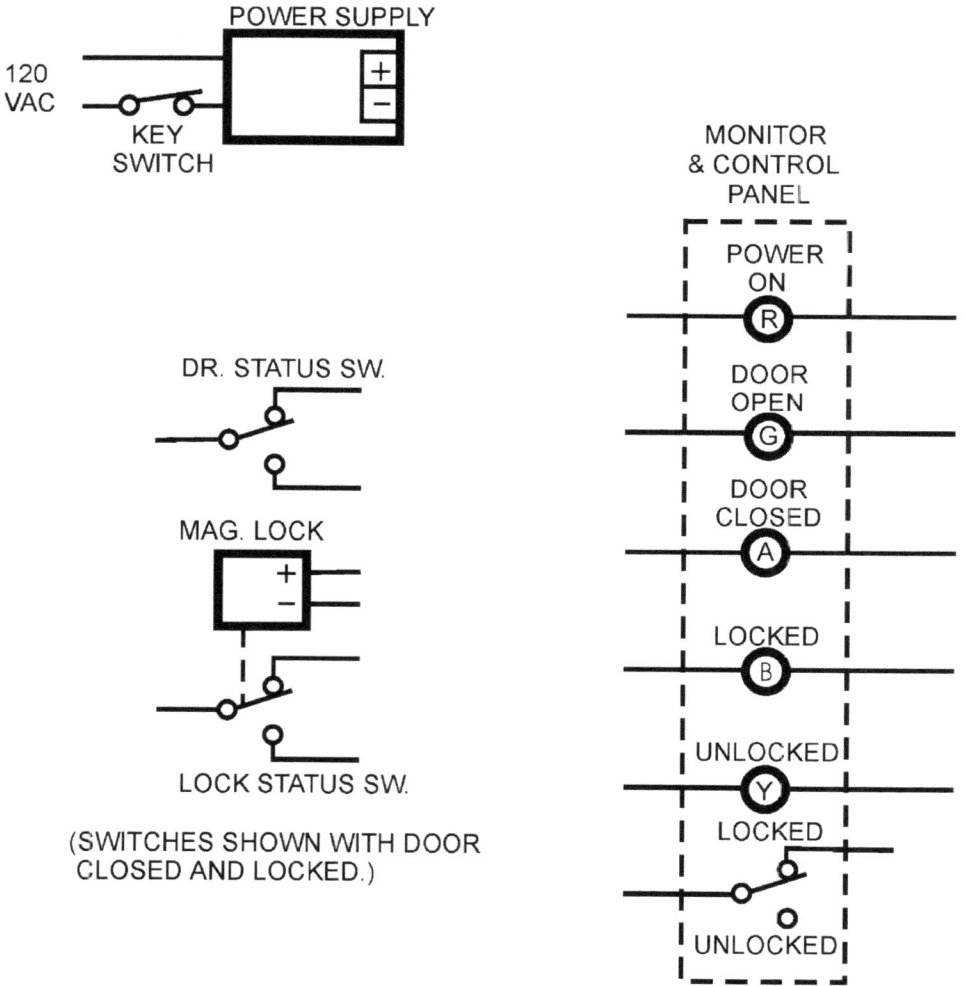

**Figure 10-7**   Monitor and control panel exercise.

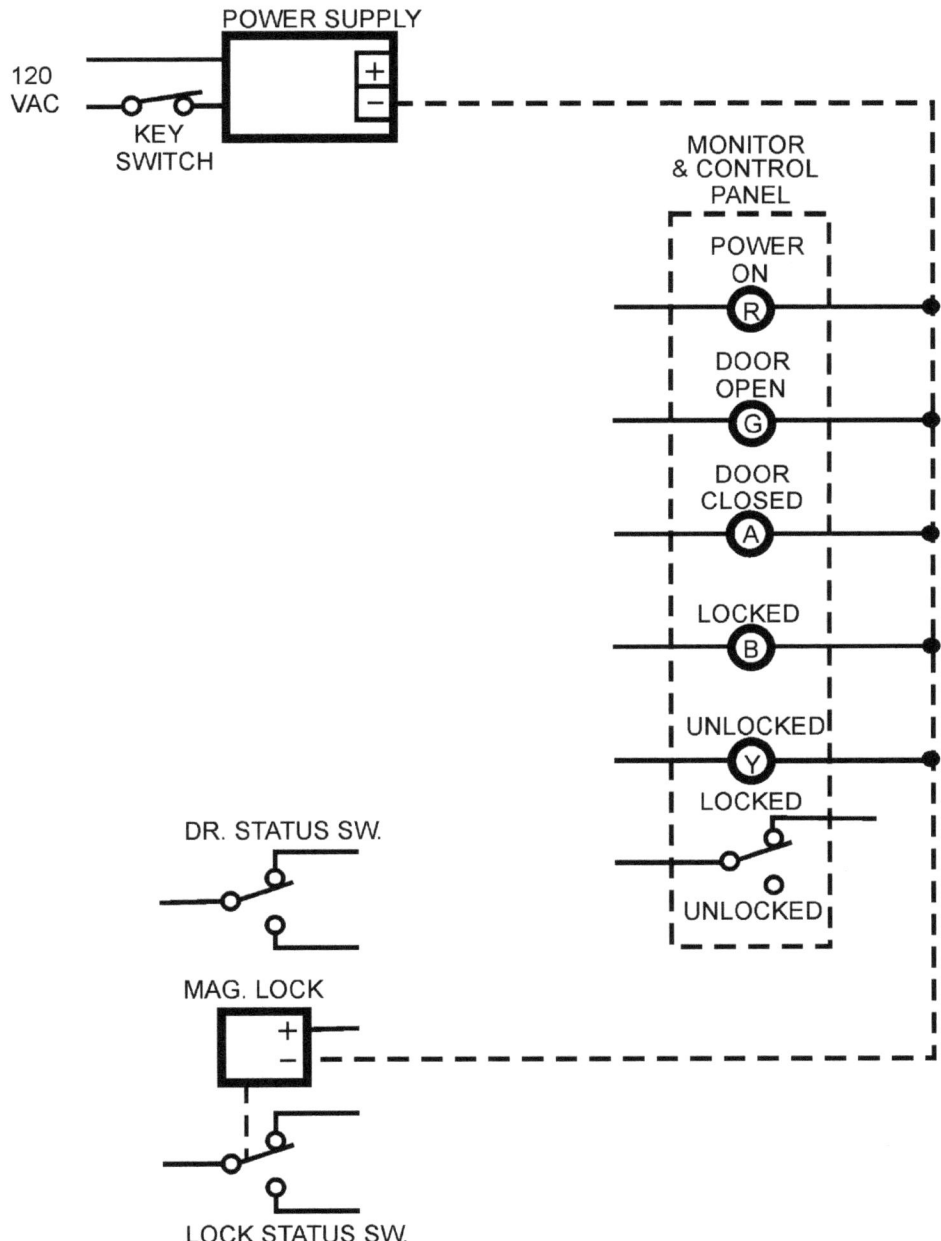

(SWITCHES SHOWN WITH DOOR
CLOSED AND LOCKED.)

**Figure 10-8**   Wire $(-)$ to all the loads in the system.

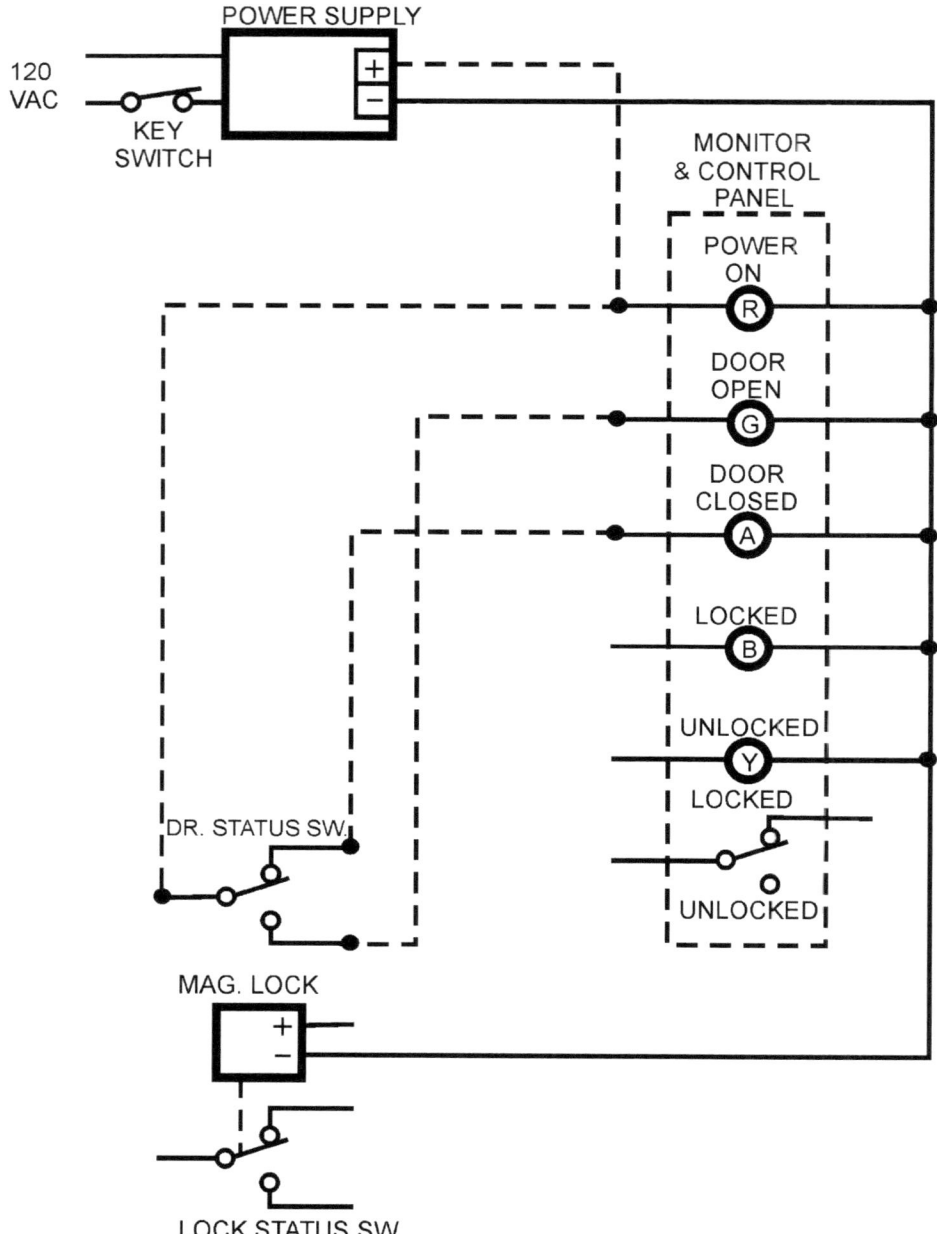

**Figure 10-9**  Wire (+) to all the loads in the system. Start by wiring (+) to the "power on" light. Also wire (+) through the door status switch to the correct "door open" and "door closed" lights.

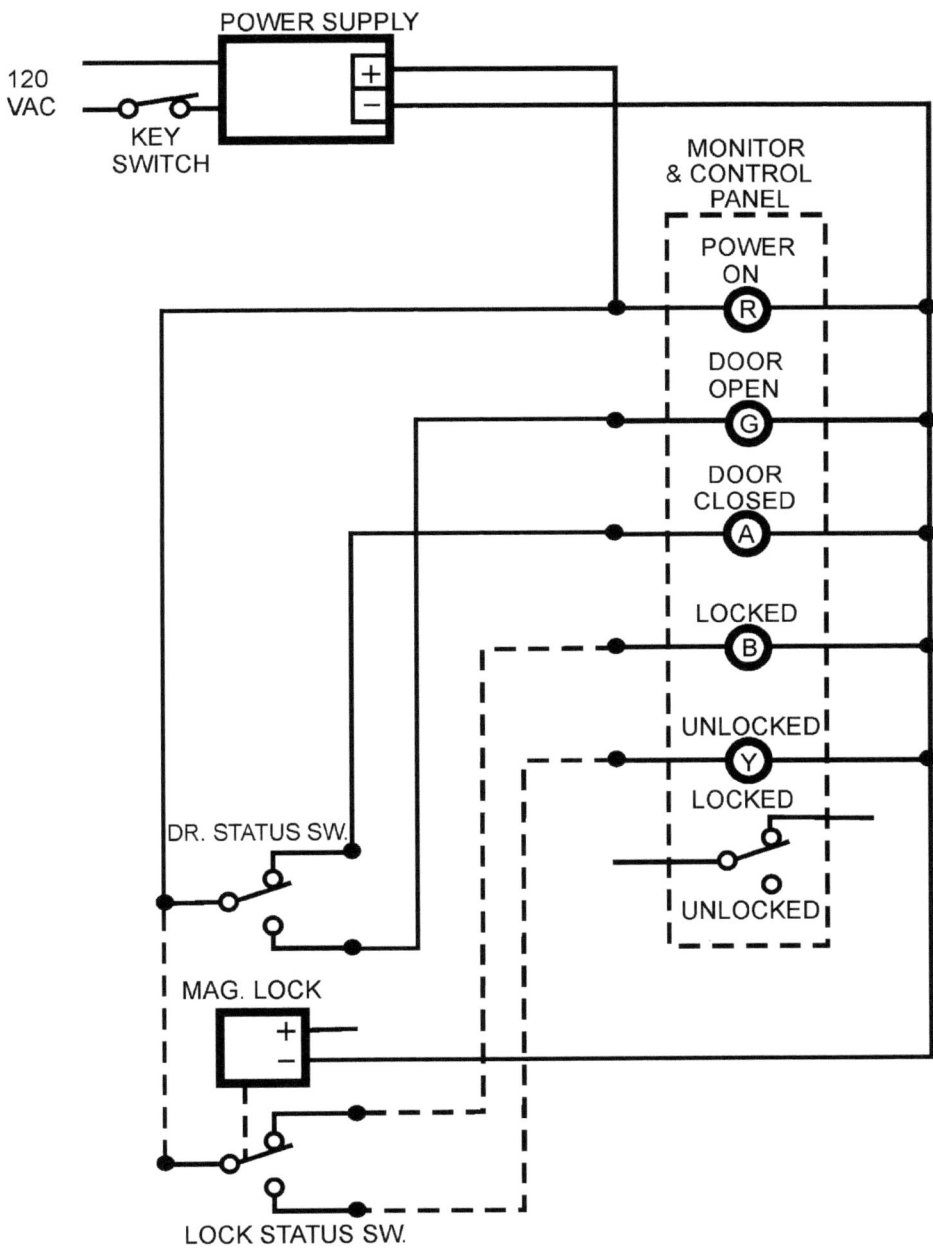

**Figure 10-10** Wire (+) through the lock status switch to the correct "locked" and "unlocked" lights.

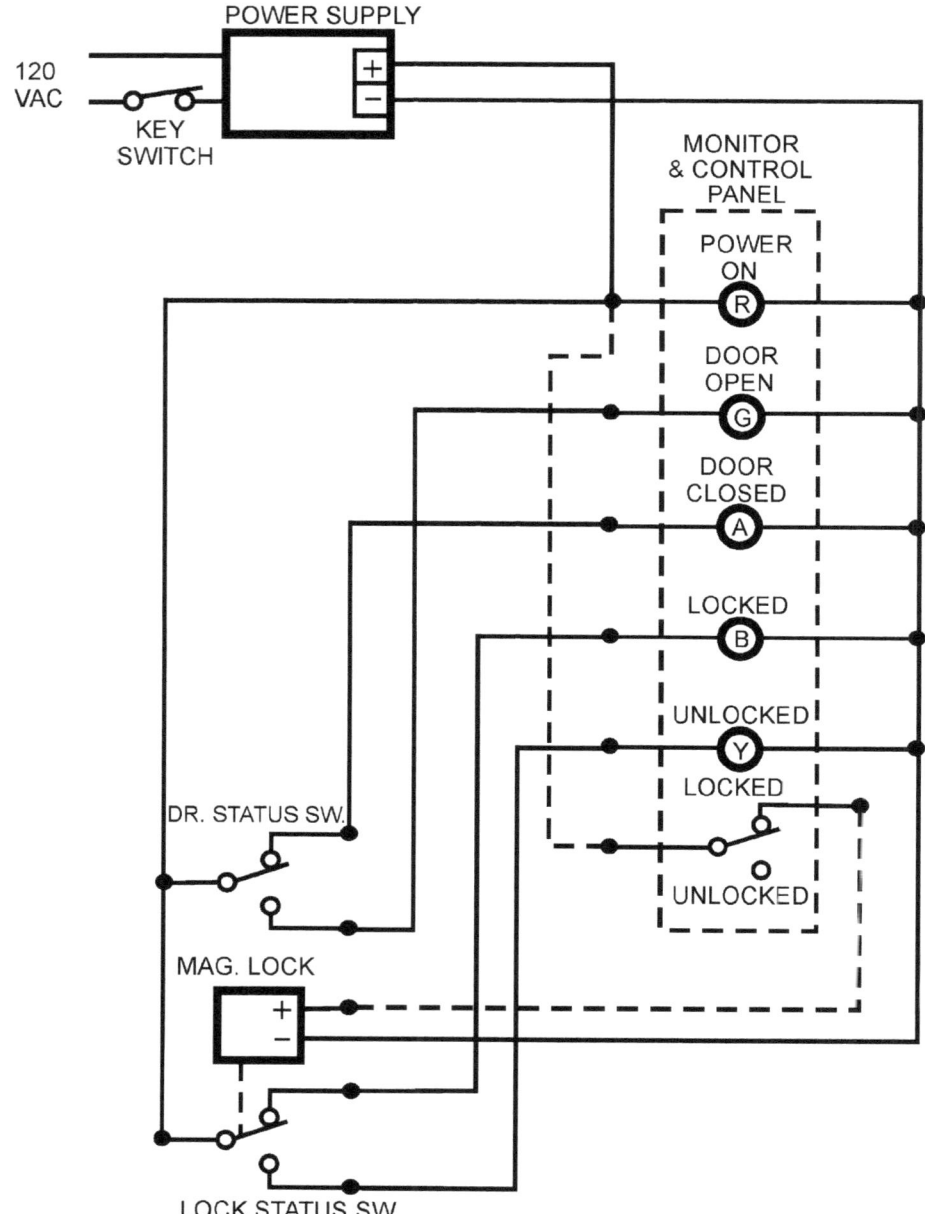

**Figure 10-11** Wire (+) through the lock control switch to the mag. lock. Monitor and control panel wiring is now completed.
Score: ☺ You should solve this—it's a crazy example, but meant to test your skill in routing (+) through the proper switches.
    ☺ Stuck? Review Chapter 7.

By following the path of (+) through the door status switch and lock status switch, you can see how power is controlled to illuminate lights to indicate certain conditions at a secured door.

The exercise is a little overdone, as it would not be necessary to monitor every condition. It was presented in this manner to accustom you to using the correct switching for specific monitoring conditions.

The wiring diagram in Figure 10-12 is included to show you an alternate way of monitoring that is more acceptable in security work.

Notice we have eliminated the "power on," "door open," and "locked" indicator lights. The two remaining indicator lights are relabeled as *door closed and locked* and *unlocked or open*. This method of monitoring is more common and more acceptable as it pretty much tells you all you need to know about a system. Try wiring this exercise (Figure 10-12); it will provide good practice in wiring switches in series and in parallel. Remember, two conditions must occur before the "door closed and locked" indicator will illuminate. Also, either of two conditions will illuminate the "unlocked or open" indicator. Solution in Figures 10-13–10-15.

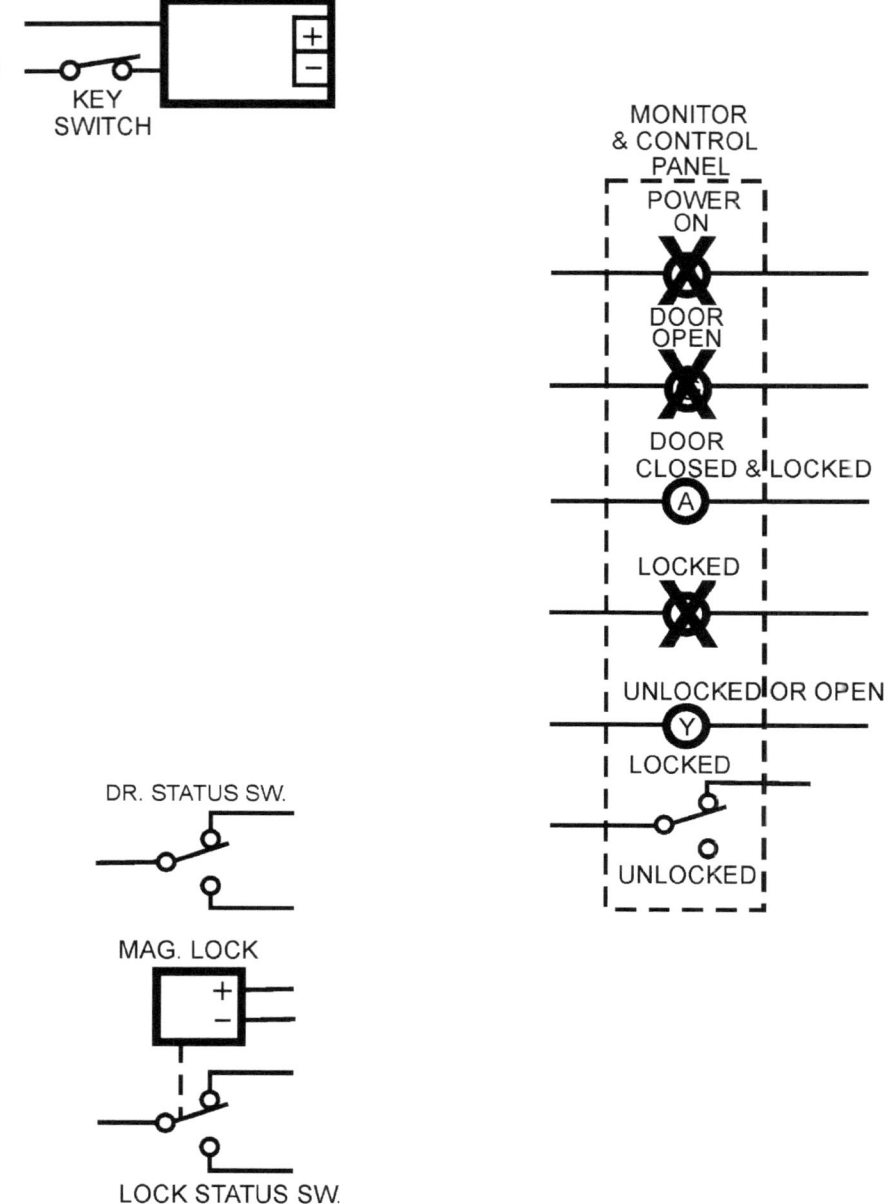

(SWITCHES SHOWN WITH DOOR
CLOSED AND LOCKED)

**Figure 10-12** Alternative layout for monitor and control panel.

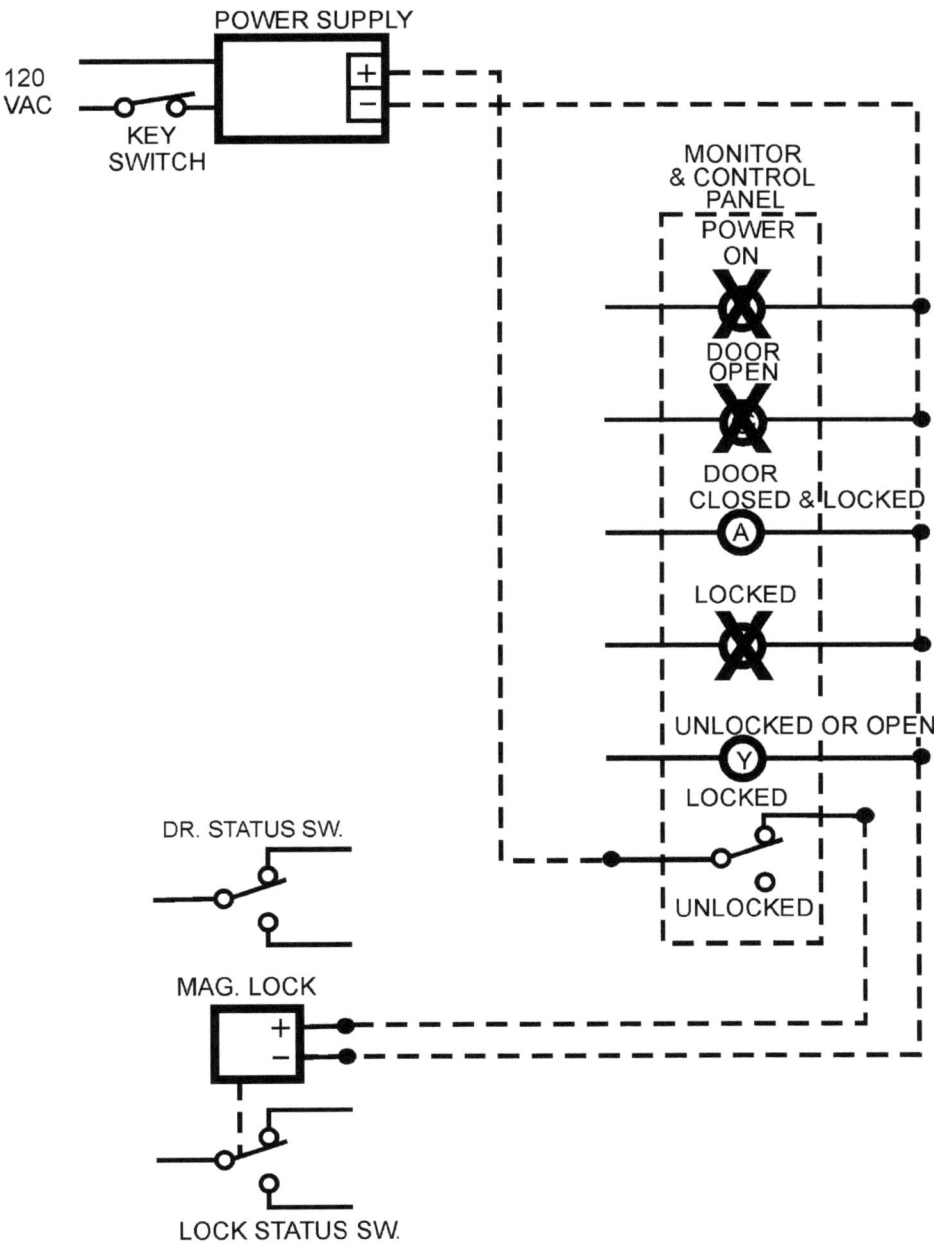

(SWITCHES SHOWN WITH DOOR
CLOSED AND LOCKED)

**Figure 10-13**  Wire (−) to all the loads in the system.
Wire (+) through the lock control switch to the mag. lock.

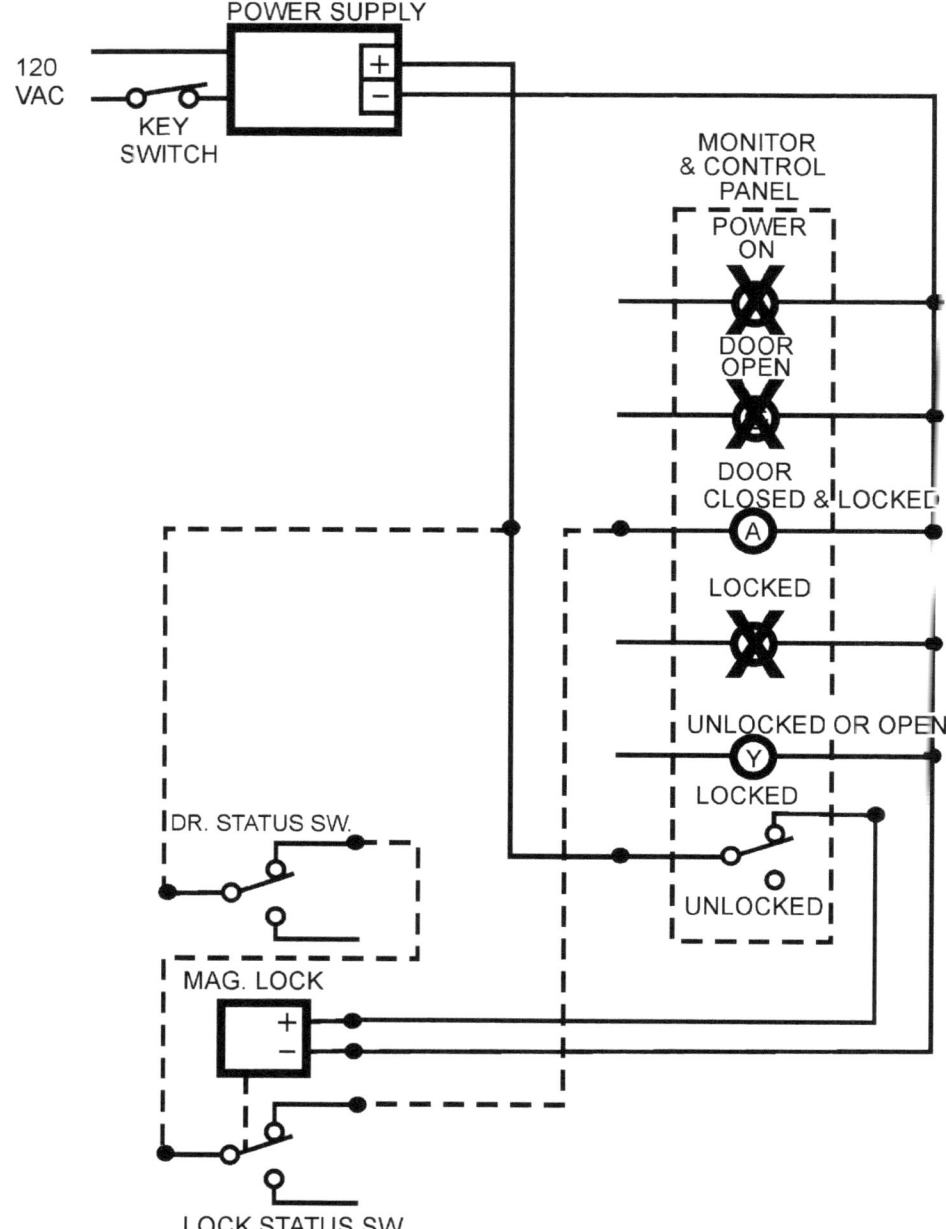

**Figure 10-14**   Wire (+) in series through the door status switch closed contact continuing through the lock status switch closed contact to the "door closed and locked" light.

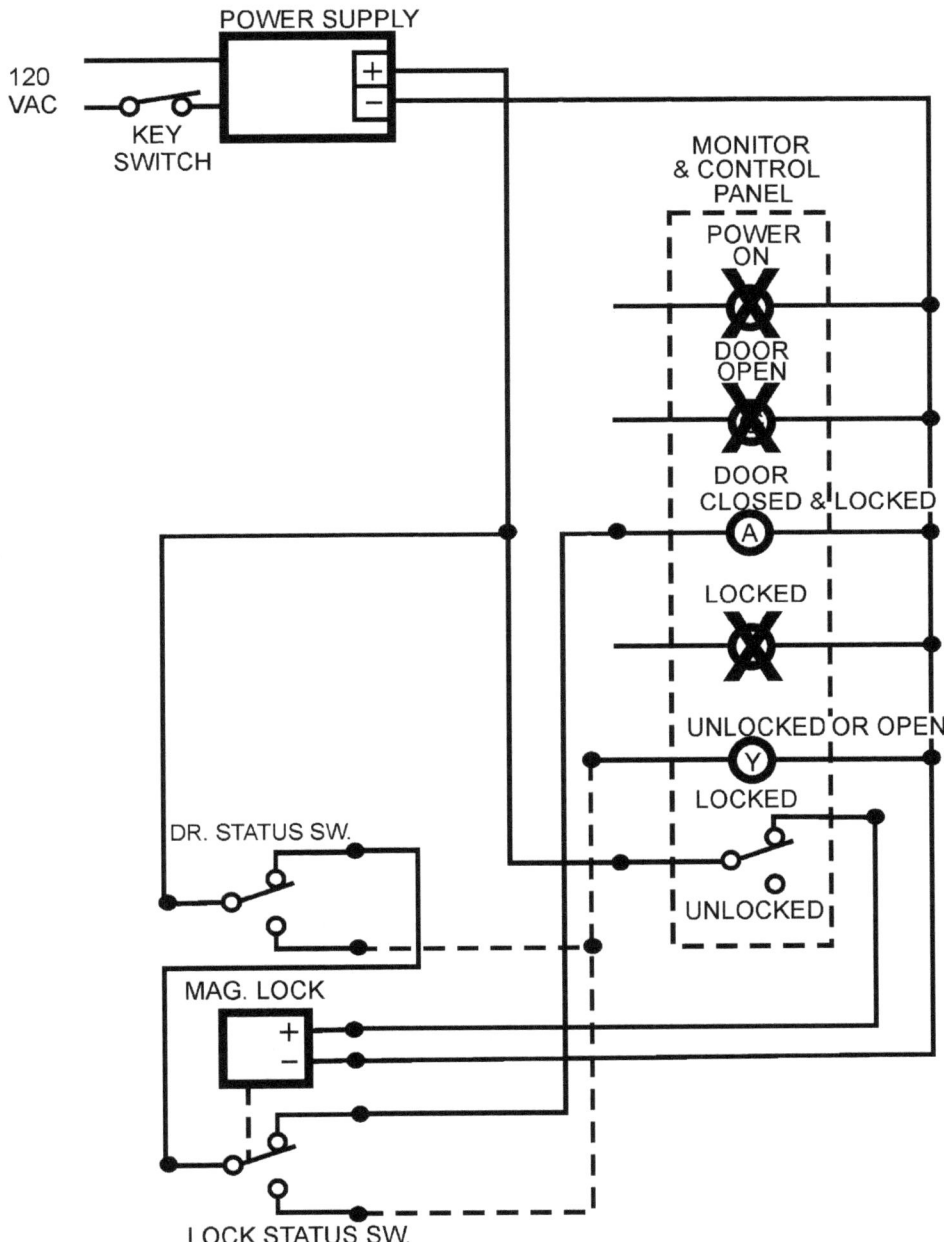

**Figure 10-15** Wire a path for (+) in parallel from the door status switch open contact and the lock status switch open contact to the "unlocked or open" light.

## Relay Logic

In Chapter 9 we did a logic diagram (Figures 9-10 through 9-14) using relay contacts. This exercise uses the same technique to create a latching relay circuit. Follow me through Figures 10-16 through 10-19 as we create a simple alarm system with a latching relay.

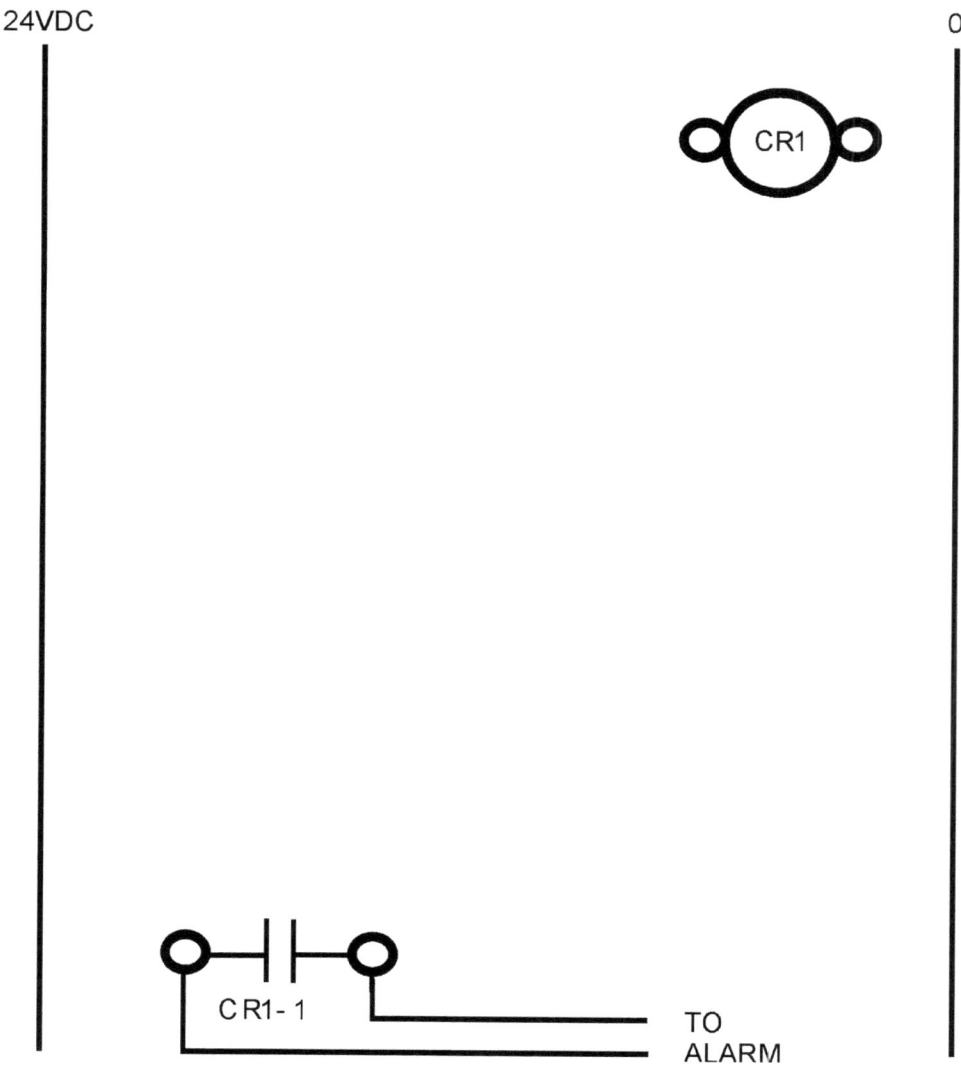

**Figure 10-16** Latching relay circuit. The point of this exercise is to send a signal (relay contact CR1-1 closure) to a remote alarm. We will use an "alarm button" to energize relay CR1, an additional set of relay contacts to latch the relay, and a "reset" switch to reset the circuit back to normal.

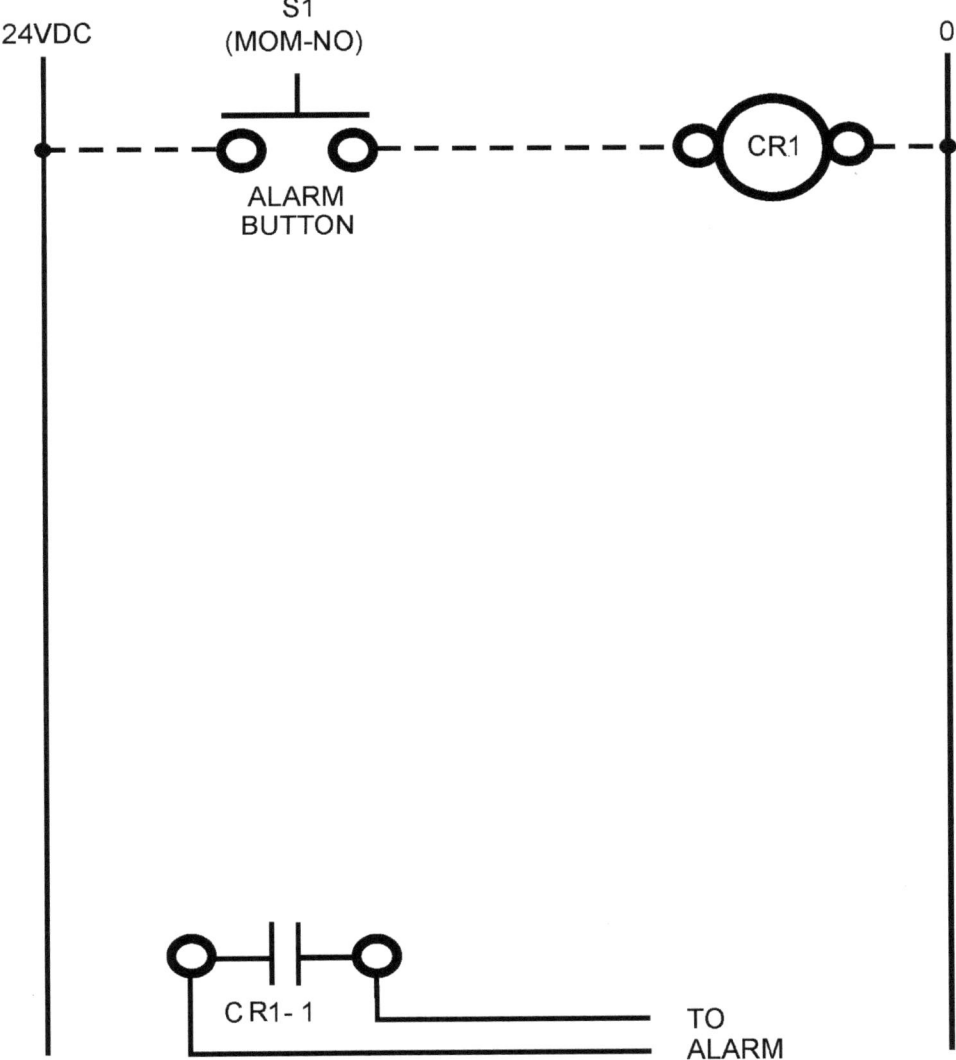

**Figure 10-17**  Alarm button S1 energizes relay CR1 causing CR1-1 contacts to close signaling alarm. Releasing button drops out CR1 and contacts CR1-1 reopen ceasing signal to alarm.

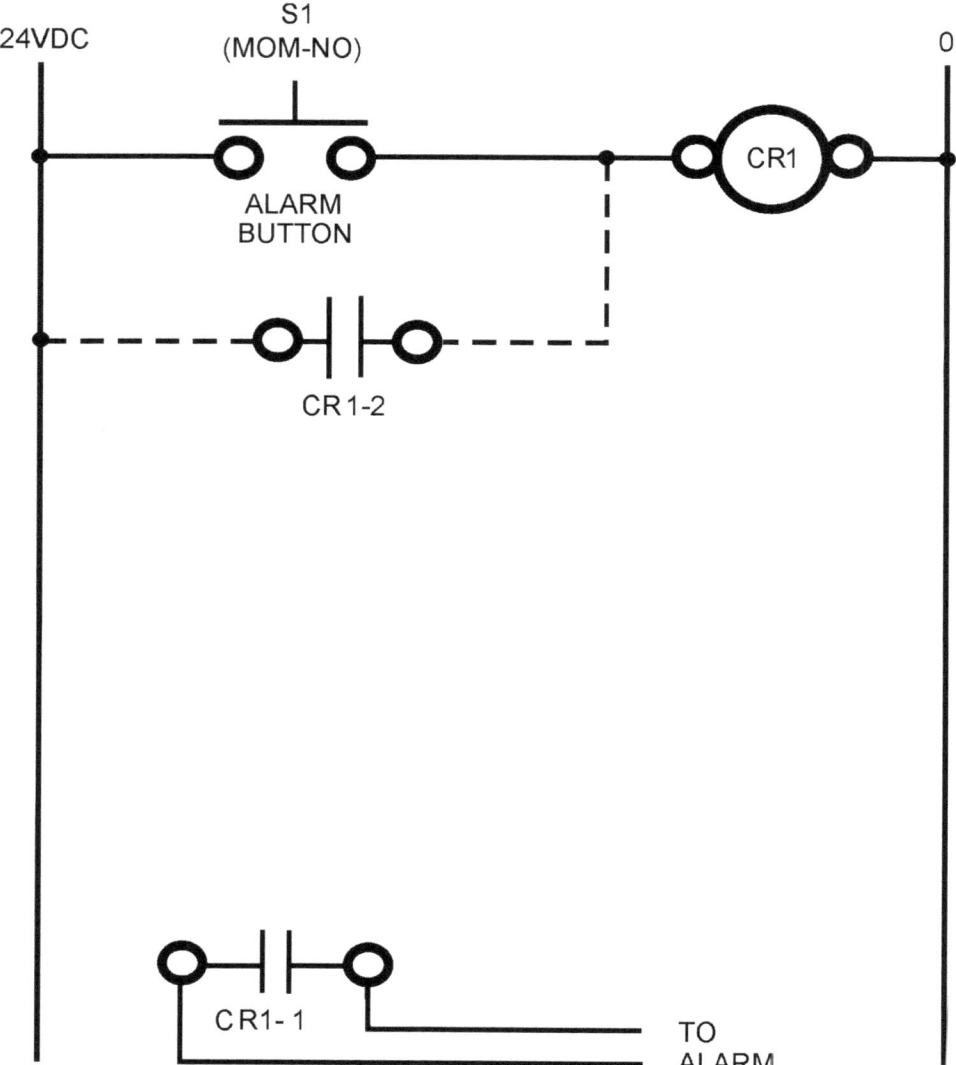

**Figure 10-18**  Alarm button S1 energizes relay CR1 causing contacts CR1-1 & -2 to close. Closed CR1-2 contacts provide new path for (+) to relay CR1 and it energizes—forever!! Closed CR1-1 contacts cause continuous alarm signal.

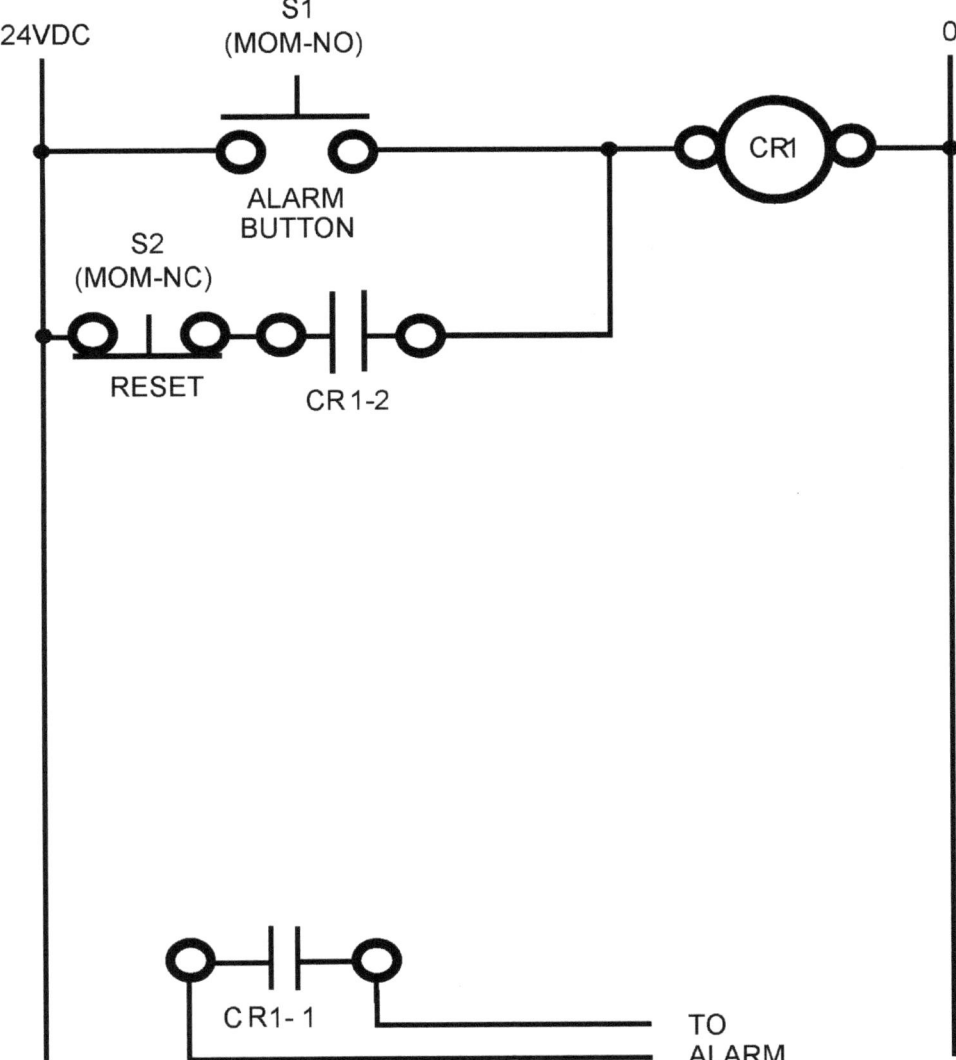

**Figure 10-19** Adding reset switch S2 will allow a way to break (+) to relay CR1 dropping out CR1, putting system back to normal.

## Latching Alarm System

Now that you have a better understanding of a latching relay, try the next exercise on your own (Figure 10-20). It is very similar to the previous circuit. The components are laid out in a manner that should make the wire runs fairly obvious. Try to solve it without looking back (or ahead at the solution) (Figure 10-21).

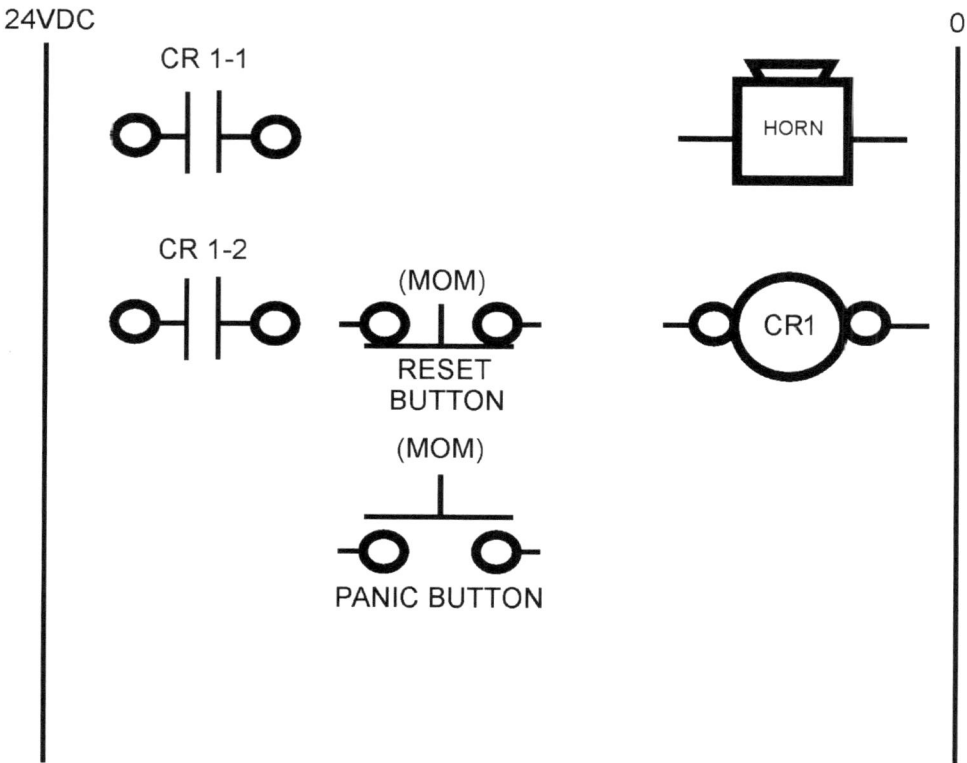

**Figure 10-20** Latching alarm system exercise.
Pushing momentary panic button latches in alarm.
Pushing momentary reset button returns system to normal.

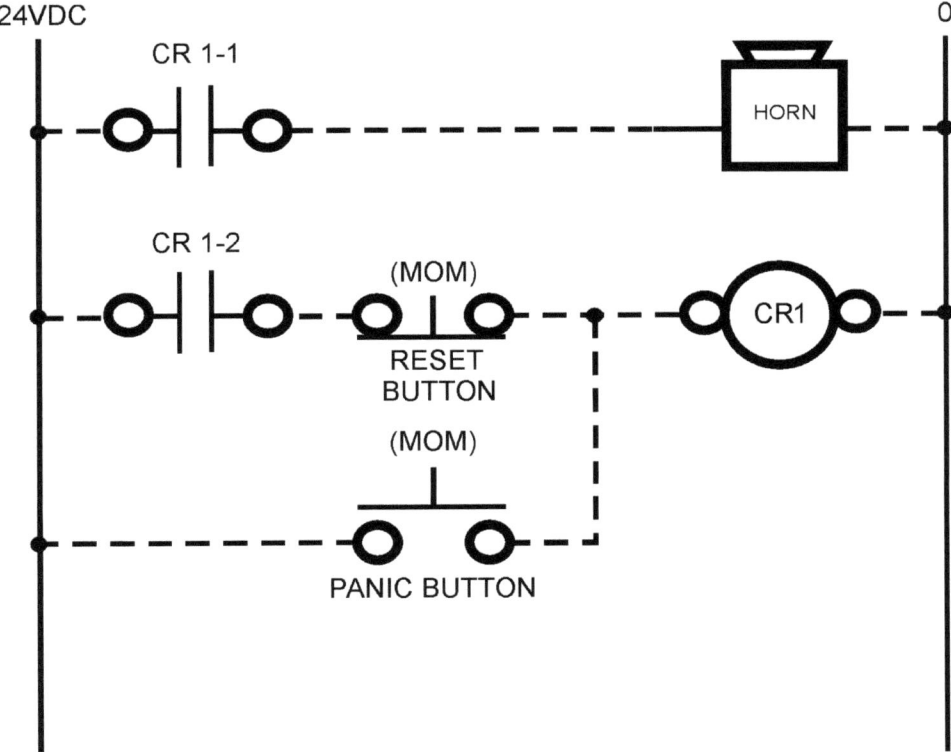

**Figure 10-21** Latching alarm system completed.
Score: ☺ This one was just for fun—but a latching relay circuit can come in handy.

## Call System with Relay

Look at the basic call system in Figure 10-22. The relay is used to allow the use of two different voltage power sources in the same system. It's a little tricky! Solution in Figures 10-23–10-25.

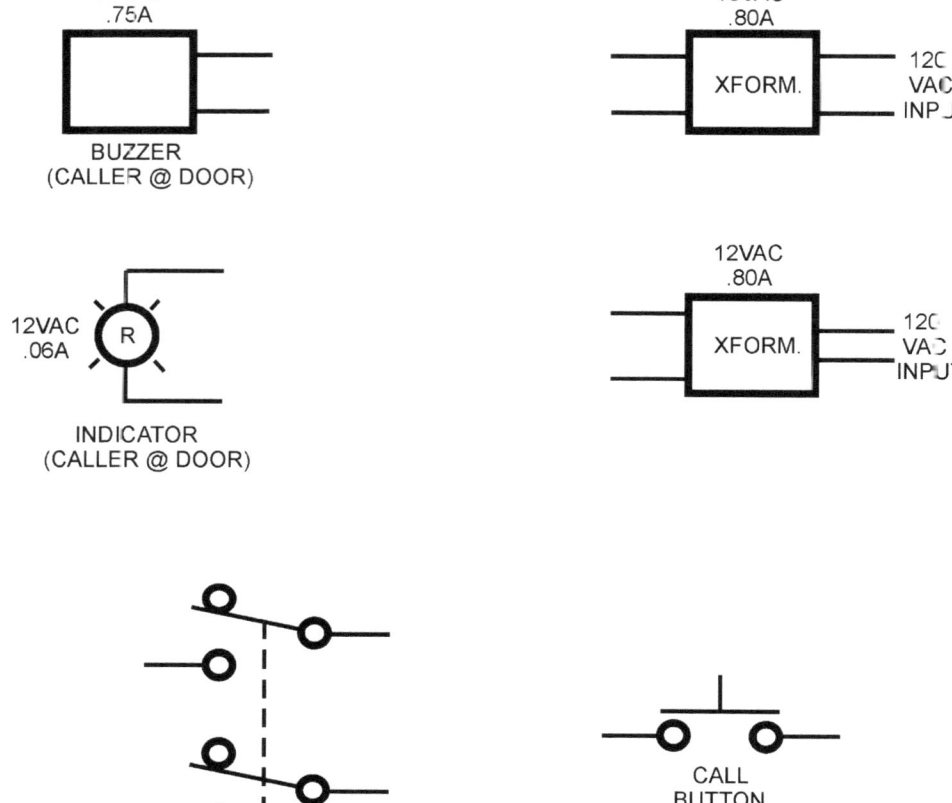

**Figure 10-22** Visual and audible call system exercise.
First determine what voltage the relay should be. Wire the call button to energize the relay.
Wire the relay contacts to route the proper power to the indicator and buzzer.

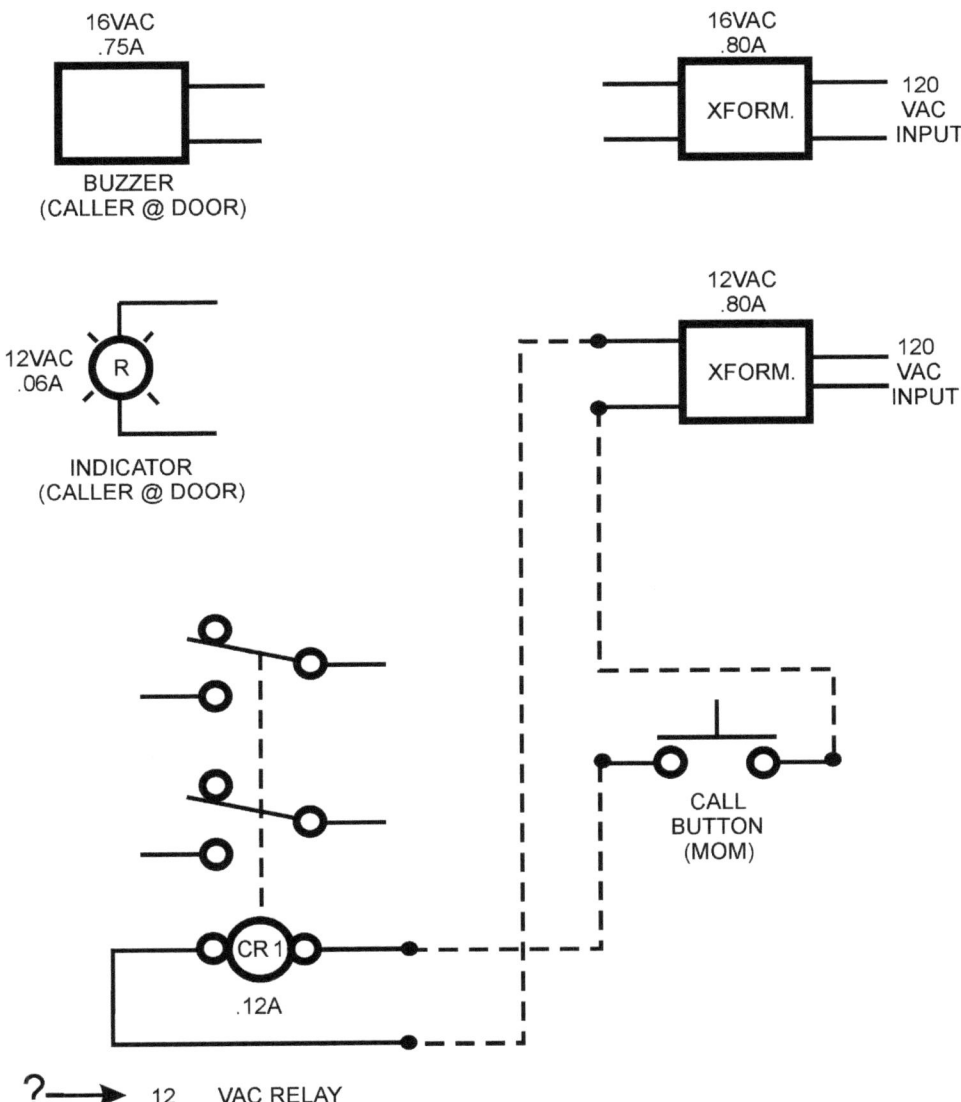

16VAC
.75A

BUZZER
(CALLER @ DOOR)

16VAC
.80A

XFORM.

120
VAC
INPUT

12VAC
.06A

R

INDICATOR
(CALLER @ DOOR)

12VAC
.80A

XFORM.

120
VAC
INPUT

CALL
BUTTON
(MOM)

CR 1

.12A

?———▶ __12__  VAC RELAY

**Figure 10-23** You first must determine what voltage the relay should be. The current draw of the 16VAC buzzer would not leave enough amperage from the 16VAC transformer to power the relay. The current draw of the 12VAC lamp leaves enough power from the 12VAC transformer to power the relay. Next run power from the 12VAC transformer through the call button to the relay coil.

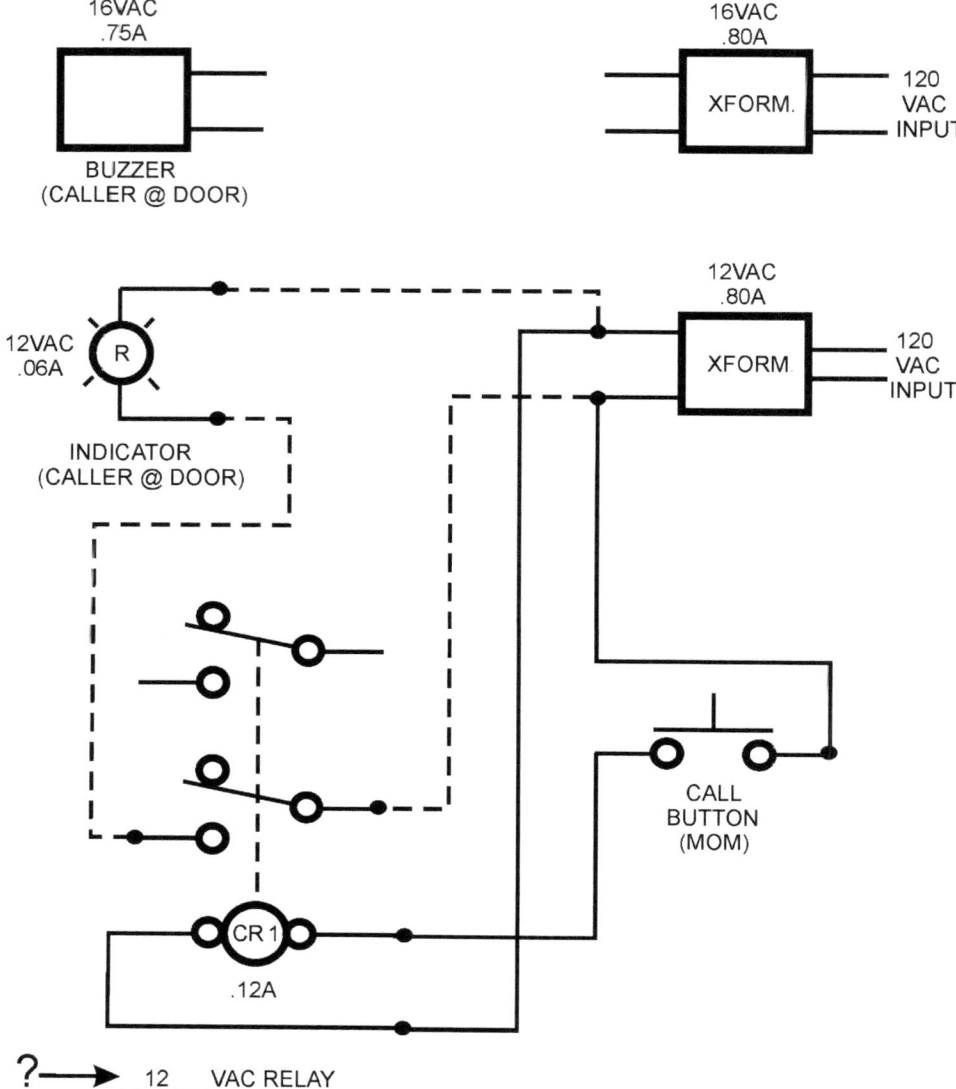

**Figure 10-24**  Wire the red lamp by running one side of the 12VAC transformer directly to one lead of the lamp. Run the other side of the 12VAC transformer through one set of open contacts of the relay.

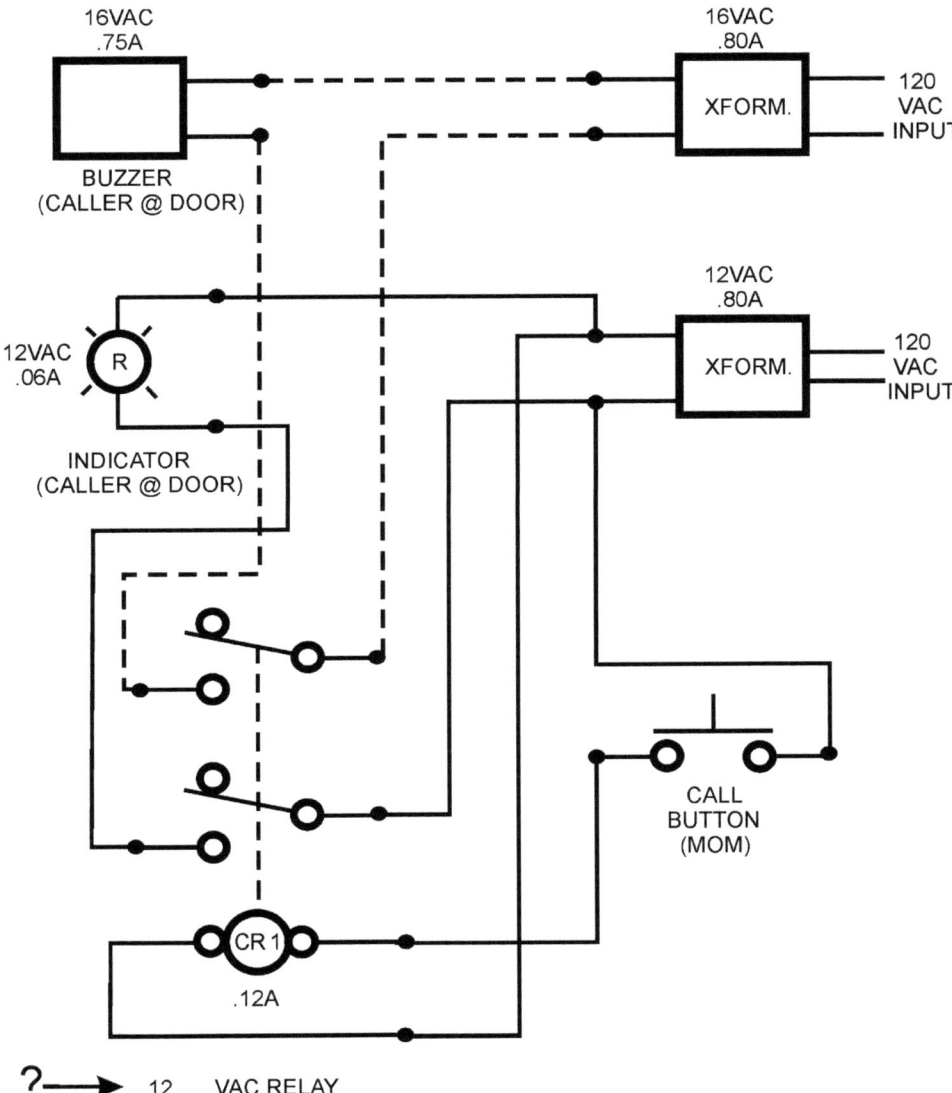

**Figure 10-25** Wire the buzzer by running one side of the 16VAC transformer directly to one lead of the buzzer. Run the other side of the 16VAC transformer through the other set of open contacts of the relay.

The system is now operational. Pushing the call button energizes the relay coil causing the open contacts to close. Power now will flow through the closed contacts illuminating the red lamp and sounding the buzzer. Releasing the call button returns the system to normal.

Score: ☺ Got it? Great, you now understand the use of switches and relay contacts.

☹ Not sure? Time to review previous diagrams.

## Security System

This exercise includes a lot of what we have learned—switches, relays, and things that need power (Figure 10-26). If you think it is too easy, don't worry; we will add a little more to it later! Solution in Figures 10-27–10-29.

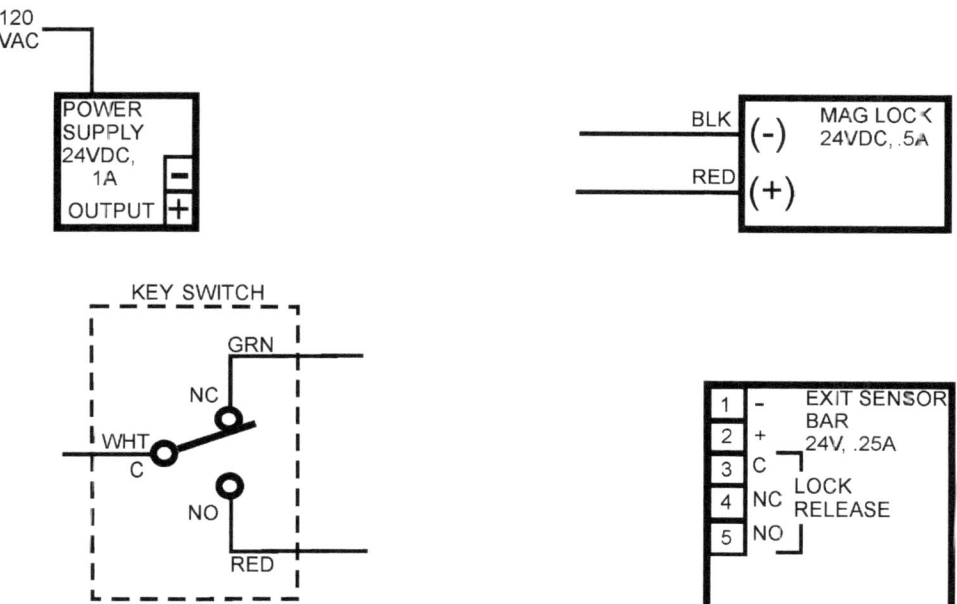

**Figure 10-26** Security system exercise.
The key switch and exit sensor bar will both release the fail-safe mag lock. The exit sensor bar contains an electronic sensor and relay. It requires its own power to operate the built-in relay. The contact labeling is shown for the relay in its deenergized state. When the sensor bar is mounted and powered the relay contacts change state.

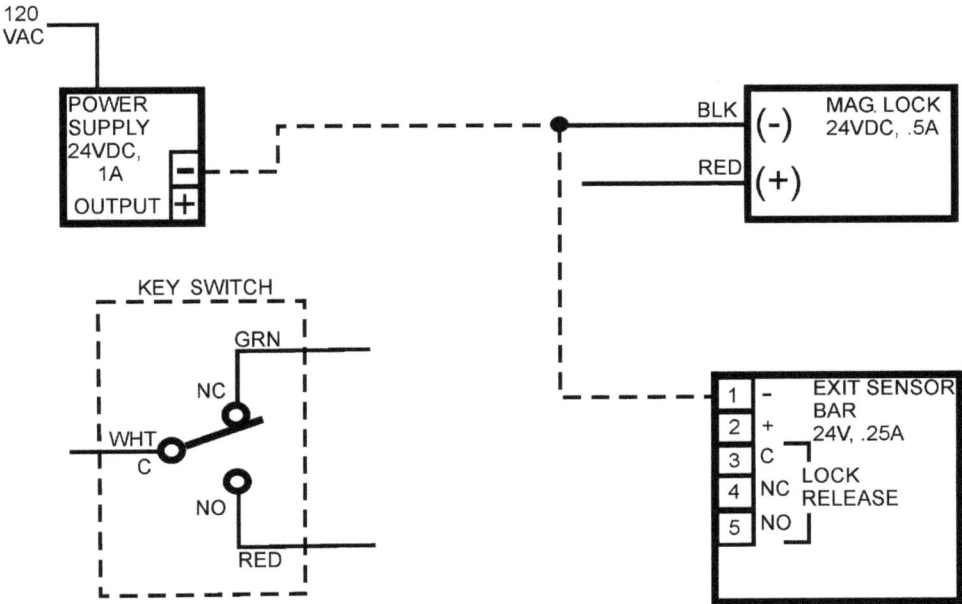

**Figure 10-27**  Wire (−) to all the loads in the system.

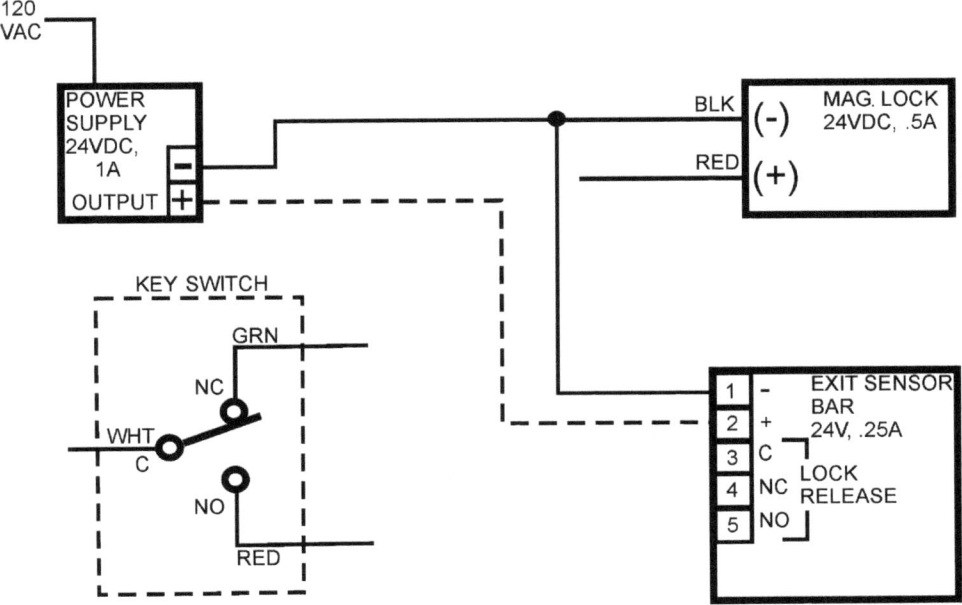

**Figure 10-28**  Wire (+) in an unbroken path to the exit sensor bar.

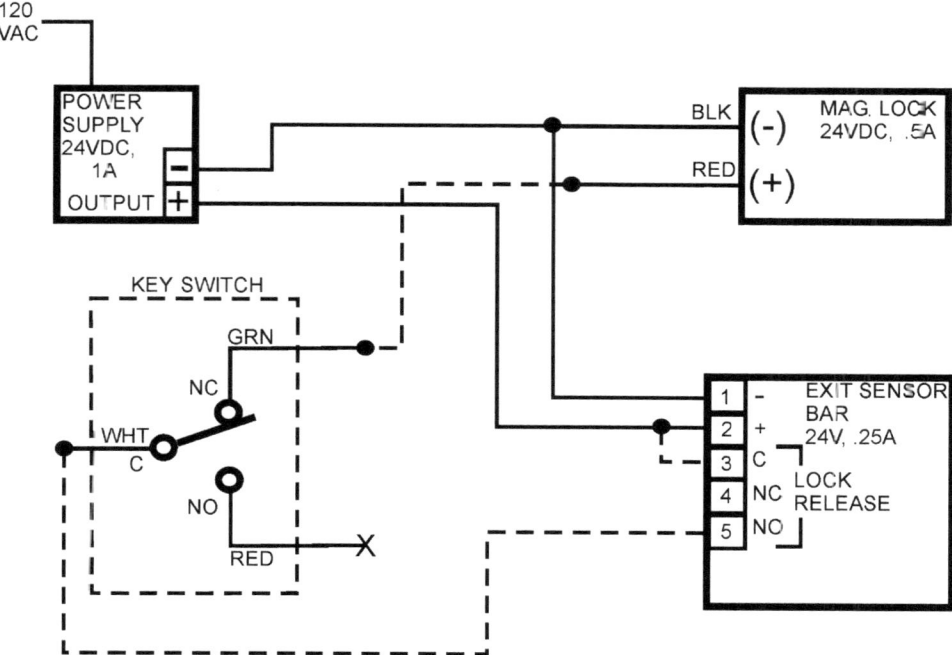

**Figure 10-29** Route (+) from the power supply to the mag lock through all the control devices. You may pick-up (+) directly from the power pupply or from any *unswitched* path from the power supply. A handy place to pick up (+) here is by simply placing a jumper from the sensor bar terminal 2 to terminal 3. Route (+) through terminal 5 (NO held closed), through the key switch to the mag lock.
Note: −x = Not used, insulate and store.
Score: ☺ Too easy? Read on and we'll add a little alarm circuit to complete this system.

Notice in Figure 10-29 that power to the magnetic lock is controlled by two "switches." One is the contact set in the exterior key switch; the other is the relay contact set built into the exit sensor bar. Also notice that power is wired to the sensor bar to operate the relay. By tracing (+) from the power supply you can see that it runs to the sensor bar, through the normally open–held closed relay output contacts, through the key switch normally closed contact, then to the magnetic lock. Opening either of these contact sets will interrupt power to the lock. *Note:* The sensor bar contacts are normally open–held closed because the sensor bar relay is normally energized in the system. This condition makes the sensor bar device as fail-safe as possible. Pushing the bar deenergizes the relay, causing the contacts to "fall back" to normally open. Also, any malfunction, for example, power loss or damaged relay, causes the relay to fail, opening the contacts controlling the lock power.

## Security System with Added Alarm Circuit

In Figure 10-30 we have added an extra set of switch contacts to the key switch. The key switch now becomes a double-pole, double-throw (DPDT) unit, readily available on the market. A second set of relay contacts, used for alarm shunt, has been added to the sensor bar. These contacts are also readily available from most sensor bar manufacturers, either standard or as an option. Both these additional contact sets will be used to control the door alarm that has been added. If we are to monitor the door, we will also need to add a door status switch.

The additions to our exercise are needed to provide an alarm when the door is opened without authorization. This condition can be caused by tampering with the lock or a forced door. Treat this as an entirely new circuit. The door status switch is the control for the new load—the alarm device. The additional key switch contacts and sensor bar alarm shunt contacts are used to prevent (shunt) the alarm from sounding during an authorized release of the lock.

See if you can complete this circuit by following the same rules you have been using. Wire (−) first to the load; route (+) through the proper switch contacts. Solution in Figure 10-31.

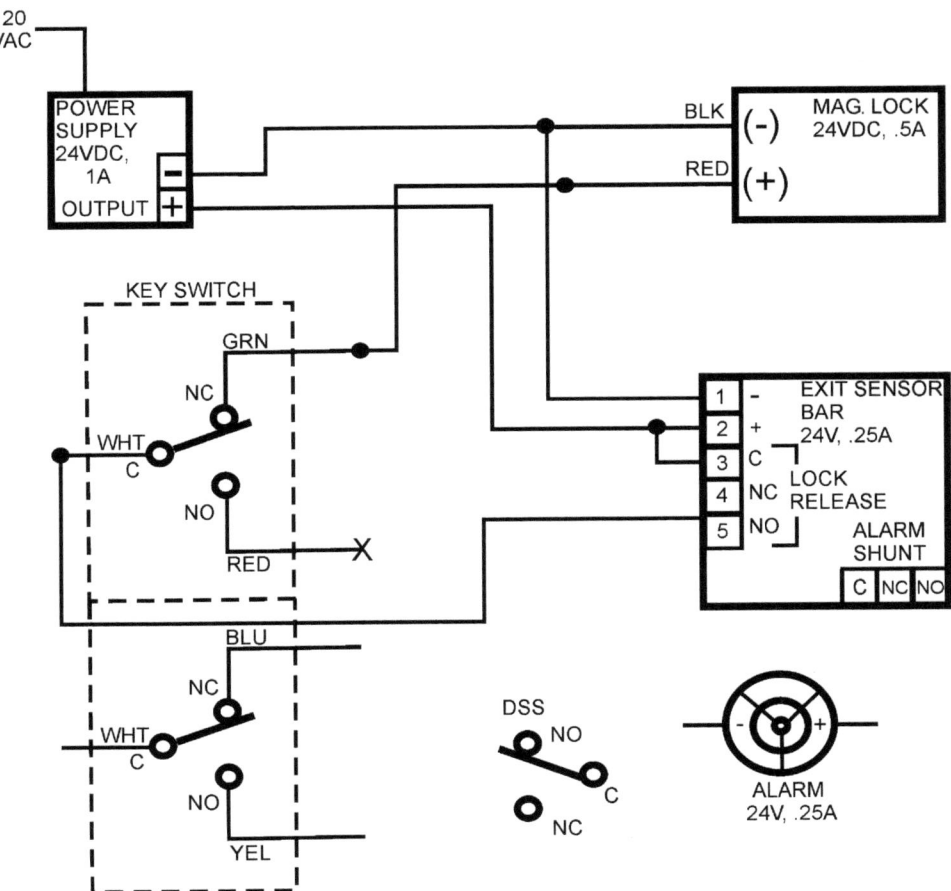

**Figure 10-30** Security system exercise with added alarm circuit.

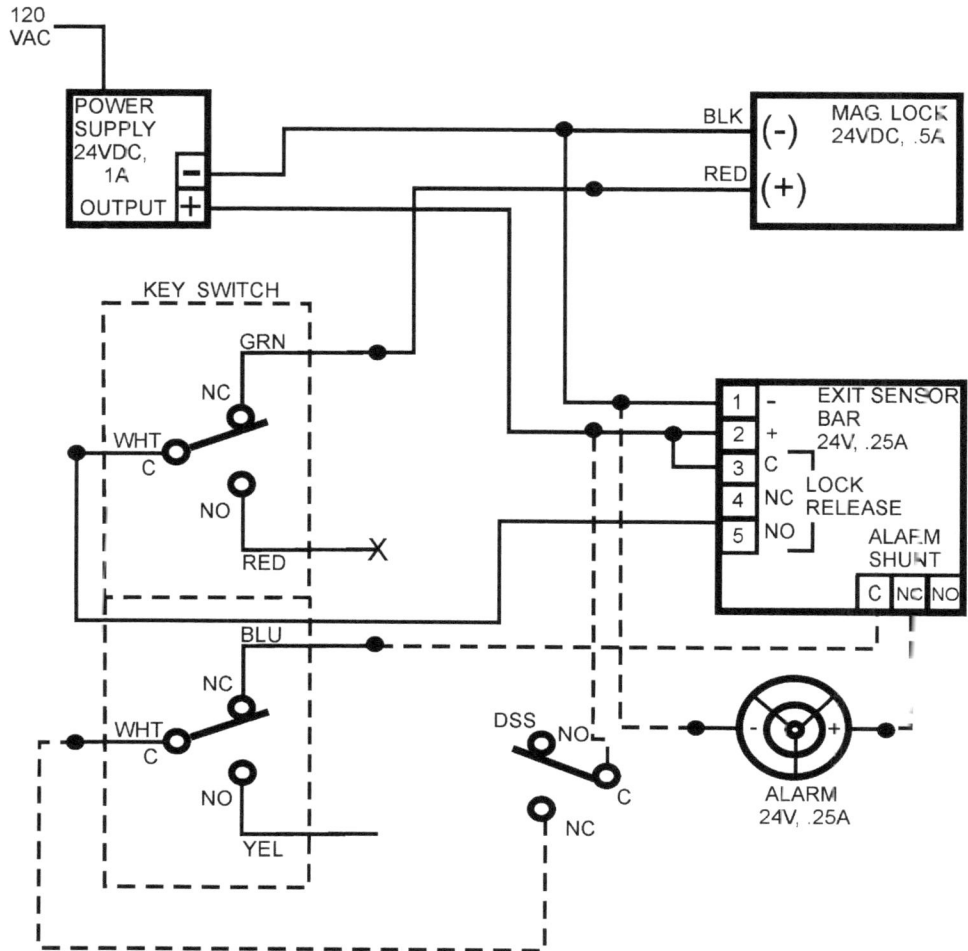

**Figure 10-31** Security system with alarm circuit completed.
Bring (−) to the alarm from any convenient place on the (−) runs. Pick up (+) from any
unbroken (+) run. Run (+) through the door status switch, through the key switch contacts,
through the sensor bar alarm shunt contacts to the alarm.
Score: ☺ Getting this entire problem passes you for the course!
☹ Having problems? Review applicable sections of the book.

Notice the way the alarm circuit is shown in Figure 10-31. If the door is opened
without using the key switch or sensor bar first, the door status switch will "fall"
closed, allowing (+) to flow to the alarm. If the key switch is first used to turn off
the magnetic lock, its second set of normally closed contacts opens. This prevents
(+) from flowing through to the alarm when the door opens. The same condition
occurs if the sensor bar is used, opening its alarm shunt contacts.

Of course, this type of circuit would probably be a little more elaborate in an
actual system. There would most likely be some associated timer devices
included to properly sequence this operation.

By now you should have a pretty good grasp on how to make and read a wiring diagram. Everything presented so far has been pretty basic, but you should be able to create more elaborate systems with a little practice. Do not worry about laying out a wiring diagram five, six, even ten times! This is normal. I also find that using multicolor highlighters to trace out different areas of a diagram really helps.

The next few chapters will take us into more advanced areas of system design and troubleshooting.

# Interlocks

In designing electronic security systems and wiring diagrams, at some point you will be confronted with interlocks. This security feature requires communication and control between various components of a system. Electric locks, access controls, and even equipment not directly associated with the security system all may have to interface with one another.

As we have previously learned, all system components can provide an output signal that monitors their status. Doors can be provided with door status switches, locks with lock status switches. Access and egress controls can provide separate output signals indicating that an authorized release is in progress. Outputs from other equipment, for example, burglar alarms, motion detectors, elevators, and X-ray machines, can all be used to control specific sequences in a security system. These outputs are commonly open and closed dry contacts, shown on wiring diagrams exactly as we have seen in previous chapters. The trick lies in knowing how to place these outputs on the wiring diagram to achieve the desired interlock sequence.

This chapter will describe several commonly used interlocks and their wiring diagrams. Once you understand the concept, you will see that an endless variety of interlock designs are possible.

## Types of Interlocks

To understand the concept of interlocks, first we look at descriptions of common types of interlocks. The terms used to describe these interlocks are not necessarily "standard" but do provide a good generic label for their function. In security work it is common to be presented with just a description of an interlock with no particular name or label. The problem is that most descriptions never contain enough information. It is important to describe everything going on in an interlock. This should include how operating each door or lock affects all other doors and locks in the interlock. You should also mentally walk yourself through the interlock, noting what is required for access, egress, and emergency controls. On large multidoor interlocks, you may have to make a chart of how each door operation affects each other door in the interlock. We will cover this a little later in the chapter.

Another important feature is whether the interlock is controlled by door status switches or lock status switches. The use of door status switches is more common. This may be due in part to the fact that many designers are not aware of the availability of lock status switches. Lock status switches provide a quicker reaction and more secure interlock. As an example of the difference, let's look at two doors, each equipped with an electric lock and door status switch. The interlock might be described as follows:

> Both doors are normally closed and locked. Unlocking and *opening* one door disables the release device for the other door until the first door closes.

This arrangement works fine for most interlocks, especially low- to medium-security systems. For high-security systems this design does present one minor problem. There exists a short time lag between the unlocking of the door and the actual opening of the door. During this time lag it would be possible to unlock, and even open, the second door.

Now let's substitute *lock* status switches for the door status switches. This interlock might be described as follows:

> Both doors are normally closed and locked. *Unlocking* one door disables the release device for the other door until the first door *relocks*.

This arrangement provides an immediate interlock activated by simply *unlocking* a door. It is a little more expensive, but offers a faster, more secure interlock.

Let's look at a description of several common interlocks. We will follow this by several basic wiring diagrams for these systems.

## The mantrap

We will start with the mantrap, as it is one of the most misused terms for describing an interlock. A true *mantrap* is most commonly found in prison and detention work. It is also sometimes referred to as a *hard interlock*. It is not as automatic as other interlocks because it requires certain manual operations as well. The mantrap provides for the secure movement of prisoners or violent psychiatric patients from one area to another. It has also been used in banking, casino, and other facilities where high-value items are moved around. A typical floor plan is shown in Figure 11-1.

A typical problem for this type of interlock might be written as follows:

> An authorized security guard is to move an internee into a secured area. The guard must not carry any key or access device and must be visually identified before access to the secure area is granted. Access and egress to the area must be secured by a hard interlock and controlled by a guard station.

The description of operation for the solution to this problem might be as follows:

Two-door mantrap has both doors normally closed and locked by fail-secure high-security electric locks. Corridor door is to be unlocked by guard station after visual identification of entering personnel. Upon unlocking of corridor door, a lock status switch shall immediately disable *any* release device for dormitory door. Corridor door shall lock immediately on reclosing. Guard station may unlock dormitory door after visual verification within the mantrap. Upon unlocking of dormitory door, its lock status switch shall disable *any* release of corridor door until dormitory door is closed and relocked. Mantrap operation shall operate in same manner for return to corridor. Battery-powered backup and a restricted emergency release switch are to be provided at remote guard station.

This is a pretty elaborate interlock, but effective control for a high-security area. The following descriptions are for three of the most basic two-door interlocks.

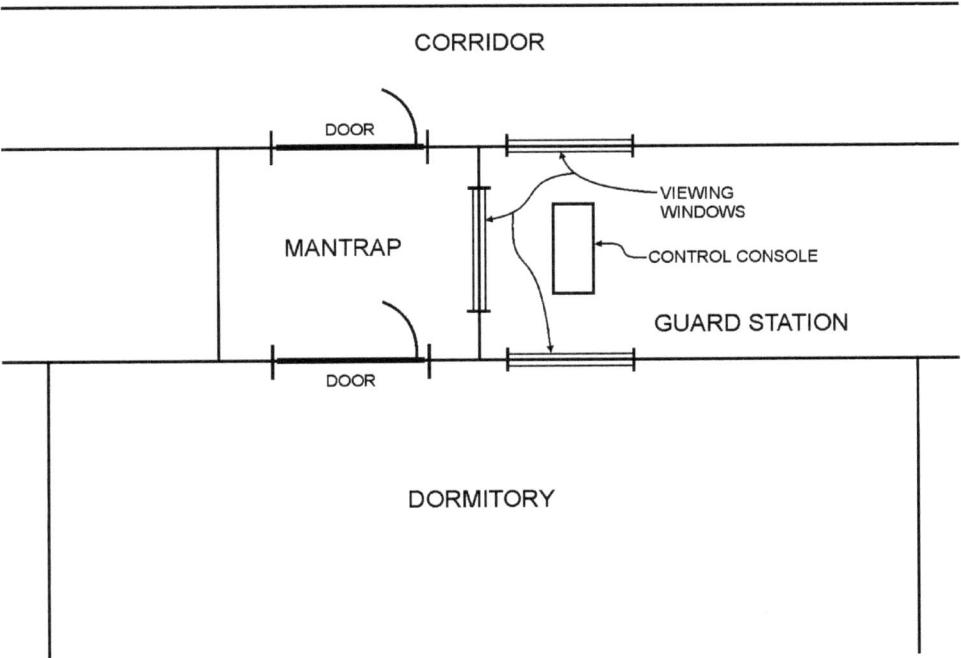

**Figure 11-1** Typical mantrap floor plan.

### Two-door safety interlock

Both doors are normally closed and locked. Opening one door locks the other door until the first door is reclosed.

### Two-door security interlock

Both doors are normally closed and locked. Unlocking and opening one door disables the release device for the other door until the first door is reclosed.

### Two-door safety/security interlock

Door A is normally closed and unlocked. Door B is normally closed and locked. Opening door A disables the release device for door B until door A is reclosed. Unlocking and opening door B locks door A until door B is reclosed.

Interlocking systems can consist of any number of doors and a variety of operating modes. The most basic systems can be designed by the correct wiring of simple door status or lock status switches. The next section describes several of these systems and their operation. More complex systems will require additional electronics, such as multicontact relays, and will be covered later.

## Two-Door Interlock Systems

The systems shown in this section use door status switches to perform the basic logic required to operate the interlock. These switches must be selected to carry the current required by the locking device(s). They usually have contacts that carry very small current loads. For this reason, only two-door systems are diagrammed. These systems could be expanded to include more doors, but a single-door status switch would then have to carry the load of several locks simultaneously. These systems would be better served by using relays with higher-rated contacts.

For each example I have provided a brief problem and description of operation for the type of system diagrammed. It is important that the description of operation be included on the wiring diagram. The wiring diagram tells the installer how to assemble the system. The description of operation tells the installer how the system should operate when completed.

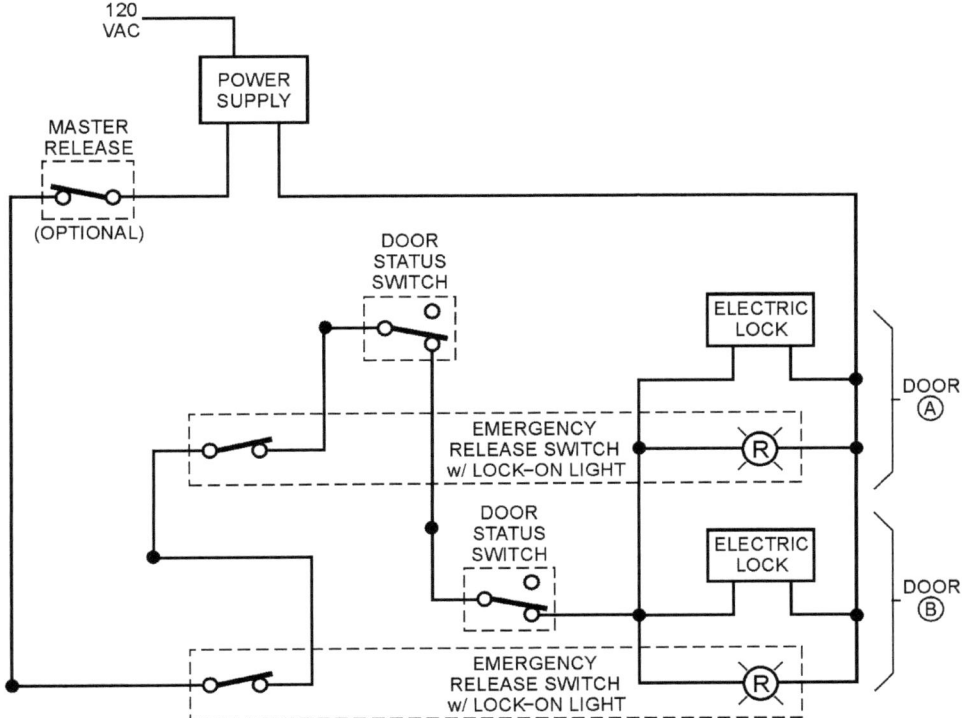

**Figure 11-2** Communicating bathroom electric locking system wiring diagram.

## Communicating bathroom system (Figure 11-2)

**Problem.** A fail-safe electric locking system is to provide privacy and convenience to the occupants of a common bathroom between two rooms. Both doors must be closed before locking can occur. An emergency release switch or power failure releases both doors. An indicator light at each exterior emergency switch will acknowledge that the bathroom is in use. Common applications would be found in hospitals, nursing homes, and dormitories.

**Description of operation.** Both doors are normally closed and unlocked. The occupant enters the bathroom and operates the master switch to lock both doors. Both doors must be closed to initiate locking. An emergency release switch outside each door will unlock both doors. An associated light indicates when both doors are in a locked condition.

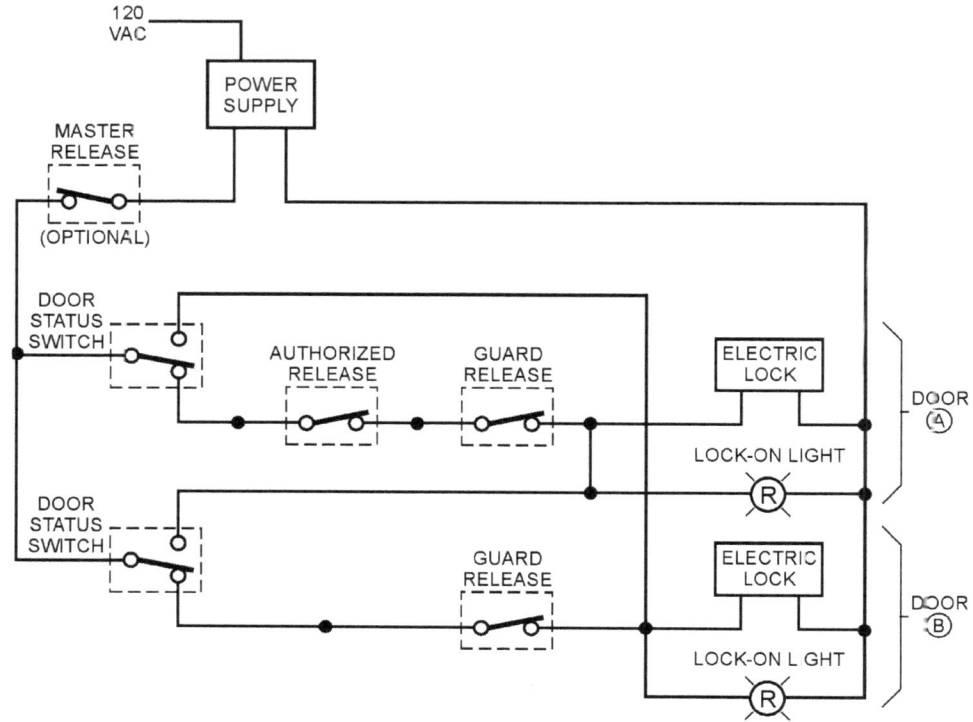

**Figure 11-3**  Mantrap electric locking system wiring diagram.

## Mantrap (Figure 11-3)

**Problem.**    A fail-safe electric locking system is to provide restricted and highly controlled access to a specific area. The mantrap system is to control access through two doors, which cannot be opened at the same time, in a pass-through area. The system allows "trapping" of a person within the pass-through area until identification is made. Common applications would be for jewelry store entrances, money-counting rooms, and high-security areas.

**Description of operation**.    Both doors are normally closed and locked. The entrance door (A) can be unlocked by authorized personnel using a key switch, card reader, or other restricted-access control. In a true mantrap, also called a hard interlock, the access control may be replaced by a call station. This requires identification of personnel before they enter the mantrap. Common application is found in detention facilities or hospital psychiatric units. Once in the mantrap, the person must call for identification, which may be performed by a closed-circuit TV camera installed inside the trap. An interior guard station may then release the interior door (B), and personnel may pass into the restricted area. If identification is not made, the guard may leave personnel secured in the trap until further action is taken. The door status switches

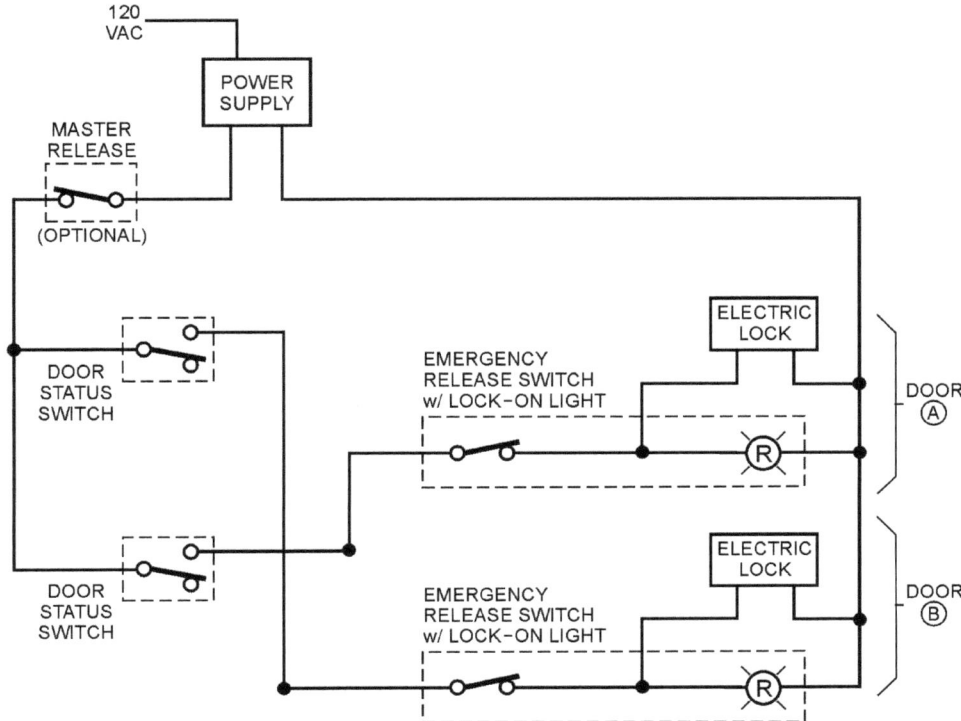

**Figure 11-4**  Two-door safety interlock electric locking system wiring diagram.

provide an override of any release switch, so that no door can be unlocked by anyone if the other door is open. The master release switch is optional and will release any locked door.

*Note:* Commercial "soft" mantraps may require fail-safe locks with fire panel tie-ins. Detention-type interlocks should be controlled by fail-secure locks with *lock* status switches rather than door status switches.

### Two-door safety interlock (Figure 11-4)

**Problem.**    A fail-safe electric locking system prevents opening of more than one door at a time. It provides emergency release switches in areas with no other means of egress. Common usage is to control or restrict the flow of light, air, or people in darkrooms, clean rooms, or similar areas in hospitals, laboratories, and industrial and military facilities.

**Description of operation.**    Both doors are normally closed and unlocked. Opening one door automatically locks the other door, which remains locked until the open door is closed. Emergency release switches may be provided in areas where personnel could be trapped if a door were left ajar. An optional master release switch will release any locked door. Fire panel tie-in may be required.

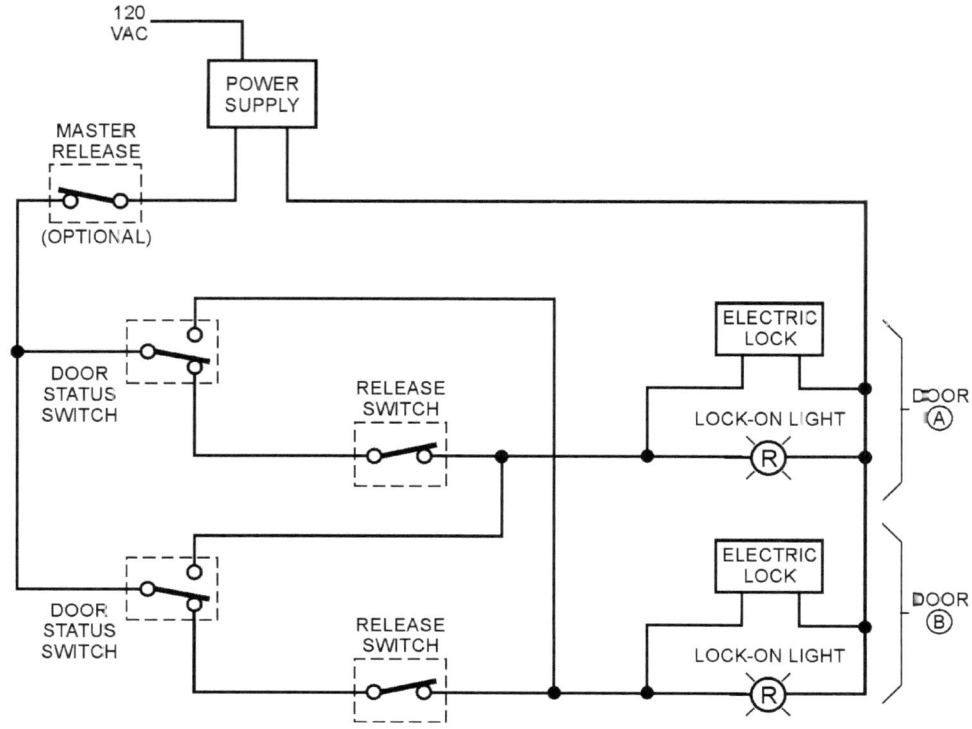

**Figure 11-5**  Two-door security interlock electric locking system wiring diagram.

## Two-door security interlock (Figure 11-5)

**Problem.**    A fail-safe electric locking system is to prevent opening of more than one door at a time. Access control for authorized unlocking may be provided on both sides of each door, if required. Common usage would be in money-counting rooms and drug storage rooms where there is no supervisor or guard control. For higher security, lock status switches may be included to override release switches (see Figure 11-6).

**Description of operation**.    Both doors are normally closed and locked. Unlocking and opening one door overrides the release switch for the other door. The second door cannot be unlocked until the first door is closed. A master release switch is optional and will release any locked door. Fire panel tie-in may be required.

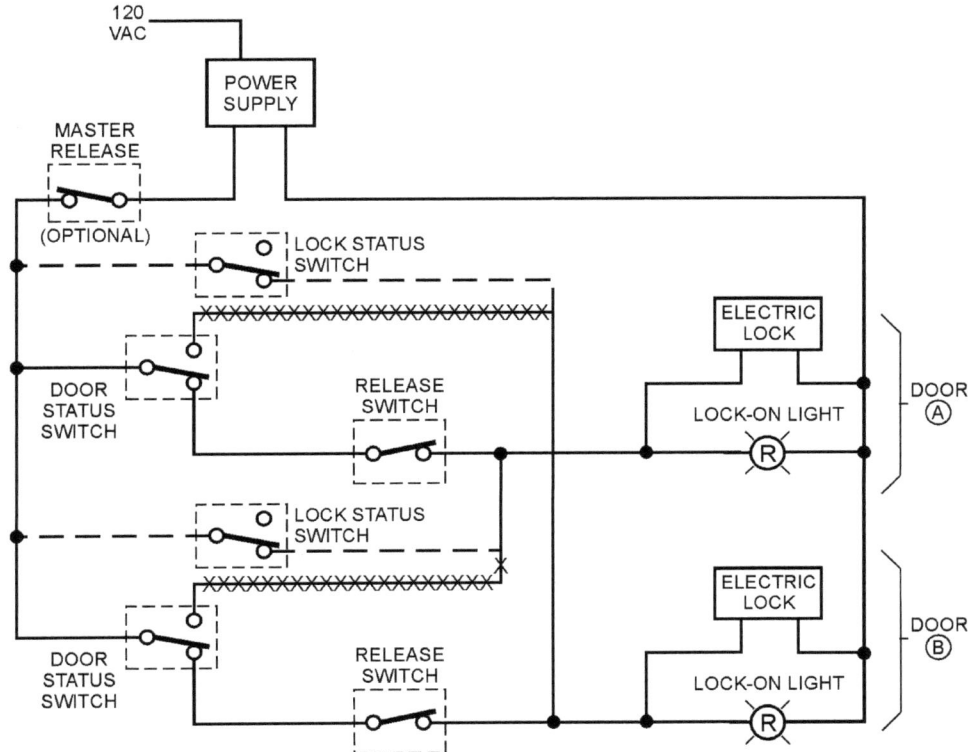

**Figure 11-6**  Two-door security interlock electric locking system wiring diagram with added security.

## Two-door security interlock with added security (Figure 11-6)

The operation of this interlock is identical to that for Figure 11-5 except the door status switch is replaced by the lock status switch. This switch, rather than the door status switch, now controls the interlock, increasing security. In the prior system, the door had to be unlocked and opened before interlocking could occur. In this system unlocking one door immediately voids the release for the other door.

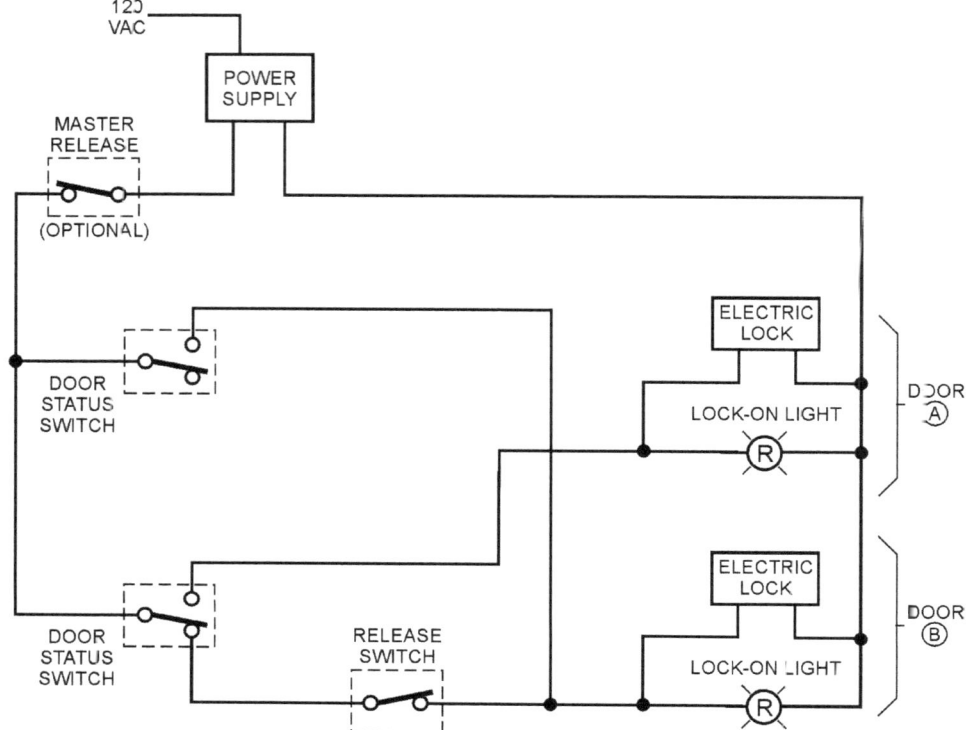

**Figure 11-7**  Two-door safety/security interlock electric locking system wiring diagram.

### Two-door safety and security interlock (Figure 11-7)

**Problem.**  A fail-safe electric locking system is to prevent opening of more than one door at a time. This system is a variation of the systems shown in Figures 11-4 and 11-5. Where the other systems specify both doors normally locked or both normally unlocked, this system requires one normally unlocked door and one normally locked door.

**Description of operation.**  Both doors are normally closed. Door A is normally unlocked; door B is normally locked. Opening door A overrides the release switch for door B until door A is closed again. Unlocking and opening door B will cause door A to lock until door B is closed again.

### Multidoor Interlock Systems

When we get into interlocks of more than two doors, things can get a little crazy. Interlocks of three or more doors may be security, safety, or security and safety interlocks, as previously described. The problem is that a door status or lock status switch can "talk" to one door, but not to two, three, or more doors. Three doors and up usually requires the use of multicontact relays. A single status switch can be used to trigger a multicontact relay. The relay contacts can then be used to communicate with multiple doors, locks, or other equipment.

Relays are also used with two-door interlocks where door status contacts are not rated to handle the current of the locking device. Many security manufacturers offer circuit boards with multiple relays for this use. All the components shown in the following wiring diagrams will have generic identifications for input and output terminations. In actual wiring diagrams you will use manufacturers' designations for terminals, wire colors, etc. The following examples will also include the use of a generic circuit board containing four double-pole relays.

## Using Relay Control

The relay board shown in Figure 11-8 was created for use in the multidoor interlocks presented in this section. Similar relay boards are available from several security product manufacturers. In some cases individual relays could be used, or relays may be built into other equipment, say, a power supply/controller.

The relay board shown in Figure 11-8 is a printed-circuit board with four double-pole, double-throw (DPDT) relays solder-mounted to it. A set of six terminals is provided for external wiring to each relay's output contacts. A separate set of terminals is provided to wire input power that will trigger each relay. Relay boards are available for 12- or 24-volt DC operation; some offer

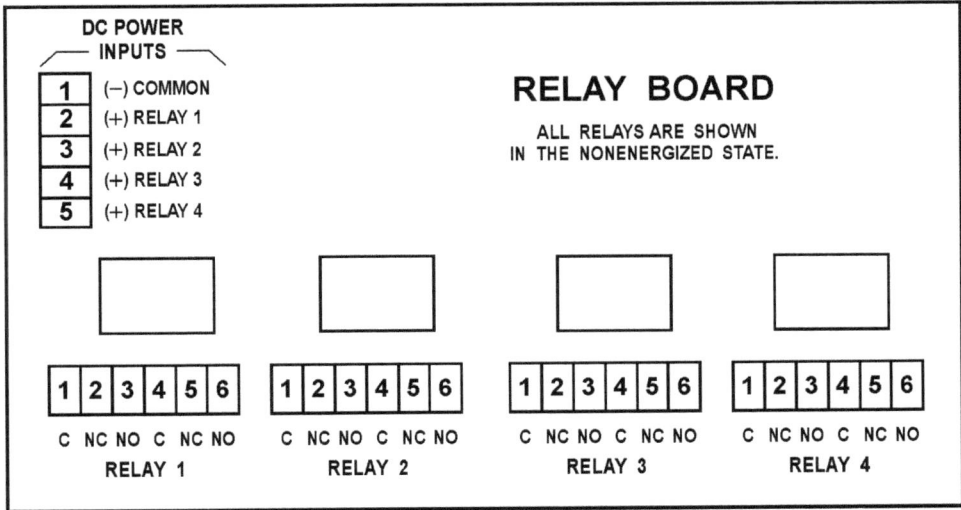

**Figure 11-8** Relay board.

field selection of operating voltage. The relays draw very little current, commonly around 35 to 40 milliamperes (abbreviated as mA). Typical relay contact ratings are 1 or 2 amperes at 30 volts DC. Figure 11-9 shows a simple example of how the relay board might be used.

In this example relays 1 and 2 are triggered by door status switches. Note that opening door 1 closes the door status switch, allowing (+) to flow from the power supply to input terminal 2. This energizes relay 1, causing the relay output contacts at terminals 1 through 6 to change state. Terminals 1 and 3, which were an open contact set, now close, sending a signal to an alarm circuit. Simultaneously the second set of contacts for relay 1 changes state. This set, wired from terminals 4 and 5, was normally closed. They now open, shunting out the use of the door 2 fail-secure lock release. The same scenario takes place if the door 2 door status switch closes, energizing relay 2. The entire wiring diagram for this system would normally show the alarm and lock wiring. This partial wiring diagram is shown only to help you understand the use of relays in a system.

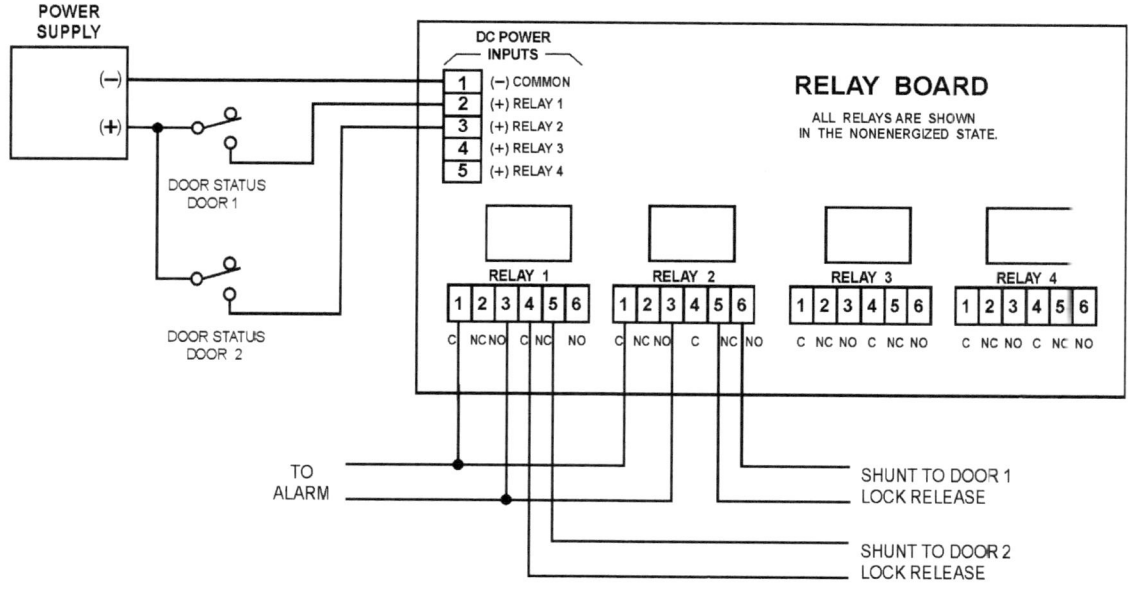

**Figure 11-9** Example of a relay board wiring.

## Interlocks with Relay Control

The following drawings show wiring diagrams for a variety of interlocks using a relay board for control. Each wiring diagram represents a common real-life scenario. However, there can be almost endless variations of a system due to the variety of products available and features required for a particular environment. By now you should be able to easily make changes to these wiring diagrams to fit a specific system. Following each figure is the description of operation for that interlock. Note that the description of operation should be included on every wiring diagram. It is shown separate in this book to allow maximum space for the drawing.

Following each description of operation is a short critique of that interlock. This narrative covers different ways you can alter the wiring diagram to meet the requirements of a particular system you may be confronted with. Some of the alterations include adding or deleting emergency release switches and fire panel interface and the use of lock sensor switches instead of door status switches, and vice versa.

The easiest way to "see" how these more complex wiring diagrams work is to use a method employed in classroom work. Get yourself three or four different-color highlighters. I always trace out the (−) runs from the power supply to the loads with one color. This gets the (−) runs out of the way so that you don't have to keep looking at them. Using other colors, trace (+) from the power supply to each load. As you move along, you will see how different contact sets control the (+) run on its way to the load. This is also the time when you will catch any errors in using open or closed contacts. When a (+) run reaches a contact set, you must also check to see what event triggers the relay to change the contact state. Sometimes this gets a little complicated, and you might want to highlight two copies. One copy would show the system in its normal state; the other would show the system after a particular event, say, opening a door or pushing a button. Read the description of operation for the following wiring diagrams. Trace out the wiring as I have mentioned. In a short time you will understand the logic used in putting one of these systems together.

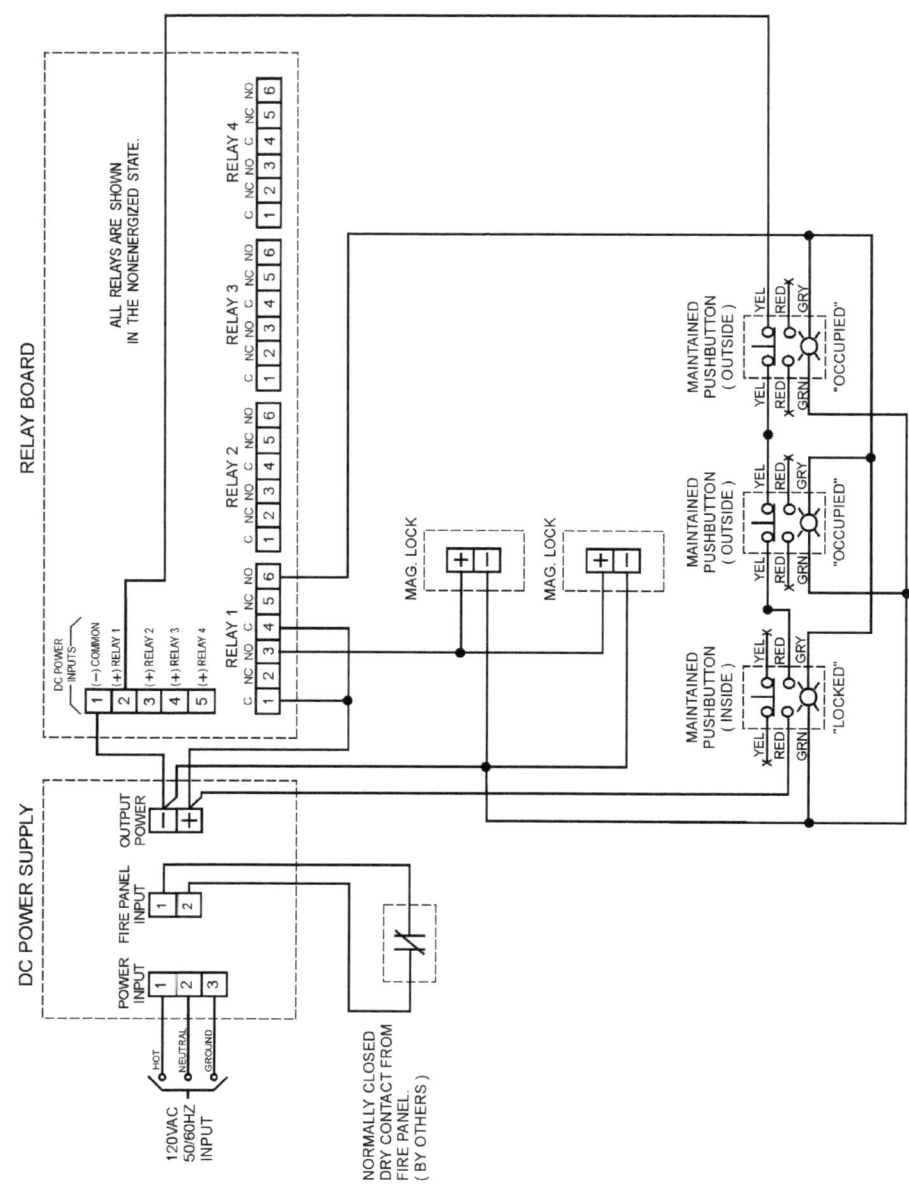

**Figure 11-10** Two-door communicating bathroom interlock wiring diagram.

154

*Description of operation:*

Both doors are normally closed and unlocked. Occupant enters bathroom and depresses inside pushbutton to lock both doors. To exit, occupant depresses inside pushbutton again to unlock both doors.

Actuating either override pushbutton located outside each door unlocks both doors for emergency access. All indicator lights are illuminated when the system is locked. Both doors will immediately unlock upon a fire panel activation.

In Figure 11-10 notice that (+) power from the power supply is routed through the normally open inside pushbutton and the two normally closed outside pushbuttons, then continues to relay 1 input power terminal 2. Also note another path (+) takes from the power supply to relay 1 common terminals 1 and 4. An occupant would enter the bathroom through either of the unlocked doors. He or she would then depress the inside pushbutton, closing its open contacts. The (+) power would then flow through the closed contacts of all three switches to relay 1 input power terminal 2. The relay coil would energize causing the normally open contacts at relay 1 terminals 3 and 6 to close.

This would now allow (+) power to flow from terminal 1 common out of the normally open–held closed terminal 3 to each of the magnetic locks. Normally open terminal 6 would simultaneously close, allowing (+) power to flow from terminal 4 common out of the now closed terminal 6 on to each indicator light. The illuminated lights show that the doors are locked and the bathroom is occupied. Should an emergency arise, depressing either outside pushbutton would break the (+) line, deenergizing relay 1. The held-closed relay contacts would "fall" back open, breaking the (+) line to the magnetic locks and releasing the locks. Opening the fire panel contacts wired to the power supply immediately kills the output power, ceasing (+) power to the magnetic locks.

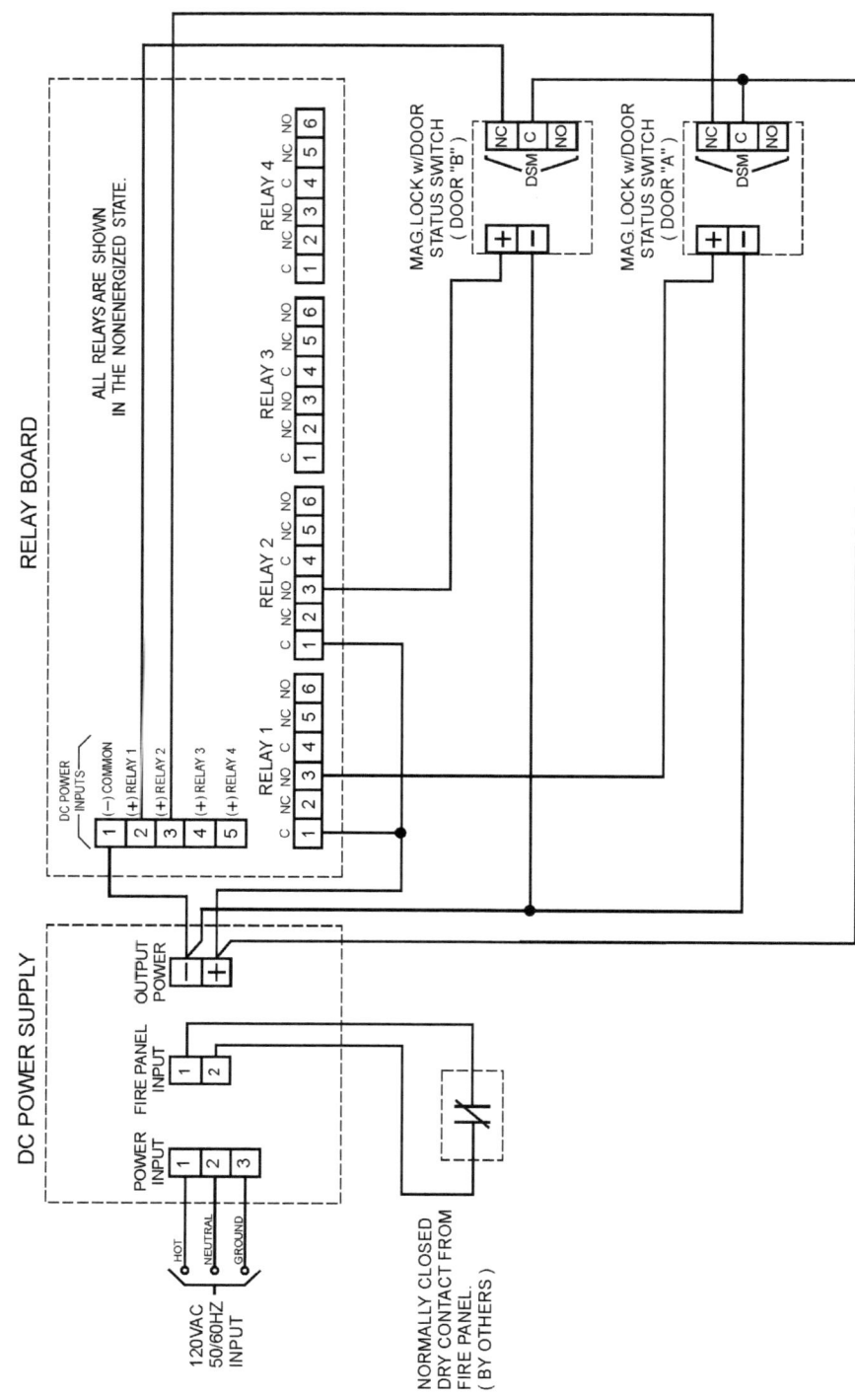

**Figure 11-11** Two-door safety interlock wiring diagram.

*Description of operation:*

Both doors are normally closed and unlocked. Opening one door locks the other until the first door is reclosed. Both doors will release immediately upon a fire panel activation.

The system in Figure 11-11 is fairly simple. The (+) power line for door A and door B is controlled by the normally open contacts of relays 1 and 2. The magnetic lock on each door is equipped with a door status switch. It is the door status switch that controls the (+) power line that runs to the relay input power terminals. While each door is closed, the door status switch normally closed contact is held open. This prevents (+) power from reaching the relay power input terminals. The relay normally open contacts remain open, preventing (+) power from reaching the magnetic locks. The doors remain unlocked. When door A is opened, its door status switch held open contact falls closed. This sends (+) power to the input power terminal for relay 2. The relay energizes, closing its normally open contact, sending (+) power to the door B magnetic lock. In this manner, opening door A locks door B. The same routine occurs if door B is opened first. Its door status switch causes the relay controlling (+) power to door A to energize, locking door A magnetic lock. The fire panel tie-in may or may not be required. If not required on this system, there would simply be a jumper across fire panel input terminals 1 and 2. Emergency releases could also be added to this system. This would be as easy as adding normally closed contacts in the (+) line between relay terminal 3 and the magnetic lock.

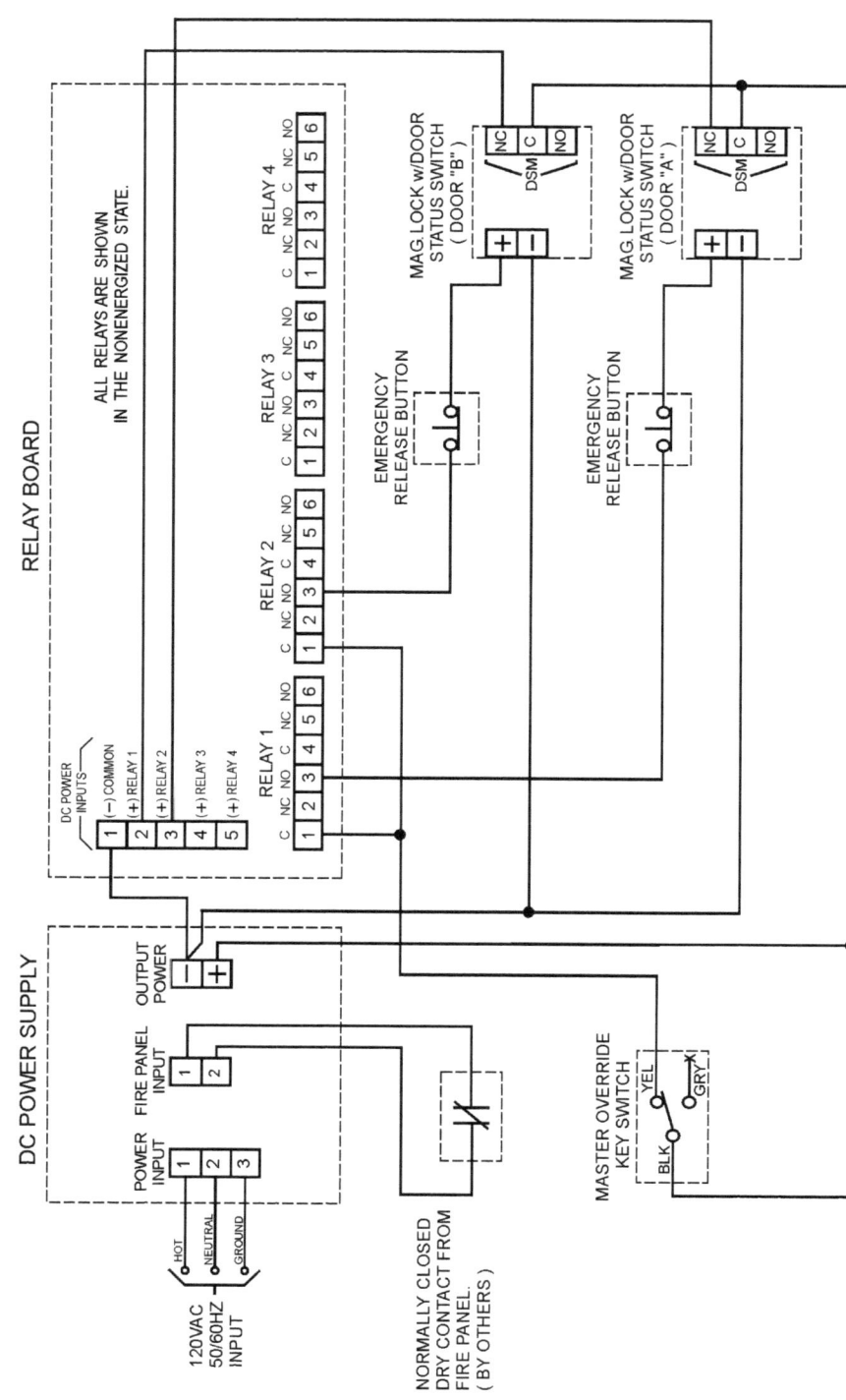

**Figure 11-12** Two-door safety interlock wiring diagram with emergency release switches.

*Description of operation:*

Both doors are normally closed and unlocked. Opening one door locks the other until the first door is reclosed. Emergency release pushbuttons unlock each door independently. The master override key switch unlocks both doors. Both doors will unlock immediately upon a fire panel activation.

The system in Figure 11-12 is almost identical to the system in Figure 11-11. The operation of the system is the same; the only difference is the addition of emergency release switches. As mentioned in the previous system, individual emergency releases are simply closed contact devices placed in the (+) power line to each magnetic lock. These devices could be simple pushbutton switches or a more restrictive device, for example, a break-glass unit that would be used less casually. This system also includes a master override key switch. It is another closed contact but placed in a (+) power line that directly affects both magnetic locks. In this case the master override switch would interrupt (+) power that is fed through both sets of relay contacts. Opening the override contacts would prevent (+) power from reaching both magnetic locks.

The use of a key switch would indicate that the owner wanted some method of turning the system on and off, possibly for maintenance work or deliveries. The master override could just as easily be a pushbutton which might be relabeled *master emergency release.* At any rate by now you should understand how these systems can be changed to suit specific needs.

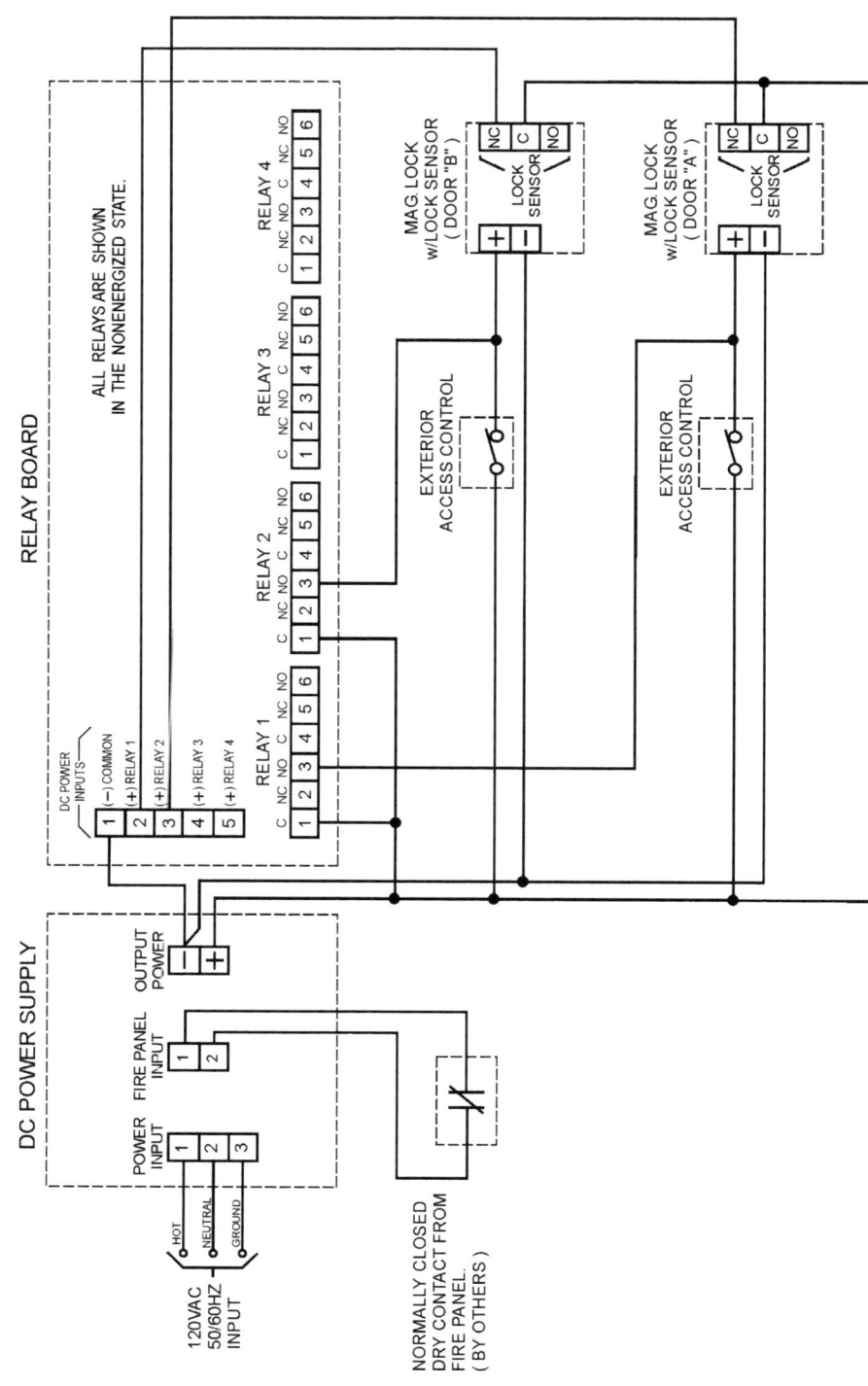

**Figure 11 13** Two-door security interlock wiring diagram.

*Description of operation:*

Both doors are normally closed and locked. Unlocking one door voids the release of the other until the first door is relocked. Both doors will unlock immediately upon a fire panel activation.

Figure 11-13 is the opposite of the two previous interlocks, where both doors were normally unlocked. In this interlock both doors are normally locked. Each magnetic lock is equipped with a lock sensor, which is a built-in switch that recognizes a locked/not locked condition. If you trace (+) power to the magnetic locks, you will find it is a direct run to the magnetic locks through normally closed access control switches. The magnetic locks are energized and locked, activating the lock sensor switch. Trace the other (+) power line from the power supply to the lock sensor common terminal. It will not go any further because the normally closed contact is now held open. If someone uses the access control for door A, its closed contact will open, breaking (+) power to the door A magnetic lock.

The door A lock sensor contact now falls closed and sends (+) power to relay input power terminal 3, activating relay 2. Relay 2 open contacts (terminals 1 and 3) now close and create a path for (+) right around the door B access control closed contacts. If someone were now to use door B access control, its contacts would open but the magnetic lock would remain energized. The (+) power would route itself through the now closed contacts of relay 2 and keep the door B magnetic lock energized. This condition would remain until the door A magnetic lock relocked, resetting the system to its normal condition. The same scenario would occur if someone used the access control for door B. The access control for door A would be disabled by the contacts of relay 1. This type of interlock would also work by using door status switches instead of lock sensor switches. This would require unlocking and opening a door to trigger the relay. It would work just as well but have a little slower activation time.

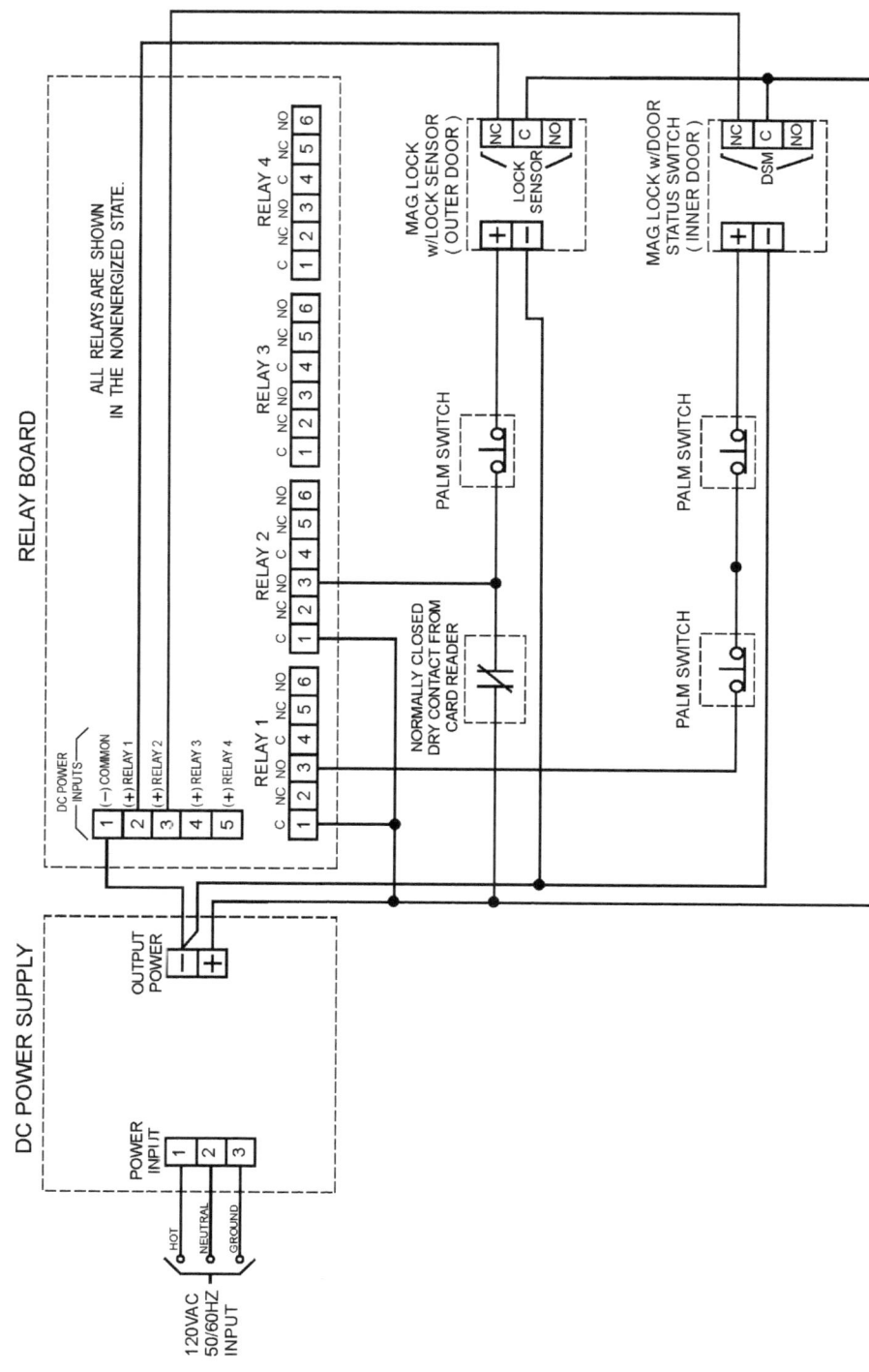

**Figure 11-14** Two door safety/security interlock wiring diagram.

*Description of operation:*

Outer door is normally closed and locked. Outside card reader provides authorized access. Palm switch inside provides emergency release.

Inner door is normally closed and unlocked. Palm switches on both sides provide emergency release when the door is locked.

Unlocking the outer door locks the inner door until the outer door is relocked. Opening the inner door disables the card reader from releasing the outer door until the inner door is reclosed.

The system shown in Figure 11-14 is sort of a combination of a safety interlock and a security interlock. In some ways it also resembles a mantrap. Notice the outer door is normally locked by a direct (+) run through the closed contacts of the controls. The inner door is normally unlocked by the break in its (+) run provided by relay 1 open contacts. If the outer door is unlocked, its magnetic lock sensor switch held open contact closes, allowing (+) to travel to relay 1 input power. This triggers relay 1, closing its NO contacts, sending (+) power to the inner door magnetic lock. The inner door remains locked until the outer door relocks. When the inner door is opened, its door status switch allows (+) to travel to relay 2 input power. This triggers relay 2, closing its NO contacts, sending (+) around the card reader contacts directly to the outer door magnetic lock. The outer door remains locked, and the card reader disabled, until the inner door is closed. The palm switches provide emergency release, and a fire panel release could easily be added.

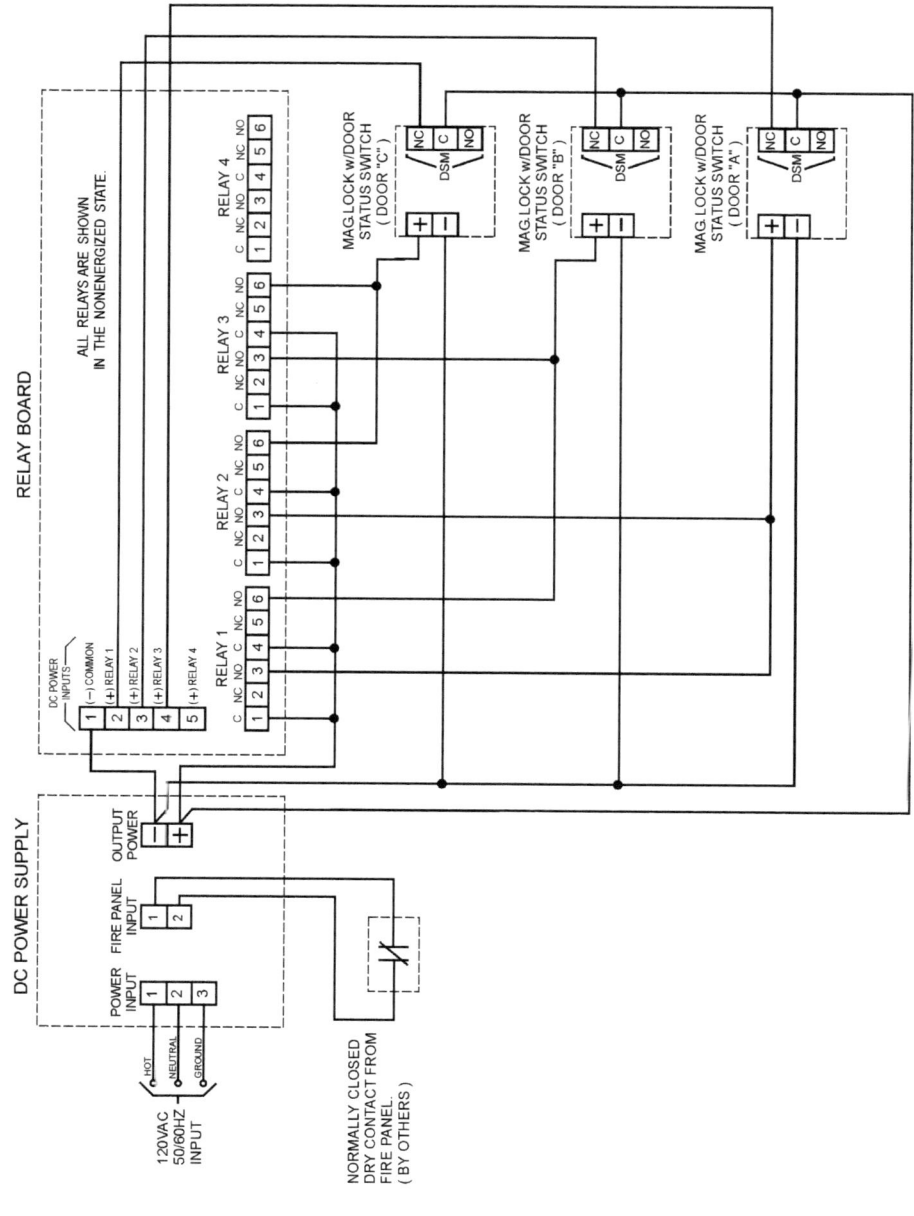

**Figure 11-15** Three-door safety interlock wiring diagram.

*Description of operation:*

All doors are normally closed and unlocked. Opening one door locks the other two until the first door is reclosed. All doors will release immediately upon a fire panel activation.

The safety interlock in Figure 11-15 is identical to the safety interlock in Figure 11-11 except for an added door. The addition of a third door requires the use of a third relay.

Opening door A trips relay 3, and its two sets of contacts lock doors B and C. Opening door B trips relay 2, and its two sets of contacts lock doors A and C. Opening door C trips relay 1, and its two sets of contacts lock doors A and B. Adding a fourth door to this system would be a problem. Notice that each relay only has two sets of contacts and can only handle two doors. The fourth door would trigger the fourth relay, but each relay could only lock two doors. What you would need is a relay board with relays that provide three sets of contacts each. Another solution would be to add a second relay board to gain more contact sets. Of course, it is possible to have a four-door interlock where opening any one door selectively locks only two of the other doors.

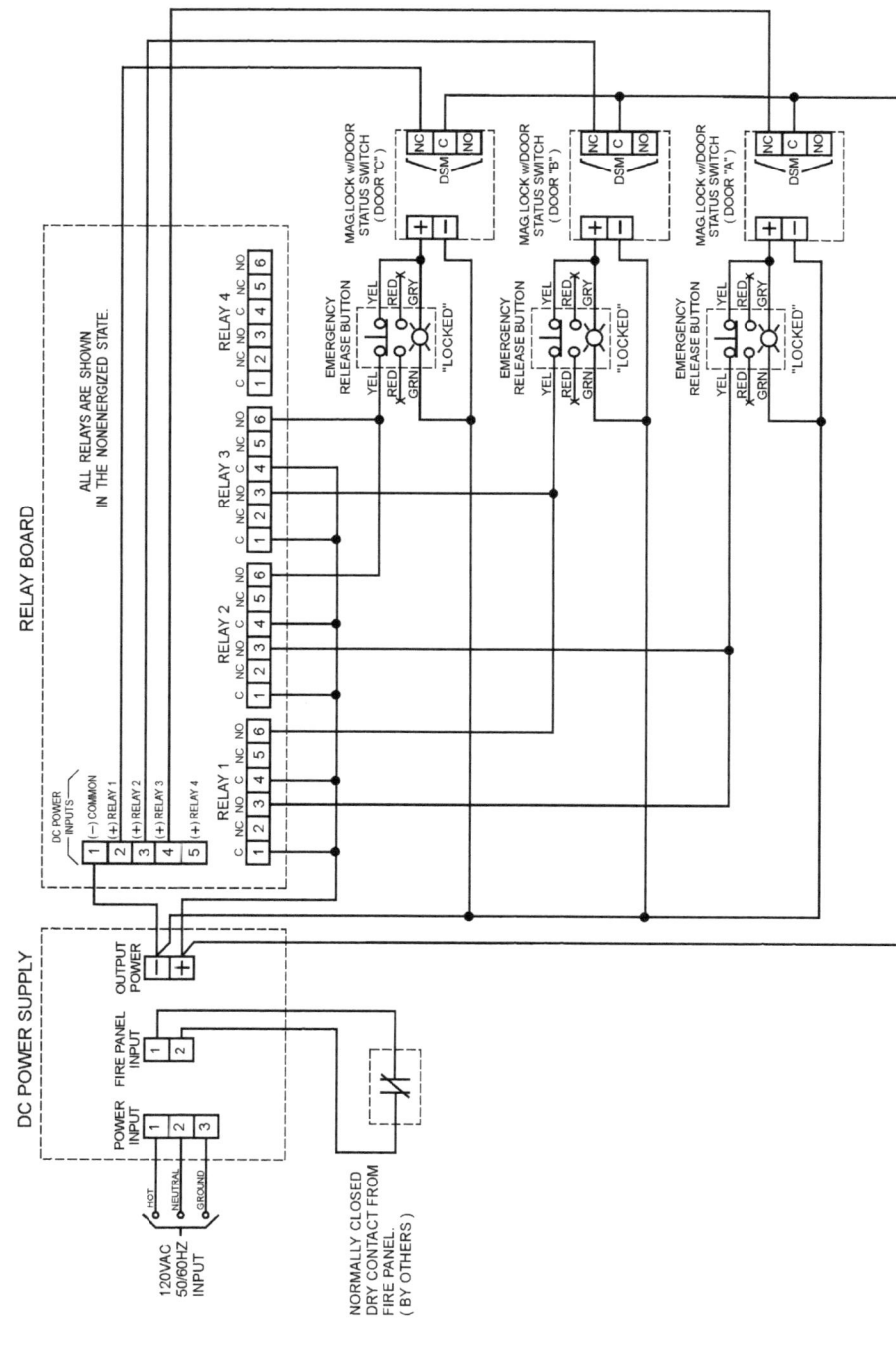

**Figure 11-16**  Three-door safety interlock wiring diagram.

*Description of operation:*

All doors are normally closed and unlocked. Opening one door locks the other two until the first door is reclosed. Emergency release pushbuttons unlock each door independently. The indicator light is illuminated when the door is locked. All doors will release immediately upon a fire panel activation.

The safety interlock in Figure 11-16 is identical to the last interlock in Figure 11-15 except that we have added emergency release buttons for each door. In this system we became a little more elaborate by adding indicator lights at each emergency release. Although the lights are labeled "locked," in reality they only indicate the fact that a lock is receiving power. This may be adequate because the assumption is that the door is closed and the magnetic lock fully bonded. If a true secure indication was required, a lock status sensor could be added to each lock. The lock sensor contacts would be used to provide a path for (+) to the light.

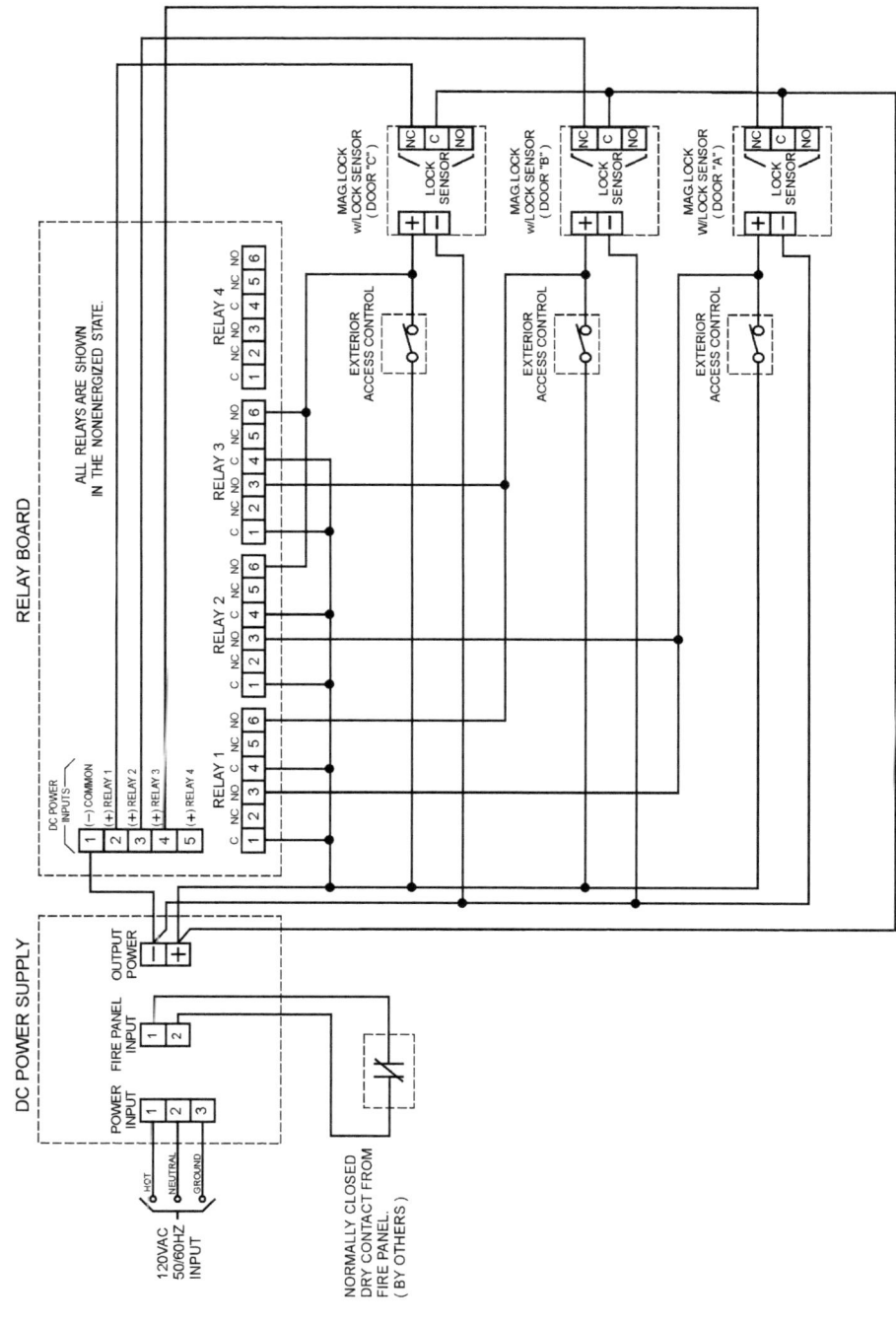

**Figure 11-17** Three-door security interlock wiring diagram.

168

*Description of operation:*

All doors are normally closed and locked. Unlocking one door voids the release of the other two until the first door is relocked. All doors will unlock immediately upon a fire panel activation.

Figure 11-17 is a three-door security interlock that is once again similar to the two door security interlock shown in Figure 11-13. Unlocking any one door triggers a relay whose contacts disable the release device of the other two doors. As noted for Figure 11-15, we have used both sets of contacts on three of the relays. Adding a fourth door would require relays with more contacts or an additional relay board.

## Summary

All the interlock systems presented in this chapter used magnetic locks as the locking device. This was the easiest way to show the variety of interlocks without the confusion of product varieties.

Other electric locks can readily be interchanged as well as access/egress devices and monitoring devices. The wiring diagrams shown should give you a good idea of how interlocks work and the logic used in their design. Two- and three-door interlocks are fairly common; but be aware, interlocks with four to eight doors occasionally come along also. These large multidoor interlocks may seem intimidating but actually can be kind of fun to design.

# 12

# Troubleshooting

In preceding chapters you learned about electronic security hardware, components, how to wire them, and the flow of power to them. This knowledge leads right into the subject of troubleshooting. Understanding how power is controlled and used, one could almost surmise where to start looking for problems in a system. This chapter will cover basic troubleshooting techniques for common problems in electronic security systems. You will also learn about an important tool used for troubleshooting electronic components and systems.

First, I will provide some brief notes on safety in working with electric circuits.

## Safety First

Most electronic hardware operates on low voltage (12 or 24 volts), and current draws are usually low. Although low voltage is generally safe to work with, you should always be aware that line power voltage may be present somewhere in the system. Line power voltage can be deadly and should not come in contact with your hands, body, or tools.

Be sure that power is shut off when you are doing any work with a system. Although high voltage can cause fatal shock, the effect of a shock is determined by the amount of current flowing through your body.

A current of one-thousandth ampere (1 milliampere) is barely perceptible. Up to 8 milliamperes can cause mild to strong surprise. An 8- to 15-milliampere current is unpleasant, but you can usually let go of the item causing the shock. A current greater than 15 milliamperes can lead to *muscular freeze,* which can prevent you from releasing whatever you are holding. Currents in excess of 75 milliamperes can be fatal.

Several factors determine the danger of electric shock. The higher the voltage of the power source, the more milliamperes flow through the body. Also, if the power source produces high current, fatal shock can be caused well below 120 volts.

Consider your home and automobile as potential sources of high-voltage, low-current shocks. Walking on dry carpeting in your home when the humidity is low causes you to pick up a static charge. If you then touch a grounded object, you feel only a mild shock or spark because the current is so small. The coil of a running automobile engine can produce at least 20,000 volts but at a very low current. Touching the coil or metal spark plug cap will give you a surprising, but nonfatal, shock.

To reduce the chance of electric shock, you should stand on a dry, nonconductive surface. If you contact 120-volt line power in this condition, you will feel little shock. While standing on a damp basement floor in your bare feet, however, you would feel severe shock. If you were standing in water, you would be electrocuted. Never stand on a wet surface when you are working with electricity!

Working safely requires an understanding of what you are doing and exactly how you will do it. Plan the job in advance, keeping loose items out of the work area. Do not defeat the purpose of protective devices, such as fuses and circuit breakers, by shorting them out or by using devices rated higher than what is specified. Exercising good judgment and care will protect you and the equipment you are working with.

## The Multimeter

If any one item can save you time and money in the field, it is the multimeter. It can save you hours in troubleshooting and often can save the cost of unnecessary consultation. The multimeter should be a troubleshooter's constant companion. Anyone involved in electronic security systems should have one and should learn the basics of its operation. Once you have learned to work with a multimeter, you will recognize its potential for use in your home and with your vehicle.

Two types of meters in use today are the analog meter [or volt-ohm meter (VOM)] and the digital multimeter (DMM). The analog meter has a pointer that indicates the reading on a calibrated faceplate. The digital meter indicates the reading by displaying the actual measurement in numerical form. Digital meters are becoming more popular because they are easier to use, have greater versatility, and are more accurate with reading errors greatly reduced.

The multimeter is a combination of an ammeter, voltmeter, and ohmmeter in one unit. Its purpose is to make voltage, current, and resistance measurements of components and circuits in an electrical system. It is designed to provide various ranges for these measurements. A labeled panel and selector switch is provided for selecting various functions and ranges. The two types of meters are shown in Figure 12-1.

The best way to learn how to use your meter is to read the operator's manual and practice a little. The manual is important because meters vary in both appearance and operating functions.

Select your meter on the basis of the features that make it most practical for your purposes. A good general-purpose multimeter will handle troubleshooting of most low-voltage systems. The basic meter should enable you to measure,

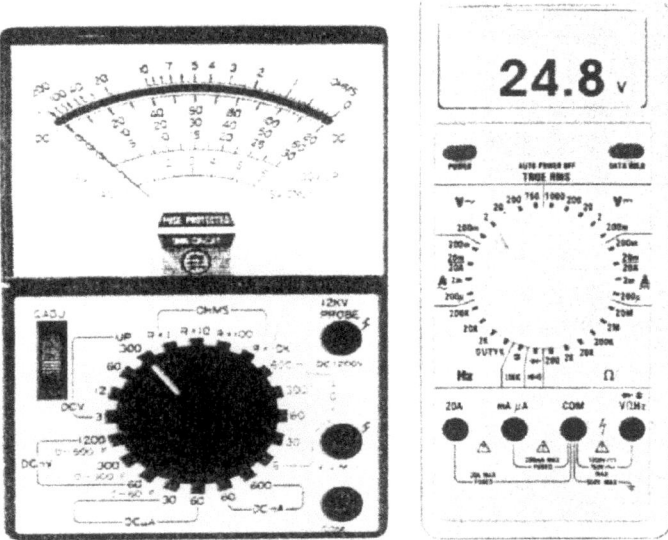

**Figure 12-1**  Two types of multimeter. Analog meter (left). Digital meter (right).

within a good range, DC and AC voltages, direct current, and DC resistance. Some digital meters include AC measurement functions. All meters are equipped with two test leads, usually one black and one red. The test ends of the leads may have probe tips or, in some cases, an alligator clip on the black lead and a probe tip on the red lead.

## Using the Meter

Meters vary in the functions that are available with them. I will outline the basics of each function in the hope that the simplicity of operating a meter will encourage you to obtain and use one.

### Voltage measurements

The purpose of the *voltmeter* function is to measure voltage across two points of a circuit. *Voltage* is defined as the difference in electrical pressure between two points. It is also called potential difference and electromotive force (emf).

The selector switch on a meter usually provides both AC and DC measurement functions, with a range selection for each. There is also a positive and a negative input terminal on the meter, normally identified by one of the methods shown in Figure 12-2.

The test leads are color-coded, one red and one black. The usual convention is to connect the red lead to the positive terminal and the black lead to the negative terminal.

| PROBE TERMINAL IDENTIFICATION ||
| POSITIVE | NEGATIVE |
| --- | --- |
| **+** | **–** |
| V-⎍ | COM |
| ⊕V· ⎍ A | ⊖ COM |

**Figure 12-2**  Typical multimeter terminal identifications. (Some meters have a separate input terminal for current measurements.)

When you are measuring DC voltage, polarity must be observed. Polarity might be identified by the marking on a component (+ or −) or by color-coded wiring. Color coding may vary, as there is no standard for low-voltage wiring and you will have to consult the manufacturer's literature. If the polarity of a DC circuit cannot be identified, one test lead probe can be held on a test point and the other probe quickly tapped on the other test point. On analog meters, the pointer will move below the 0 mark, indicating that you have the polarity reversed and that you must swap the positions of your probes to be correctly polarized. Digital meters will read the voltage either way, indicating the polarity on the display. When you are using the AC voltage function, polarity need not be observed. Either probe may be connected to either test point.

If the voltage value being measured is not known, it is advisable to set the selector to the highest range on the function (AC or DC) chosen. You may then switch down until a satisfactory reading is observed.

Applying high voltage with the selector on a low-voltage range may damage the meter. Voltage readings are always taken with the circuit power turned on. Use extreme caution in measuring voltages of 120 and above. Figures 12-3 and 12-4 show simple voltage measurements being taken.

Readings must be taken with the test probes touching uninsulated junctions or bare wire. If no uninsulated place can be found to place the probes, the following procedure is commonly used: Shut off the power to the circuit. Drive a sewing needle through the insulation into the metal wire of each conductor. Be careful that the needles do not touch each other or any grounded object. Turn the power back on, and take the measurement by placing the probes on the needles. Shut the power off and remove the needles. No damage should be done to the insulation because of the small diameter of the needles. As a precaution, tape the area where the needle entered the insulation.

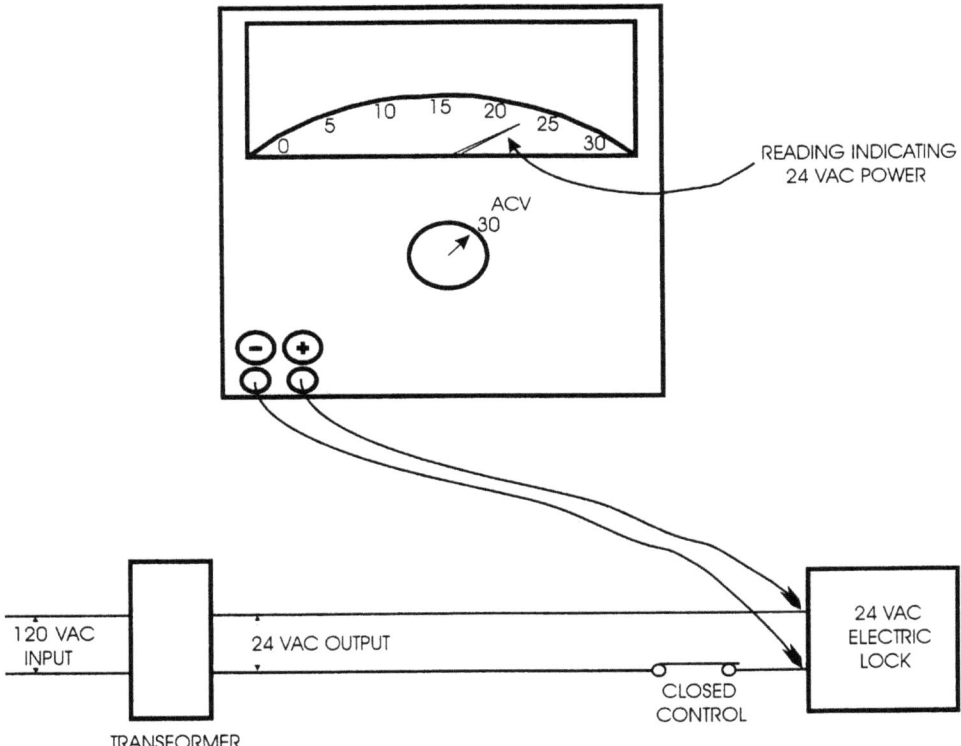

**Figure 12-3** Reading an AC circuit voltage.

## Current measurements

The *ammeter* function is used to measure the current being used by a load. *Current* can be defined as the movement of electric charges. The unit of measurement for current is the ampere (or amp).

Although most multimeters will measure direct current, many will not measure alternating current. Do not attempt to measure alternating current unless your meter has that specific function. Most digital meters have both DC and AC functions.

The test leads should be placed in an open circuit as shown in Figures 12-5 and 12-6.

Set the selector to the desired function, DCA for direct current and ACA for alternating current.

When you measure direct current with an analog meter, polarity must be observed. If polarity is not known, it may be determined with the voltmeter, as previously described in "Voltage Measurements." This will prevent possible damage to the meter. In AC circuits polarity need not be observed.

It is recommended that you start at a high range setting and switch down until a satisfactory reading is observed. When you are measuring current, the

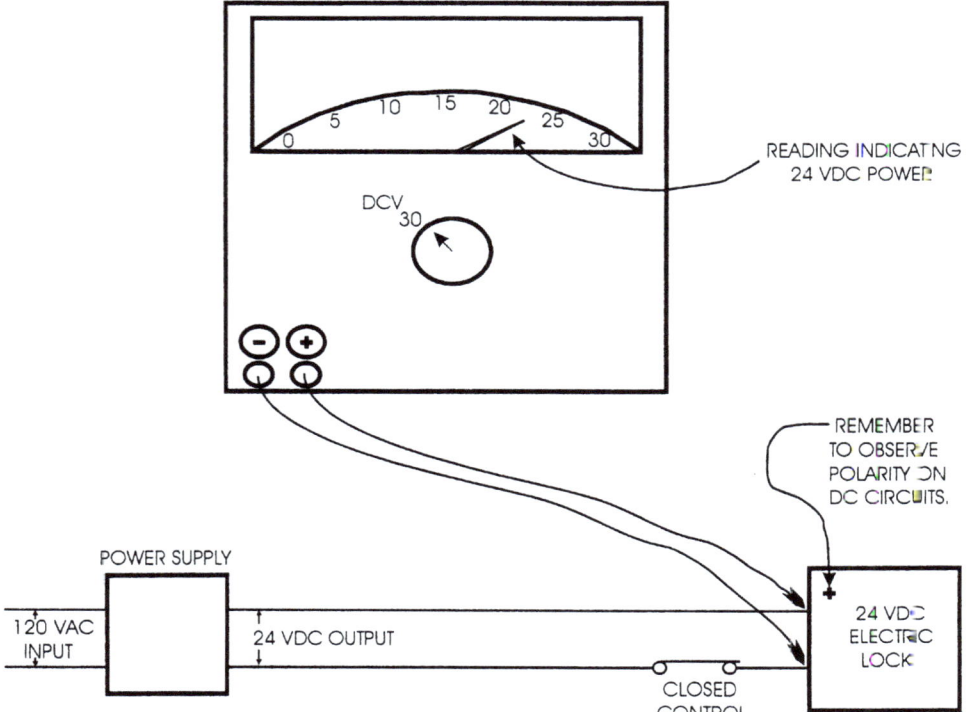

**Figure 12-4**   Reading a DC circuit voltage.

meter must be in series with the load. See Figures 12-5 and 12-6 for examples of current measurements.

Cut the power to the system, and open the circuit at a convenient place. Connect the test probes in their proper positions, and apply power to the circuit. You will now be measuring the current draw at that particular place in the circuit.

### Resistance measurements

The ohmmeter function is used to measure the resistance of a component in a circuit. *Resistance* is defined as the opposition to the flow of current. The measurement unit of resistance is the ohm, often indicated the Greek letter omega ($\Omega$). You can measure the resistance that a component offers to the current flow in a circuit.

Unlike for voltage and current measurements, the power to the circuit must be shut *off* when you are measuring resistance. It is advisable to discharge all capacitors in a circuit, remove all batteries, and unplug all line cords.

The ohmmeter function requires its own power source to provide low current to pass through the component under test. A battery located inside the meter

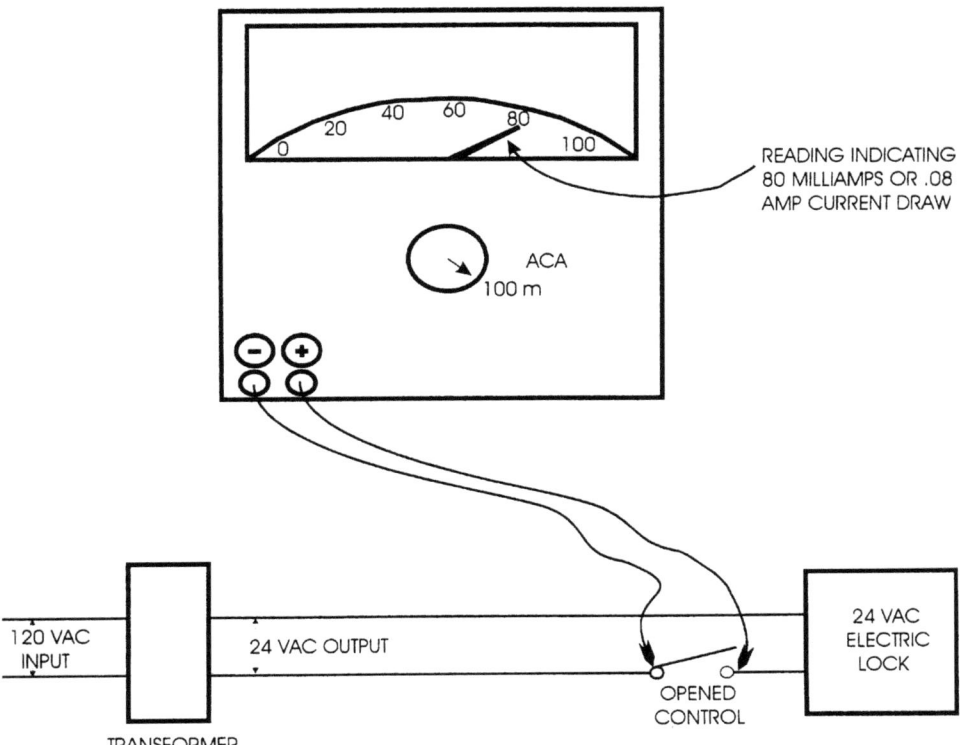

**Figure 12-5**    Reading current in an AC circuit. (Note that with the switch open the meter is completing the circuit.)

supplies this current. When not in use, meters without an on/off switch should not be left in the ohmmeter function, as it will drain the battery.

Analog meters vary as to whether the ohmmeter scale will read left to right or right to left. Consult the meter's manual for the correct procedure to "zero" your meter before starting a measurement.

With the test leads on the component to be tested, place the selector to one of the ohm range positions. It is best to select a range that allows you to read the resistance near the middle of the scale. Check to ensure that power is off in the circuit. The test probes are placed across the component under test, as shown in Figure 12-7. Polarity need not be observed in resistance measurements.

When a component has very little resistance, you will get a low reading; conversely, high resistance will give you a high reading. You may have to run through several ranges until you find a range that gives you a midscale reading. Remember to zero the meter for the range you select, to obtain an accurate reading.

If the pointer does not move in any range, you are measuring across an *open* circuit, indicating a break in the component. (This may be an "0" or "infinite"

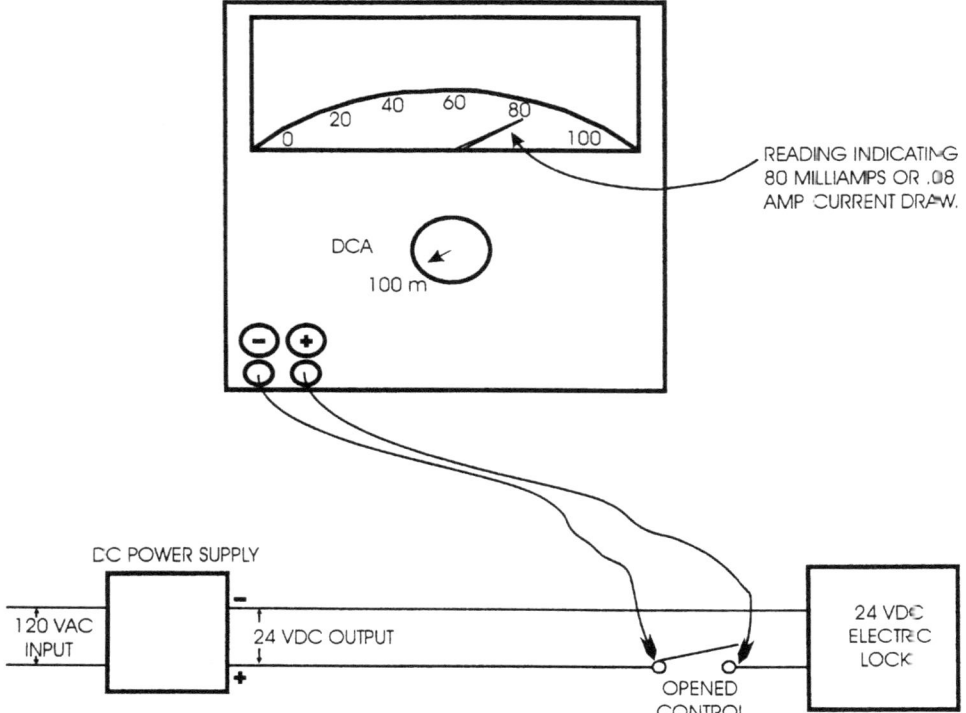

**Figure 12-6**   Reading current in a DC circuit. (Note that with the switch open the meter is completing the circuit.)

reading on your analog scale, depending on the type of meter being used.) If the pointer deflects across the entire scale, you are measuring across a *short* circuit, indicating that the component may be shorted. (This again may be an "0" or "infinite" reading on your scale.) It may be necessary to completely disconnect the component from the circuit to prevent other components in the circuit from interfering with the reading.

## Meter review

Use of a multimeter can save you time and money. A multimeter functions as (1) a voltmeter, to measure voltage (the difference in electrical pressure between two points); (2) an ammeter, to measure current (movement or flow of electric charges); and (3) an ohmmeter, to measure resistance (opposition to current flow). Select a meter that best suits your needs.

Use caution when you are working with or near high voltages. Observe polarity when measuring DC voltage and direct current. Do not attempt to measure alternating current unless your meter is capable of doing so. When measuring resistance, you must make sure that the circuit power is off.

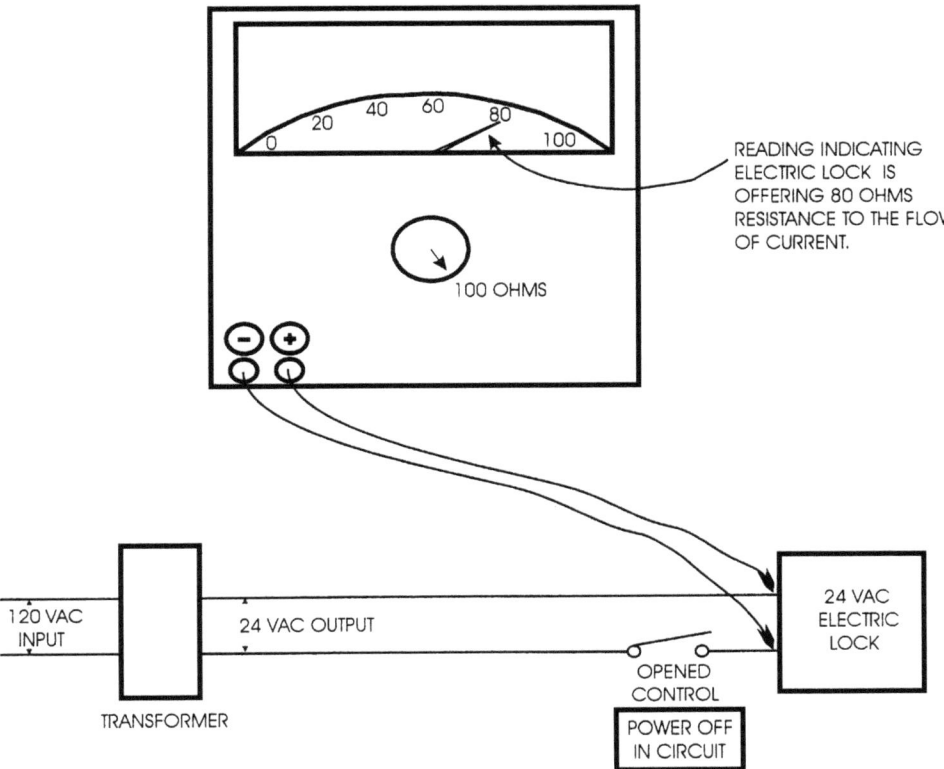

**Figure 12-7**   Reading the resistance of a circuit component.

I recommend a digital meter for most hardware trade people. It is easy to use and takes more abuse. A good unit that measures voltage and resistance is fairly inexpensive. It is not often you will need to measure current, but this feature can be included for a few dollars more.

## Solving Problems

The importance of the VOM was seen in the preceding section. Although it is the tool you will use the most, your first step will be just plain good detective work. What you are told is that the trouble will often be far from the real problem. The first two details to document are (1) a list of symptoms and (2) whether the system ever worked properly.

If a system is determined to be just installed, and it never worked or works improperly, the problem is very often miswiring. A component failure this early would be unusual, since manufacturers normally test products before shipment. Always ask for the system wiring diagram used for installation. Nine out of ten times the wiring diagram is wrong, or the actual hookup wiring is wrong.

If a system has been up and running, the problem usually lies in product failure or some event adversely affecting the system. Products can easily be tested with the VOM. Other causes require detective work. These are some of the questions to ask:

- *Was anything in the system changed or added?* Items to look for are added components incorrectly wired, added components overloading the power supply, and changes in programmable access equipment.

- *Was any part of the system abused or vandalized?* Access control equipment is the most vulnerable to vandalism or abuse by irate users. Locking devices are also often tampered with. Even abused doors and frames can interfere with proper operation of the system.

- *Has any unusual event occurred?* Earth or building movement can misalign locking devices. Power outages, brownouts, and electric storms can affect any electrical equipment, especially programmable access control units.

Answers to these questions can quickly lead you to a good starting point. If none of these items seems to have occurred, it is time to start at the component level. The power supply is usually the most accessible and logical place to start. From there you may have to move from component to component, looking for a faulty product. Last, check the wiring, looking for poor connections, breaks, and incorrect wire size.

The balance of this chapter will cover some actual troubleshooting problems and methods used to solve them.

## "It doesn't work!"

This is the most common statement you will hear when something in a system goes wrong. Usually the component that "gets the blame" is the one that most prominently displays a manufacturer's identification. It doesn't matter what or where the problem is; someone has to be called. Normally, time-consuming expensive calls can be avoided by using a little common sense (and a meter).

Let's look at Figure 12-8, a system diagram we completed in Appendix C (Figure C-1E).

A typical complaint is, "The lock doesn't lock." I always first ascertain that the lock is physically installed correctly and that a system wiring diagram is available.

Next I determine the voltage of the system and whether it is a fail-safe or fail-secure lock. In our example it is a fail-safe magnetic lock and a 24-volt DC system.

I believe the most logical place to start is to check for output power at the power supply. Figure 12-4 shows a meter setup for reading DC voltage. If we look at our exercise Figure 12-8, the meter probes should be placed at A and B. The voltage should read 24 volts DC or slightly higher. Many low-voltage supplies include a battery-recharging circuit so the output voltage may read

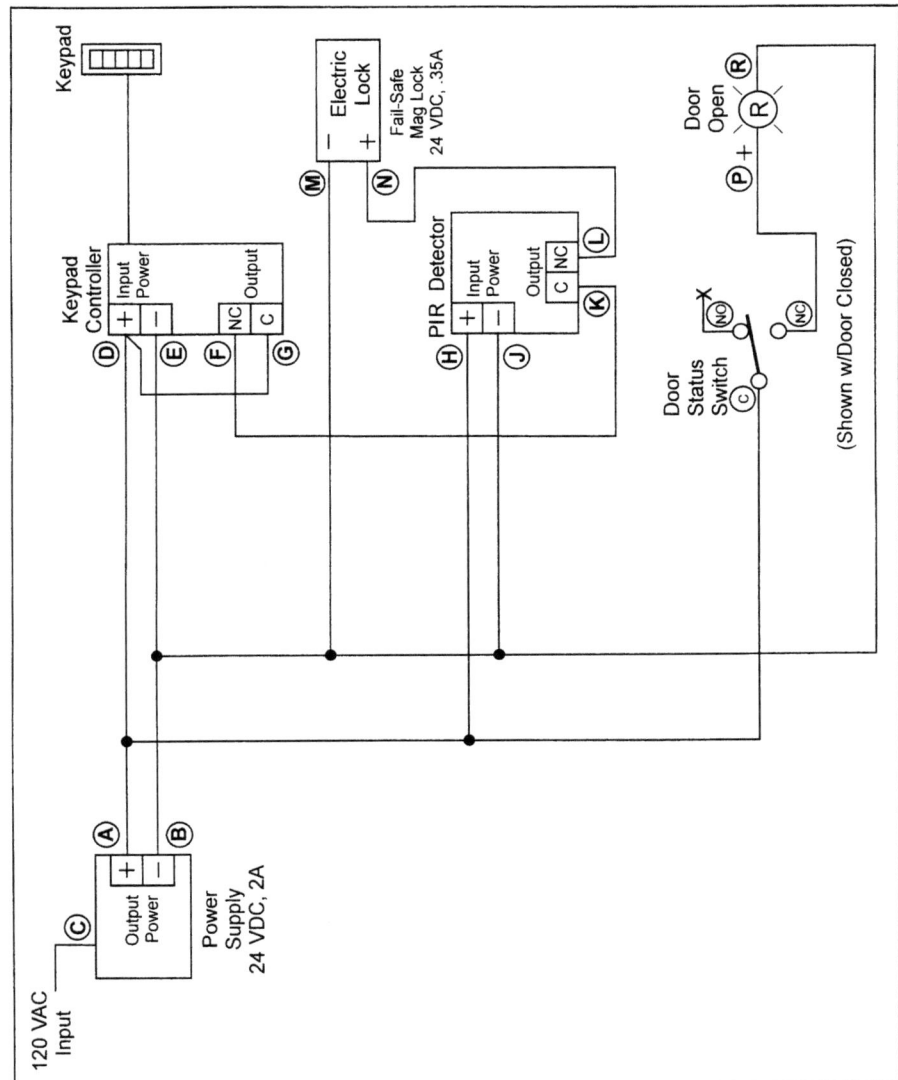

**Figure 12-8** Troubleshooting a system.

181

around 28 volts. Some power supplies are also voltage-selectable. If the reading is 12 to 14 volts, look for a voltage selector switch or jumper. If the correct voltage is present, the problem lies between the power supply and the lock. If voltage is not present, I move to the high-voltage side of the power supply. Normally this is 120 volts AC, and you must reset your meter and use extreme caution. Check the input voltage at C. There should be 120 volts AC between the hot (black wire) and neutral (white wire) input terminals. If there is any question in checking high voltage, call an electrician. If 120 volts is not present, this is the problem. I have seen job sites where the security system is completed before the electricians have installed circuit breakers or even high-voltage service. If 120 volts is present, the power supply may have a blown fuse. If any fuse is blown, this may indicate a short circuit somewhere in the system. If a fuse is replaced and blows again, you must check all wiring and components between the lock and power supply for other problems. Components that use power, for example, locks, access controls, etc., may be defective, and we will cover that a little later. If all fuses are okay, line power is present, and there is no output power, then it would be safe to say the power supply is defective.

If the power supply has the correct output power, move to any component that switches power to the lock. Starting at the keypad controller, there should be 24 volts DC on the input power terminals marked D and E. If there is no power, the wiring either is wrong or has a break in it. If input power is 22 volts or lower, the wire gauge is too small for the wire run. If input power is okay, move to the output contacts at points F and G. These should be closed contacts, and you should read 24 volts DC here also. If no voltage is present, check that the wiring isn't connected to an open contact in error. If power is present, you can also check the controller operation by entering a valid code at the keypad. The voltage reading should drop to zero until the unlock time setting expires. If everything checks out okay, move to the PIR detector.

The checks here are the same as those for the keypad controller. There should be power at the input terminals marked H and J. There should be 24 volts at the closed output contacts marked K and L. If there is no voltage at the closed output terminals, it may be faulty wiring. It may also be interference in the PIR detection area. Be sure the detection area is clear of any human or animal. Allow a minute or two for the PIR detector to "settle in." There should be 24 volts DC at the output terminals. That voltage will drop to zero when a person enters the PIR detection area. If all equipment controlling the lock checks out okay, move to the magnetic lock.

The magnetic lock should have 24 volts DC right at the lock, whether it is at input terminals or lock lead wires marked M and N. If power is present and the lock is not secure, check the lock and armature mounting. The armature must be able to flex somewhat when it is properly mounted to the door. When the door is closed, the armature must be up against the face of the lock. There must be no foreign matter, for example, paint, tape, or dirt, between the mating faces. The magnetic lock may also be checked to see if it is defective. Shut off the system power, and disconnect the lock from the (+) input wiring. Change your meter to read direct current, as shown in Figure 12-6. Place one meter probe on the magnetic lock input terminal or wire. Place the other probe on the disconnected system wire. Turn the system power back on. The meter should read 0.35-ampere current draw. A high or low reading may indicate a defective lock or some defect in the circuit wiring.

You can further check the lock by completely disconnecting it from the input wiring. The magnetic lock is simply a big coil of wire. You can check the ohm value of the coil, as shown in Figure 12-7. Set your meter to the ohms ($\Omega$) setting and place the probes across the magnetic lock leads. The reading should be around 60 to 75 ohms. You might want to contact the lock manufacturer to find out what the exact reading should be. You may also use Ohm's law; volts ÷ amperes = resistance (ohms). In this case the coil resistance is 24 ÷ 0.35 = 68 ohms. A very low or high reading would indicate an open or shorted coil—a defective lock.

Note that this reading must be taken directly across the lock coil leads. Some magnetic locks will have a circuit board which must be disconnected. The lock leads may have a plug-type connector. Slip a small-diameter piece of bare wire into the sockets. This will give you a connection to the coil for the meter. The lock may also have four lead wires, indicating it has a dual winding coil. It would be best to call the manufacturer to find out the resistance values and how to check the coil.

This system also includes a basic monitoring circuit to indicate the position of the door. One common complaint is that the light indicates door closed instead of door open. This is a simple problem to solve: The (+) wire is on the wrong contact of the door status switch. Another problem is welded contacts. Door status switch contacts are fairly delicate and will not tolerate high voltage or high current. Contact operation can be checked with the switch in or out of the circuit. With the switch in an active circuit, as shown in Figure 12-8, set your meter to read DC voltage. With the door closed, meter probes on C and NO should read 24 volts. Meter probes on C and NC should read 0 volts. If both readings are 24 volts, the contacts are welded. The door status switch must be replaced. The door status switch may be single-pole, single-throw (one set of contacts). Check this by placing the meter probes on C and NC. With the door closed you should read zero voltage; with the door open you should read 24 volts.

The door status switch may also be checked when completely disconnected from the system. Most meters have a setting marked with a horn symbol or diode symbol (check your manual).

Connect the probes across C and either of the other contacts. If the contacts are closed, you will hear a tone signal from the meter. The contacts can be checked for opening and closing by manually operating the switch with the permanent magnet half of the switch set.

Checking an indicator light, or audible device, is easy. In Figure 12-8, check for voltage across P and R. If voltage is present and the device doesn't work, it is defective. Some installers carry a separate power supply with them. Any component that uses power can be disconnected from the system and tested with the separate power supply.

## Summary

All the troubleshooting steps previously described should enable you to solve basic problems within a system. The following is a summary of the checkpoints we have covered in this chapter. It could also serve as a checklist for troubleshooting a system.

- *Is all hardware correctly installed?* Sometimes a site survey is required first. I would not agree to troubleshoot a system that had poorly installed hardware. Insist corrections be made first.

- *Is the system wired correctly?* A system wiring diagram must be available before troubleshooting can begin.

- *Has the proper gauge wire been used?* Too often a smaller than required wire gauge is used. This creates voltage drops and incorrect operation of electronic components. The system may have to be rewired. An alternate solution may be to move the power supply closer to the load.

- *Are wire connections secure?* Many times a simple loose wire connection is the culprit. It is worth checking first. One problem that can go unnoticed is a single strand of stranded wire touching an adjacent terminal. This can cause some really erratic troubles in a system.

- *Is the power supply sufficient to handle the load(s)?* Add up all the current draws in the system. The power supply should have at least the same amperage output, and somewhat higher is better.

- *Is any of the equipment defective?* Problems may immediately point to a particular piece of equipment. Removing the suspect item and checking it with a separate power supply may save a lot of time.

# 13

# Practical Applications

By now you have been supplied with sufficient information to create your own wiring diagrams. You have been supplied with drawing techniques, the use of a variety of electronic hardware, and the logic involved in designing a wiring diagram for specific problems. The information presented was organized in a manner to help you understand the procedures involved in designing wiring diagrams from basic to complex systems.

Although we have covered a multitude of systems, it would take another entire book to cover all the system variations you will be confronted with. This last chapter was assembled to present a reference section of commonly specified electric hardware applications. It provides an outline of a variety of common applications and a generic list of selected products. The scope of this information is limited, as other factors may influence the selection and application of the components shown. Unusual operating conditions, the local environment, and life safety codes must all be considered in the design of these systems.

And notice another thing: There are no wiring diagrams! The intent of this chapter is to provide a typical solution for specific applications. Select an outline closest to your particular situation and modify it as necessary. Then choose the manufacturers' products and gather the technical information for each item. Lay out each item, using the symbols you have learned, and identify all terminations and hookup points. Then complete your wiring diagram. This chapter is divided into sections covering general applications, specific types of locking devices, and specific types of entrance and exit doors.

## General Applications

The following section provides system outlines utilizing several different types of electric locking devices. It is meant to show that a variety of electric hardware devices are available for system solutions. Each elevation drawing includes a brief description of the locking device.

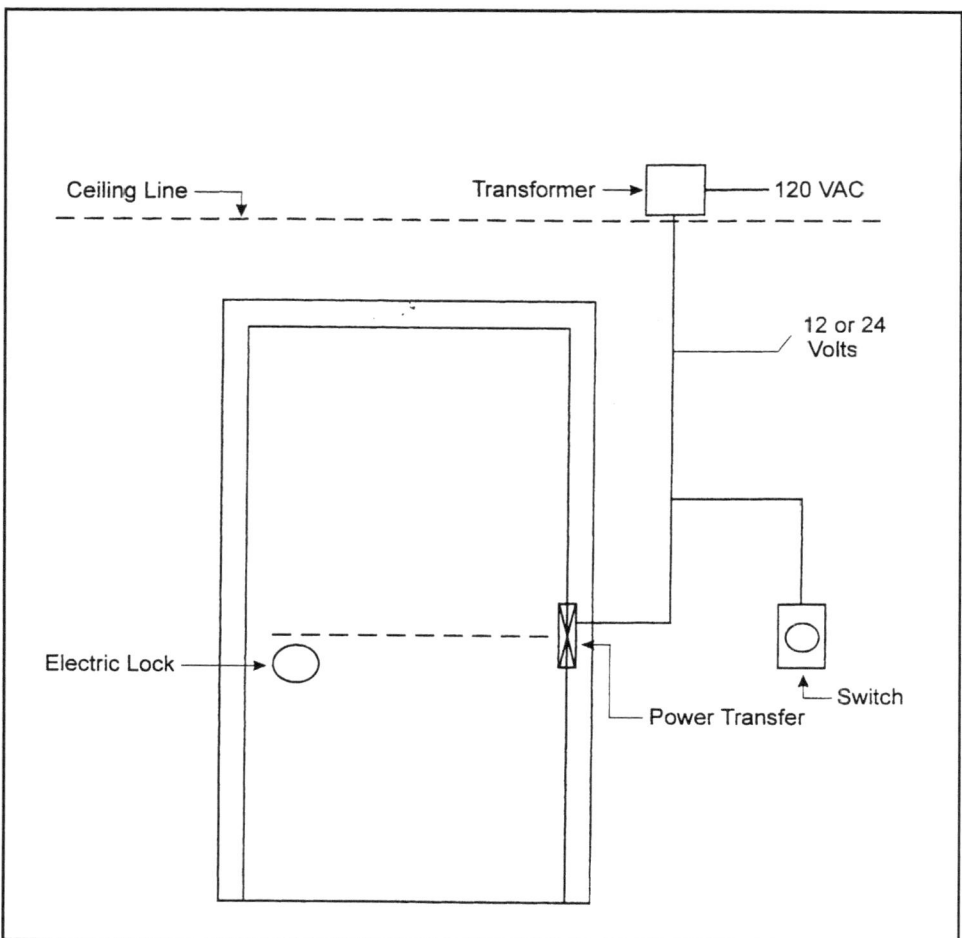

**Figure 13-1**    Electrified lock set application.

The electric lock set may be selected as fail-safe or fail-secure.

**Electrically locked (fail-safe).**    Outside knob or lever is continuously electrically locked until unlocked by key, switch, or power failure. Inside knob or lever is always free for immediate exit.

**Electrically unlocked (fail-secure).**    Outside knob or lever is continuously locked until unlocked by key or electric current. Inside knob or lever is always free for immediate exit.

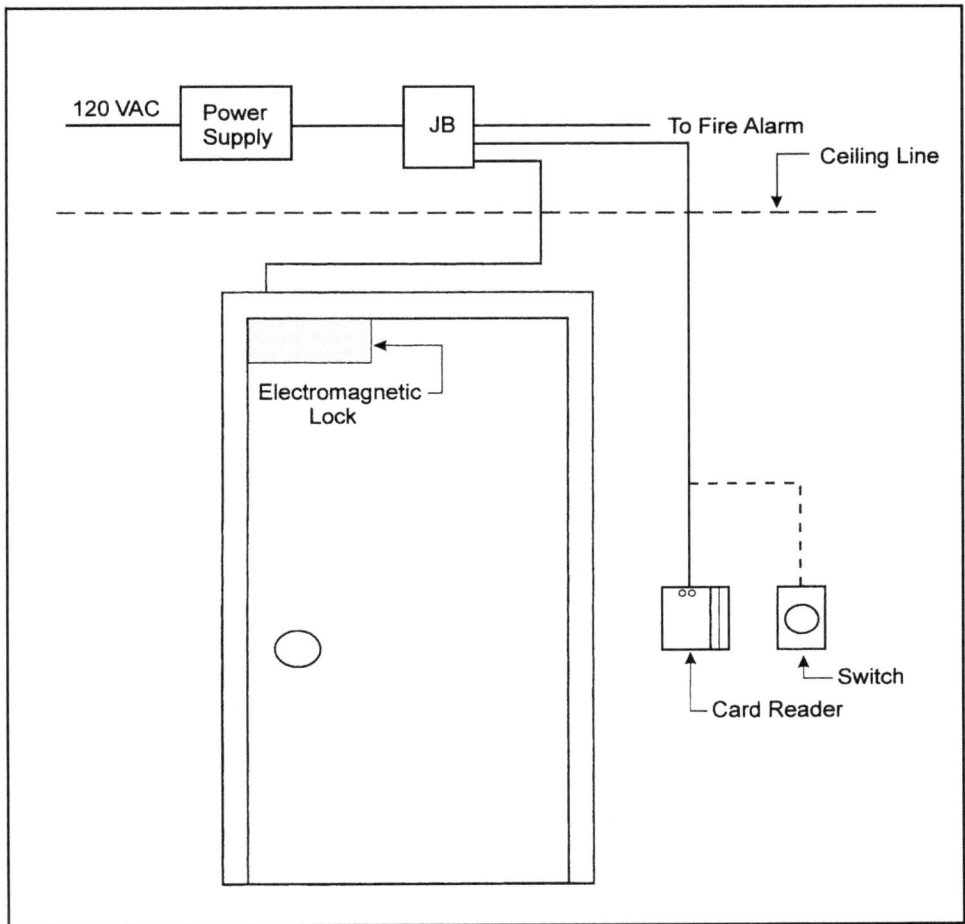

**Figure 13-2**  Electromagnetic lock application.

Single door enters a secured area. The electromagnetic lock is controlled by an exterior card reader. Exit is allowed by a signal from the interior pushbutton, or by loss of power. (*Note:* This is a nonegress opening.)

Electromagnetic locks are inherently fail-safe and can be tied into a building's fire alarm system or other hazard-sensing system, which causes the release of the lock upon activation of the emergency system.

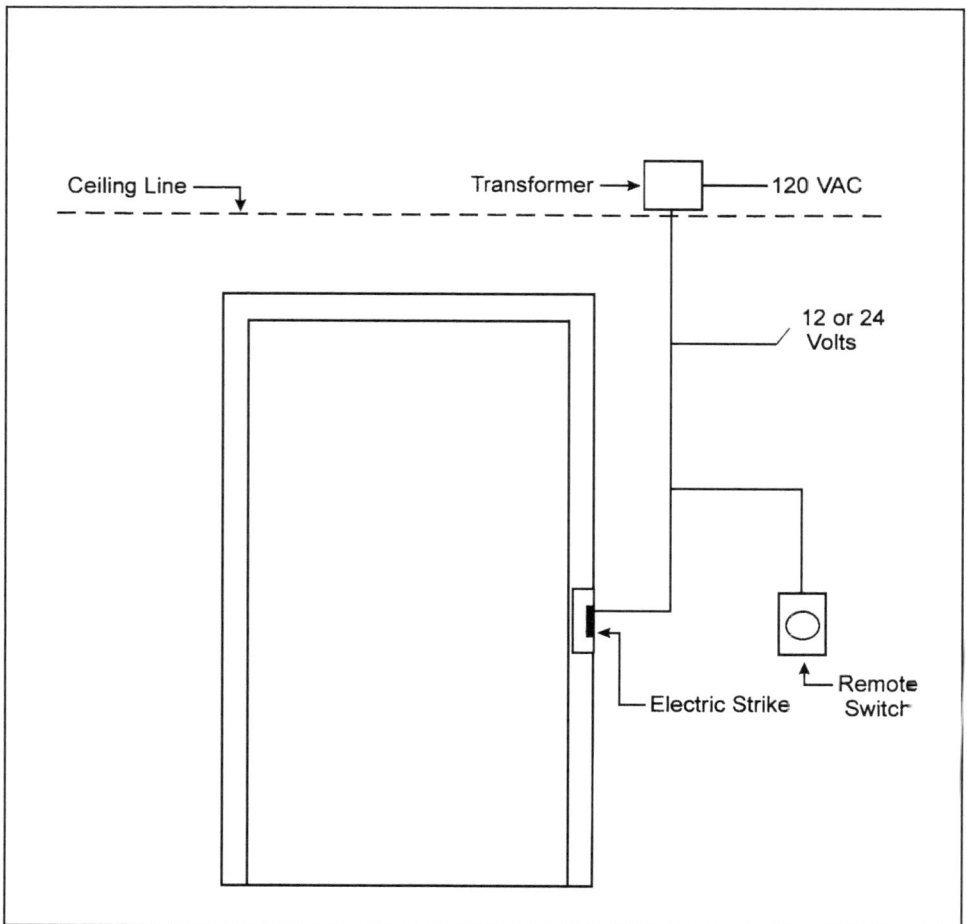

**Figure 13-3**   Electric strike application.

Electric strikes provide remote release of a locked door. They allow the door to be opened without retracting the latchbolt. This occurs by the releasing of the electric strike lip (sometimes called the keeper or gate).

Strikes may be selected as fail-secure or fail-safe and are available in 12 or 24 volts, AC or DC.

**Fail-secure.**   It requires power to be applied to unlock the strike lip. On loss of power the strike is locked.

**Fail-safe.**   It requires power to be applied to lock the strike lip. On loss of power the strike is unlocked. Most building codes prohibit the use of fail-safe strikes on labeled openings.

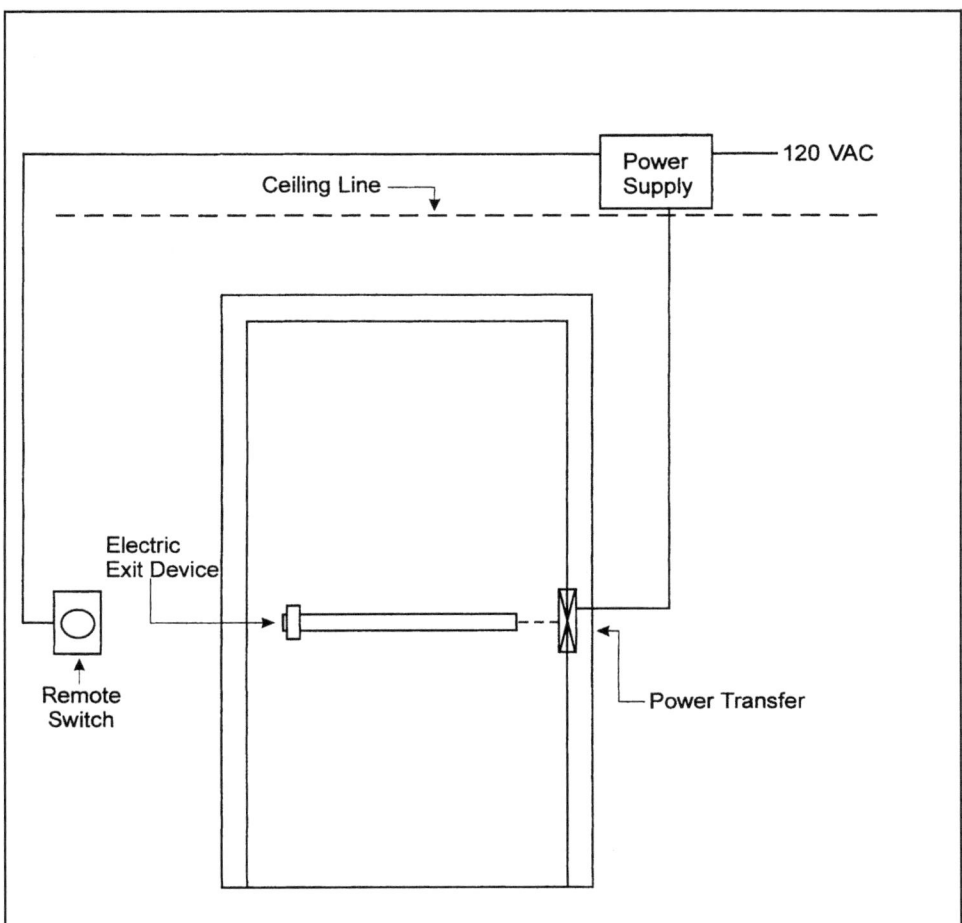

**Figure 13-4**  Electric exit device application.

**Electric latch retraction.**    Electric latch retraction provides remote latch bolt retracting ability for exit devices. This feature is available for rim, mortise, and vertical rod devices. Devices are also available that provide remote locking or unlocking of the outside trim.

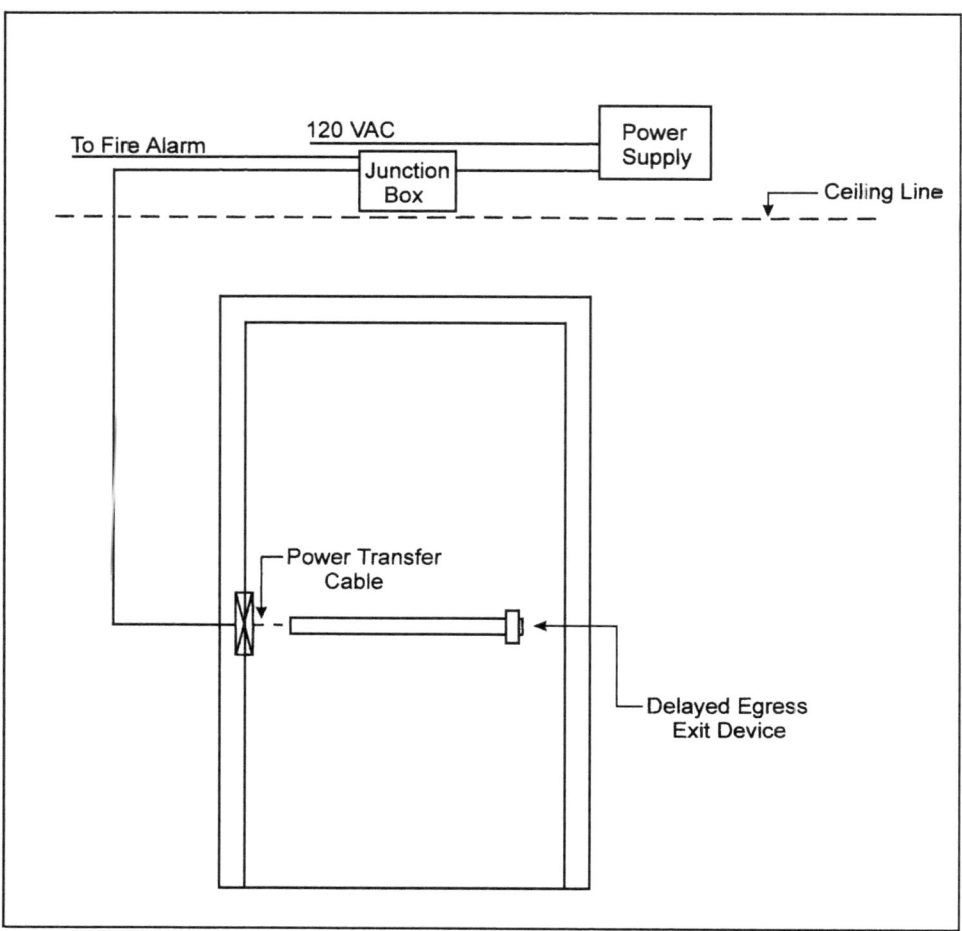

**Figure 13-5**    Delayed egress electric exit device application.

This device provides a degree of security without compromising life safety for egress openings.

**Secure mode with alarm and delayed release.**   The secure mode will operate whenever the door is latched and the exit device is activated to the armed state. When the exit device pad is depressed, a horn will sound along with a rapid flashing indicator light at the device. The exit device will remain locked for the preset period of delay of 15 or 30 seconds. After the delay period, the push pad will retract the latchbolt, permitting egress from the area.

**Authorized exit mode.**   For authorized exit while the device is armed, the door may be released for a preset period of time without activating the alarm by turning the key cylinder counterclockwise to the "off" position and resetting in the clockwise position. The system can remain disarmed in the counterclockwise position.

**Emergency exit mode.**   Whenever an emergency detector (smoke, fire, water flow, etc.) signals that an emergency condition is present, the device will unlock instantaneously, sound a horn, and flash an indicator light rapidly. The door may be opened immediately by depressing the push pad.

In the event of a power loss, the electronics become inactive, allowing the device to operate as a normal exit device.

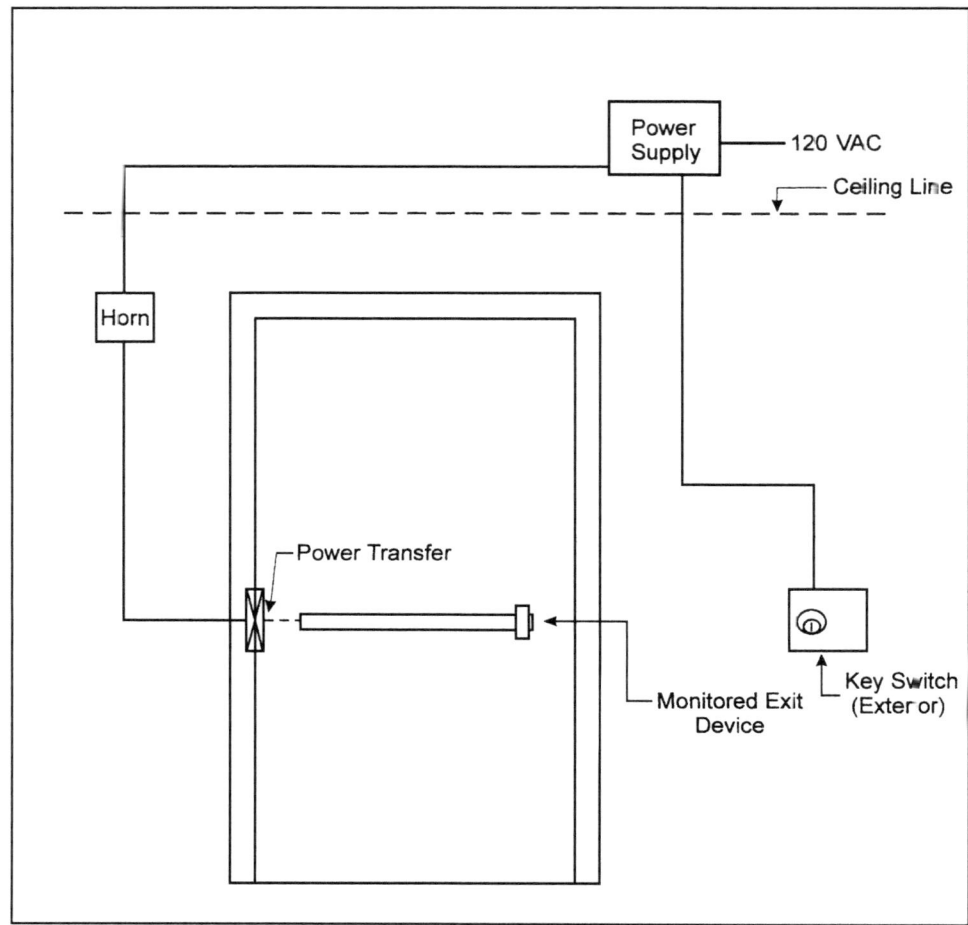

**Figure 13-6**  Monitored exit device application.

A signal switch feature is used to report the unauthorized use of an opening. Available in rim, mortise, and vertical rod, these devices are equipped with two internal single-pole, double-throw (SPDT) switches. One switch monitors both the touch bar and the latch bolt assembly, making the latch bolt tamper-resistant for positive security. The other SPDT switch is connected to the mortise cylinder for alarm "bypass." The device can be used as a local door alarm as shown or may be wired to a remote alarm or security console.

## Magnetic Lock Applications

The following section provides a variety of specific types of openings and their treatment using electromagnetic locks. The popularity of magnetic locks is due to several factors, including these:

- It is a true fail-safe locking device (there are no moving parts to bind, jam, or prevent release).

- It is easy to install.

- It requires low maintenance.

- It has low current draw.

- It has a high holding force.

Each drawing is an elevation diagram for a particular system. Included with each drawing is a system description of operation that could also be used as a system specification. These systems are actual solutions for the types of openings shown in the examples. But as mentioned earlier, there are many ways to design a particular system. The systems shown are guidelines, and you may change the equipment used to suit your needs.

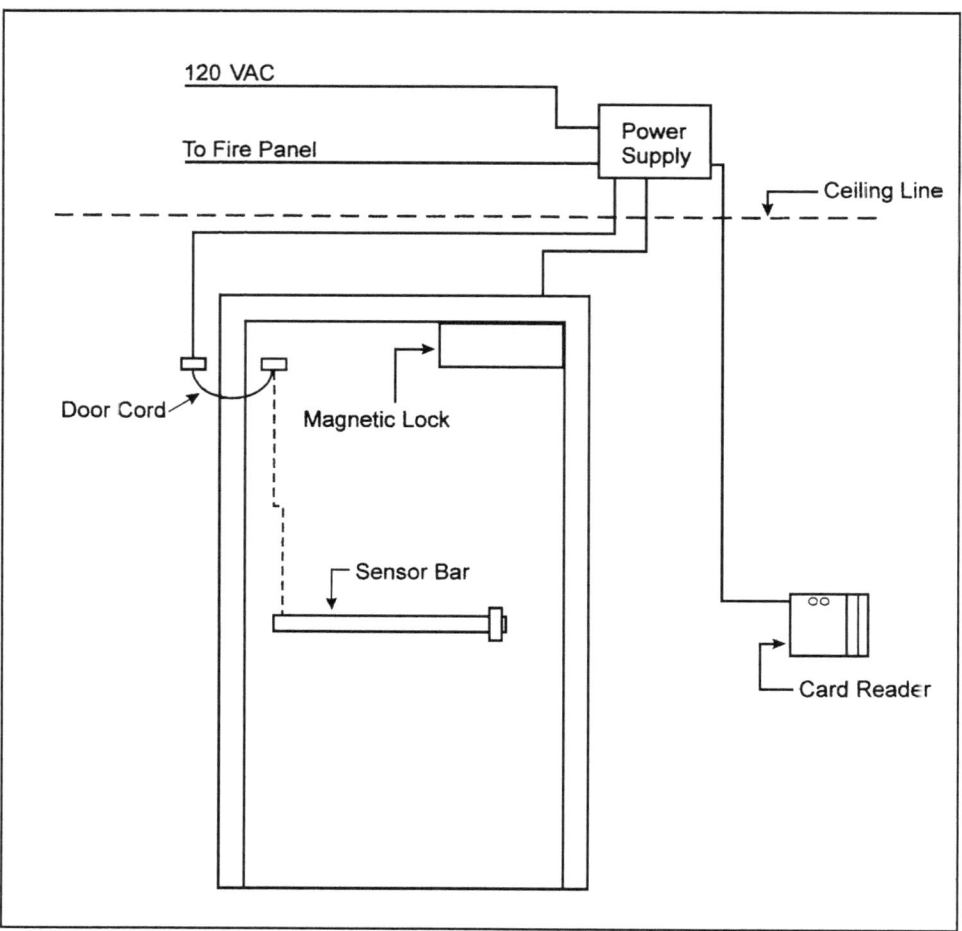

**Figure 13-7**   Front entrance.

## System description of operation

The door shall be normally closed and secured by an electromagnetic lock. Egress shall be allowed immediately, at any time, by activation of a fail-safe sensor type of exit bar. Ingress shall be allowed by card reader. The card reader shall be programmable and shall allow audit trail retrieval by means of a portable computer. The system shall interface with an approved supervised fire system for immediate release upon fire panel activation.

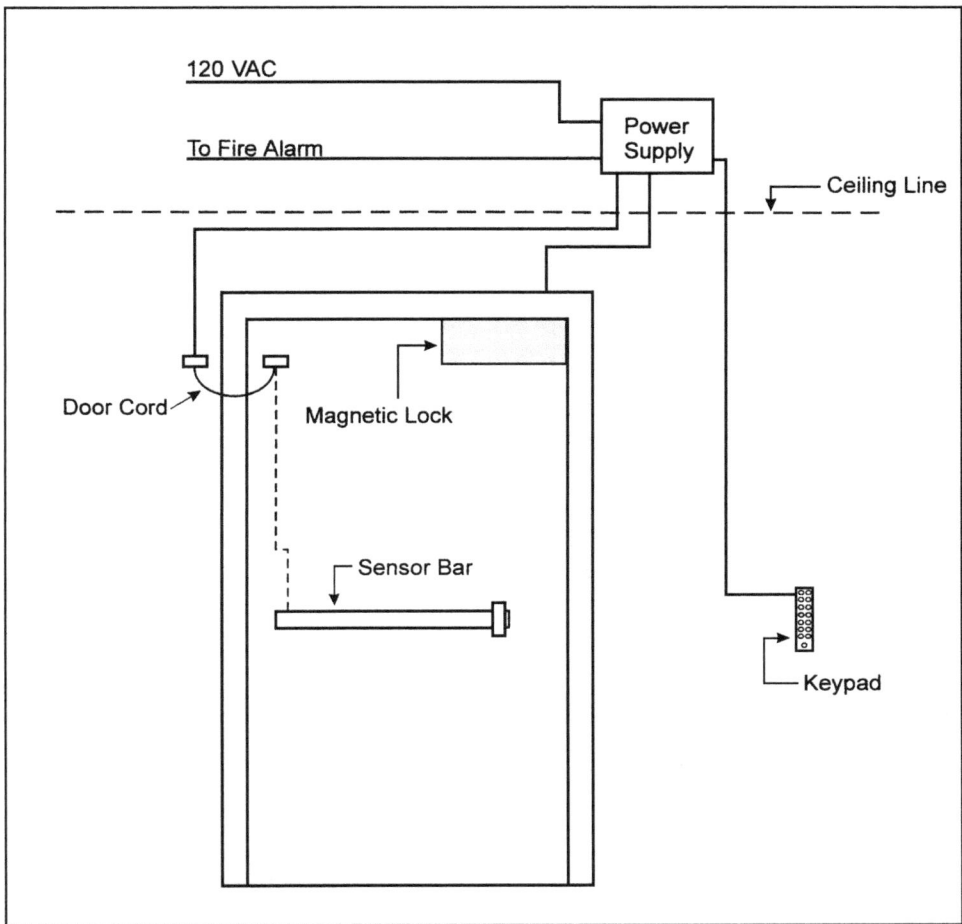

120 VAC

To Fire Alarm

Power Supply

Ceiling Line

Door Cord

Magnetic Lock

Sensor Bar

Keypad

**Figure 13-8**    Employee entrance.

## System description of operation

The door shall be normally closed and secured by an electromagnetic lock. Egress shall be allowed immediately, at any time, by activation of a fail-safe sensor type of exit bar. Ingress shall be allowed by entering a valid code into a digital keypad. The keypad shall be wired to a controller contained within the power supply housing. The controller shall be programmable and shall allow for 60 individual codes. The system shall interface with an approved supervised fire system for immediate release upon fire panel activation.

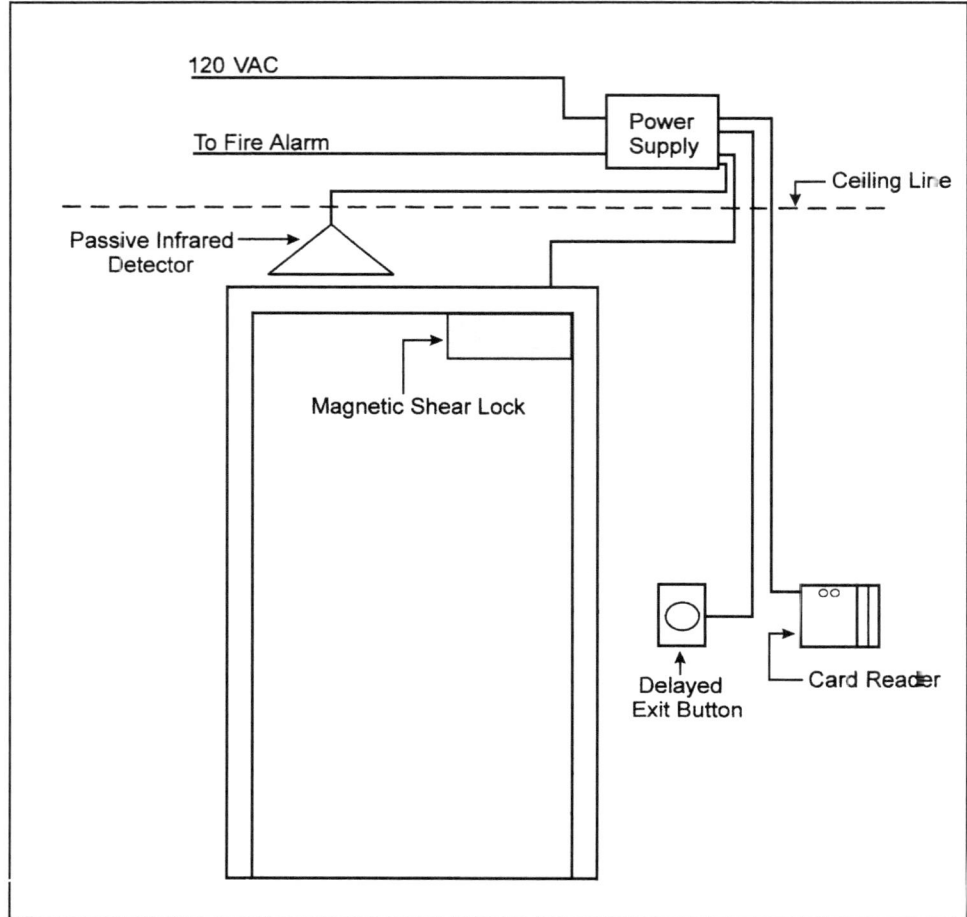

**Figure 13-9**  Office building entrance.

## System description of operation

The door shall be normally closed and secured by a concealed electromagnetic lock. Egress shall be allowed immediately by activation of a passive infrared detector. A delayed-action pushbutton shall be provided for redundant egress release. Ingress shall be provided by a card access system. The lock shall interface with an approved supervised fire system for immediate release upon fire panel activation.

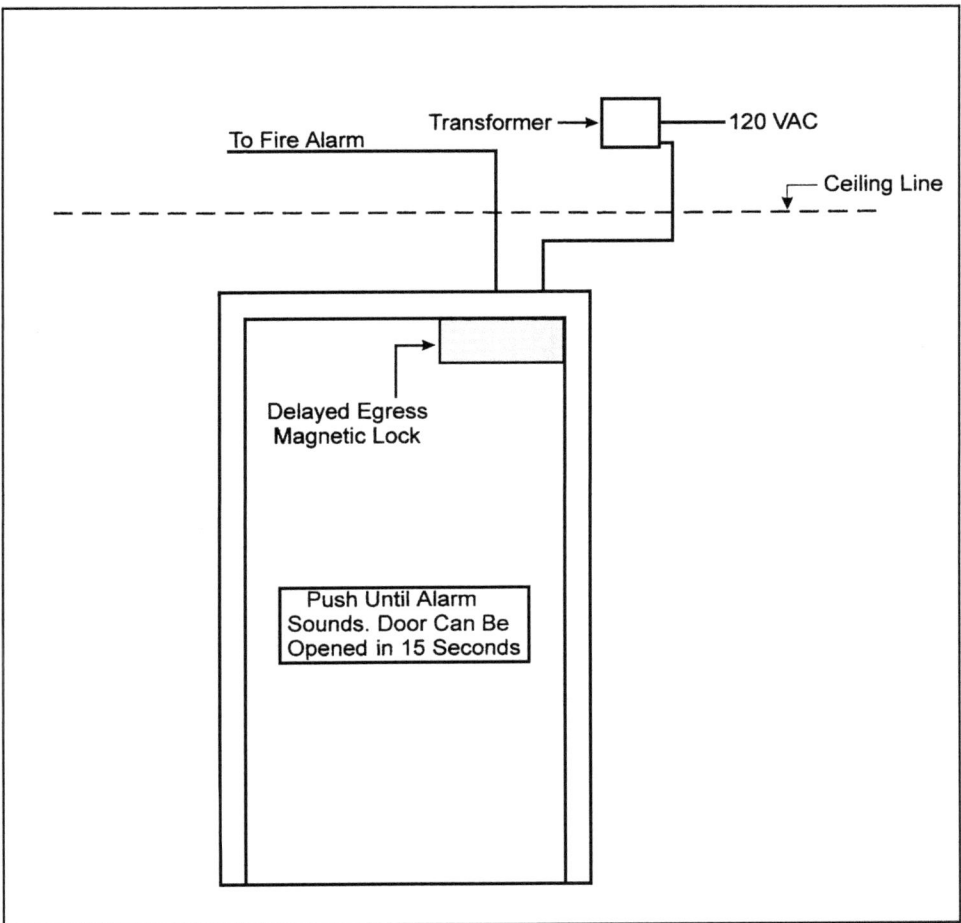

**Figure 13-10**   Emergency exit door.

## System description of operation

The door shall be normally closed and secured by an electromagnetic lock. Any attempt to egress during nonemergency situations shall initiate an irreversible 15-second delay before the locking device releases. The irreversible delay shall be initiated by no more than 15-pound force continuously applied to the release hardware for a period not to exceed 3 seconds. Initiation of the delayed-egress cycle shall activate a signal that the system is functional. The system shall include a mandatory sign for application to the egress side of the door. The sign shall read: Push Until Alarm Sounds. Door Can Be Opened in 15 Seconds.

Authorized egress and system reset shall be by a key switch integral to the lock. The lock shall interface with an approved supervised fire system for immediate release upon fire panel activation.

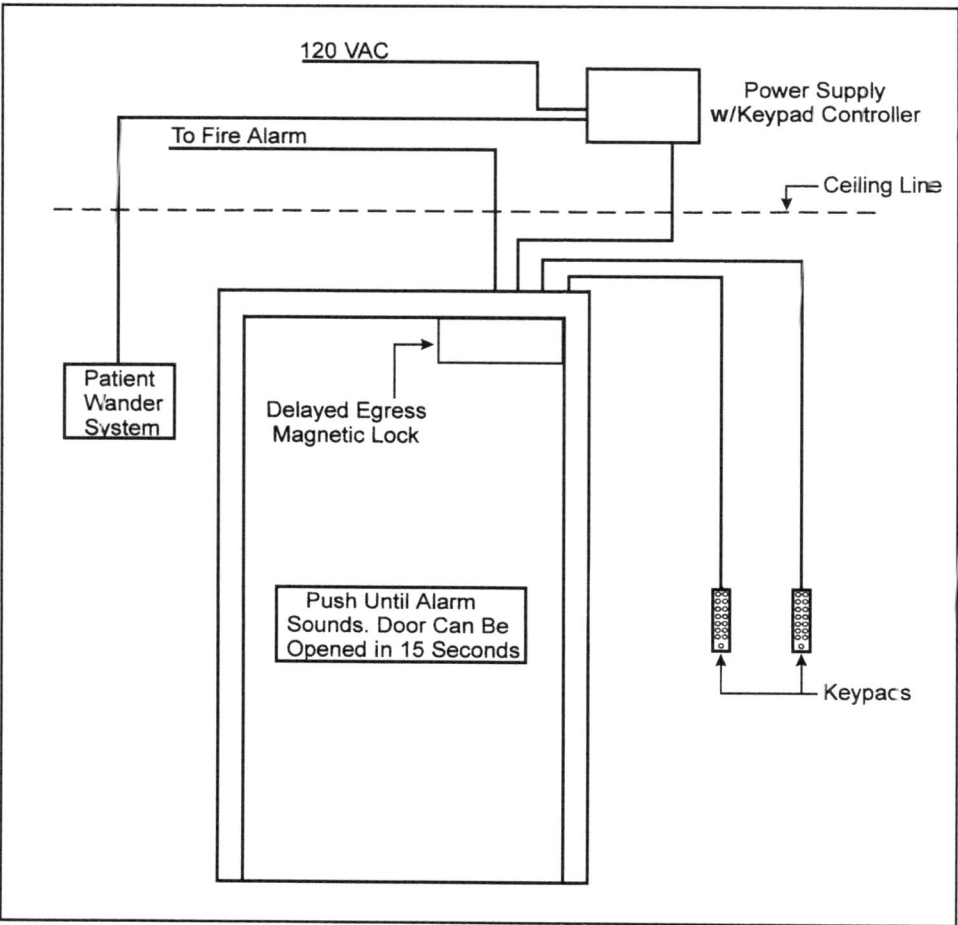

**Figure 13-11** Front entrance—healthcare center.

**System description of operation**

The door shall be normally closed and unlocked. The opening shall be equipped with a delayed-egress magnetic lock and a patient wander system. A patient wearing a proximity bracelet, upon approaching the door, shall trigger the wander system alarm and cause the magnetic lock to energize. An attempt to egress during the locked condition shall initiate an irreversible 15-second delay before the locking device releases. The irreversible delay shall be initiated by no more than a 15-pound force continuously applied to the release hardware for a period not to exceed 3 seconds. Initiation of the delayed-egress cycle shall activate a signal that the system is functional. The system shall include a mandatory sign for application to the egress side of the door. The sign shall read: Push Until Alarm Sounds. Door Can Be Opened in 15 Seconds.

Authorized egress and ingress during a locked condition and system reset shall be by programmable keypads. The lock shall include an audible alarm that will sound if the door is held opened longer than 10 seconds. Alarm will silence upon closing the door. The lock shall interface with an approved supervised fire system for immediate release upon fire panel activation.

## Electric Exit Device Applications

The electrified exit device is one of the most versatile locking devices in use today. Not only does it satisfy life safety codes for means of egress, but also it allows control over both egress and access. The mechanical exit device originated to provide immediate unrestricted egress at emergency exit doors. Some of the features that electrification of the exit device later provided are

- Electric latch retraction—for remote control of access and electric dogging for use with automatic door operators.
- Built-in electronics for delayed-egress mode—to provide a degree of security to egress openings.
- Built-in switches—to provide monitoring and alarm shunt features.

The following section provides a variety of opening types and control systems.

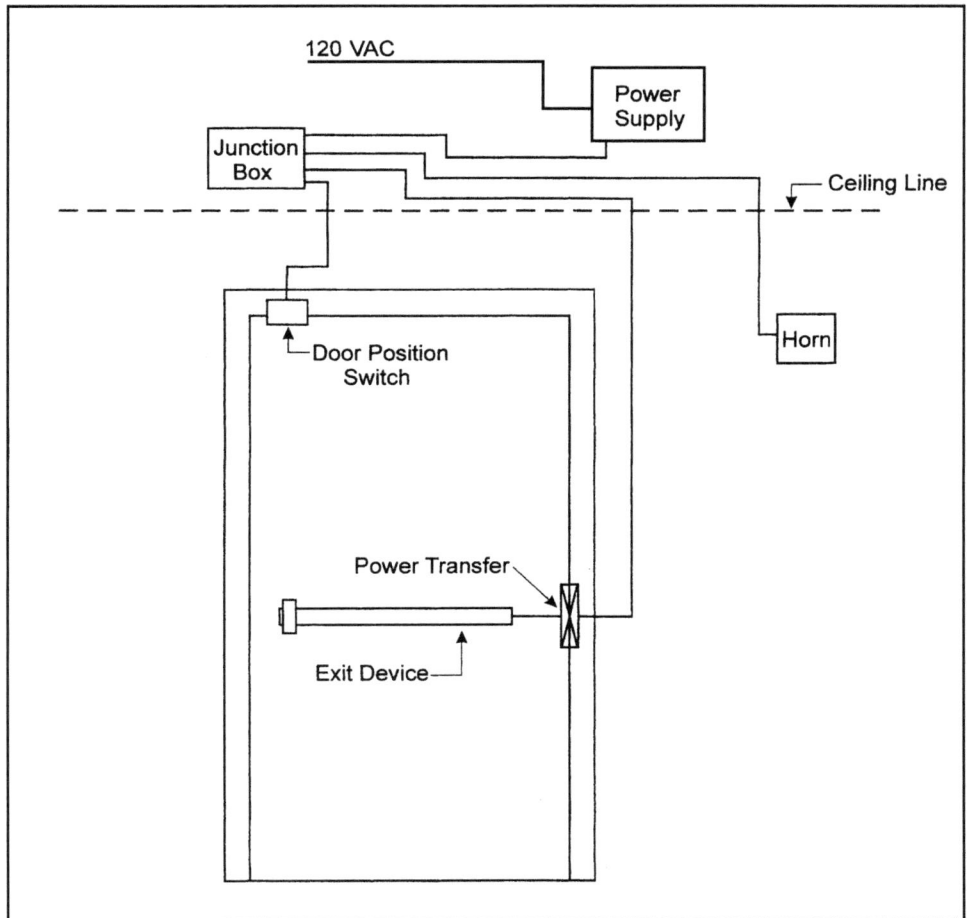

**Figure 13-12**   Exit door—egress only with alarm.

Alarmed exit door provides an audible alarm for unauthorized use of the opening.

## System description of operation

**Interior operation.**   Pushing the touchpad will sound the alarm. It will continue to sound until the door is resecured and a key is used to reset the system.

**Exterior operation.**   There is no valid entry from the exterior. If the door is forced open, the alarm will sound and a key must be used on the interior side to reset the system.

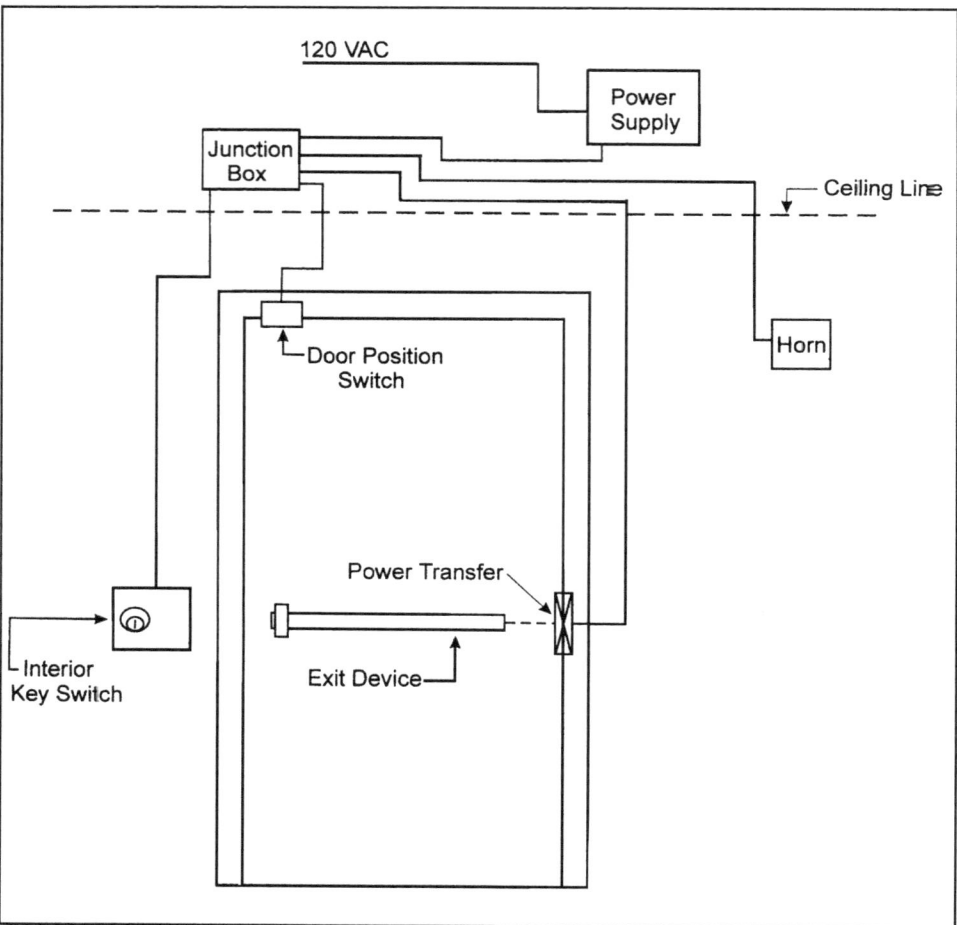

**Figure 13-13**   Exit door—egress only with alarm and disarm function.

Alarmed exit door provides an audible alarm for unauthorized use of the opening.

**System description of operation**

**Interior day operation.**   A key may be used in the exit device to inhibit the alarm for up to 60 seconds, allowing an authorized person to leave the area. To disarm the system for daytime use, the user must turn the key in the counterclockwise direction and then back to center.

**Interior night operation.**   To rearm the system, the key switch must be turned in the clockwise direction and then back to center. If the door is in the armed mode when opened, the alarm will sound until reset with the key.

**Exterior day/night operation.**   There is no valid entry from the exterior. If the door is armed and forced open, the alarm will sound. A key must be used on the interior side to reset the system.

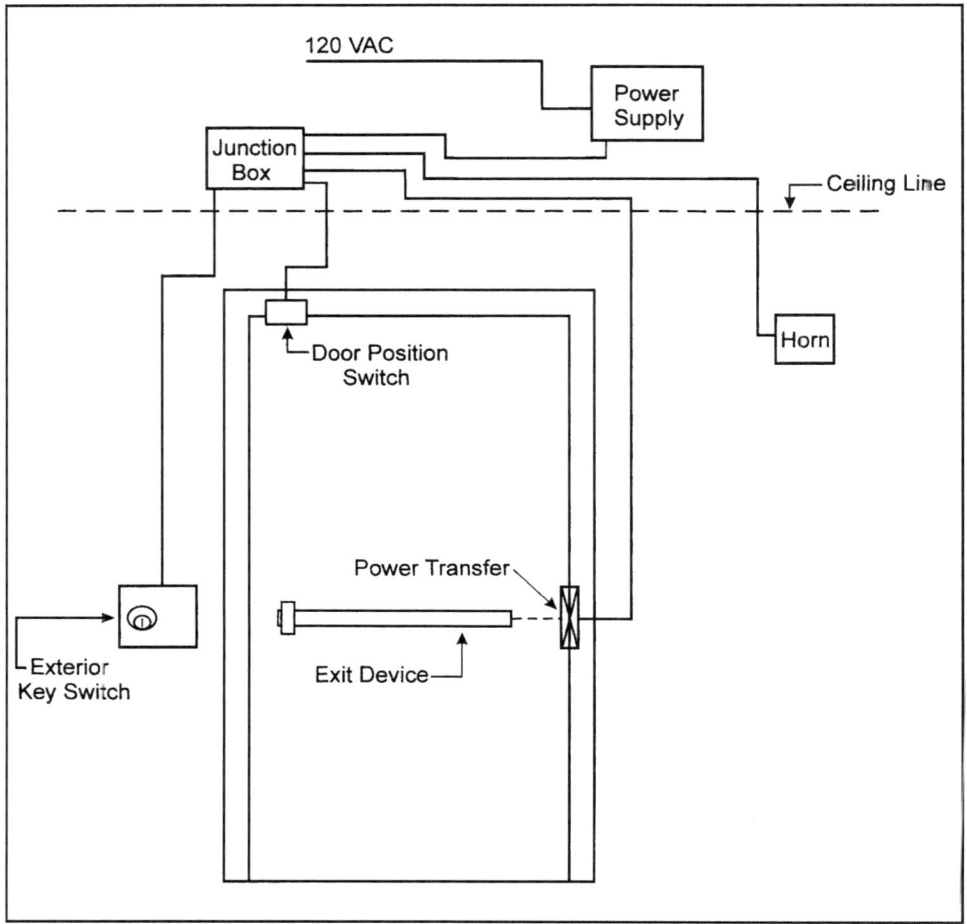

**Figure 13-14** Exit/Entry door—with alarm and disarm function.

Alarmed exit/entry door provides an audible alarm for unauthorized use of the opening. For service or delivery purposes, a key switch allows use of the door from the exterior.

## System description of operation

**Interior day operation.** Turning the key in the exit device will disarm the system for momentary use. If the door is held open (depending on the time set), the alarm will sound. If the door is in the armed mode when opened, the alarm will sound.

**Exterior day operation.** To disarm the system for daytime use, the user must turn the key switch in the counterclockwise direction and then back to center. Once this has occurred, the door may be open for an indefinite amount of time. To resecure the opening, the key must be turned in the clockwise direction and returned to center.

If the door remains propped open, the alarm will sound. If the door is forced open when in the armed mode, the alarm will sound.

**Exterior night operation.** When the user turns the key switch in the clockwise direction and back to center, the alarm system will disarm for a preset time. A key must be inserted in the trim to retract the latch, allowing the user to enter the door. If the door is held or propped open, the alarm will sound.

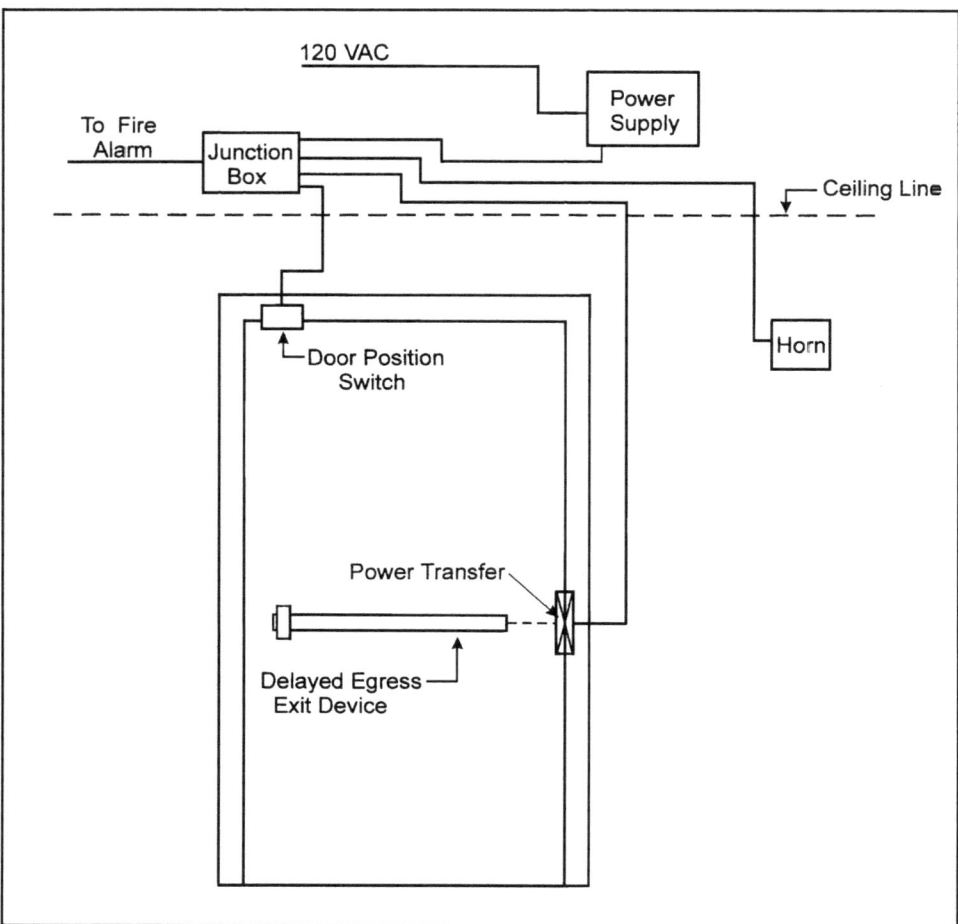

**Figure 13-15**   Exit door—delayed egress with alarm and disarm function.

Exit control provides a delay in opening in nonemergency situations. It is commonly used to limit employee pilferage or customer theft in high-security installations or retail stores.

## System description of operation

**Interior day operation.**  Without a key, exit is delayed for 15 seconds. Any attempt to exit will cause the local horn to sound. At the end of 15 seconds, the user may exit. The alarm will continue to sound until reset.

If immediate egress is required, the device may be disarmed using the key. The key is required to rearm the system.

If the fire alarm is triggered, the local horn sounds and immediate egress is available. To rearm the device, the fire alarm must first be cleared, then the key used to reset.

**Interior night operation.**  Without a key, exit is delayed for 15 seconds. Any attempt to exit will cause the horn to sound. At the end of 15 seconds the user may exit. The alarm will continue to sound until reset.

**Exterior day operation.**  When the exit device is disarmed from the interior and the exterior cylinder is unlocked, entry may be gained by use of the exterior trim.

The opening is equipped with a door position switch. If the door is not reclosed when rearmed, the system will sound its alarm and the device will not relock.

**Exterior night operation.**  The exterior trim is locked, entry is not allowed. If the door is forced open, the alarm will sound.

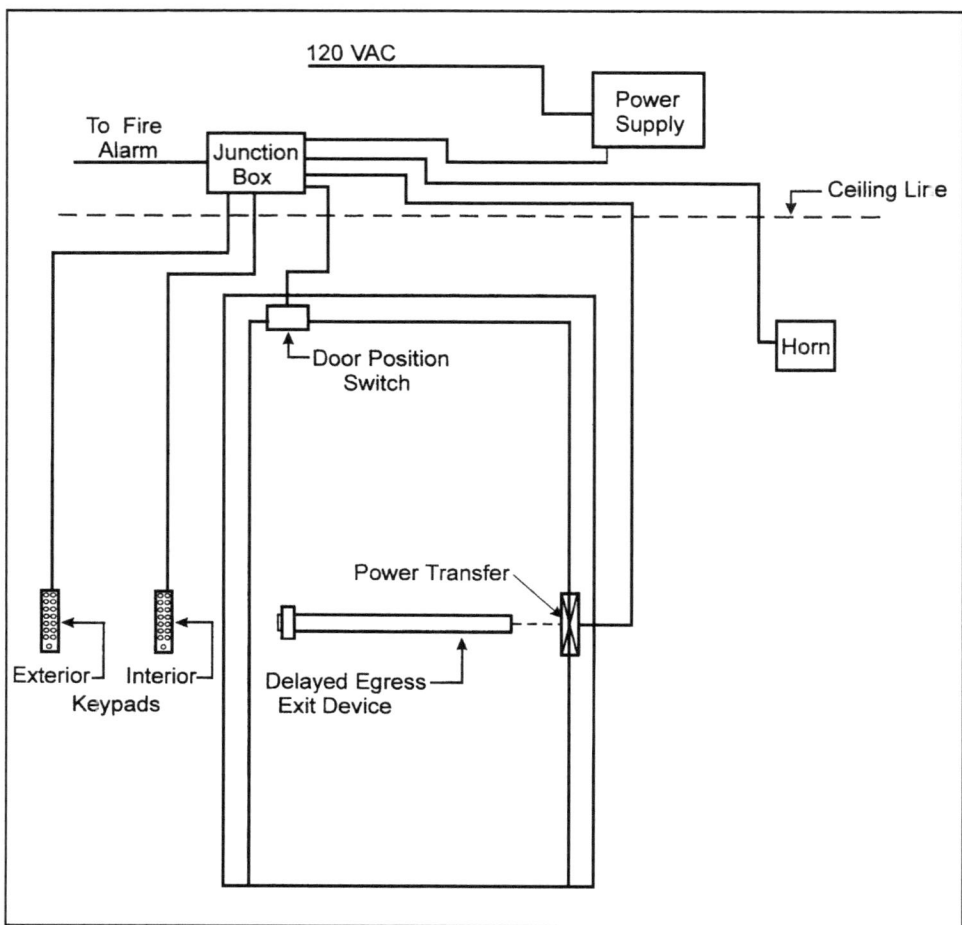

**Figure 13-16**  Exit/entry door—delayed egress with access control.

Exit control provides a delay in opening a door in nonauthorized and non-emergency situations. Access control permits authorized entry and exit without delay or alarm. Monitoring ensures the opening has been resecured after use.

## System description of operation

**Interior day operation.**   The opening is equipped with digital keypads. Only authorized personnel are able to activate the device. The keypad can be programmed to unlock the device for 5 to 60 seconds. Without valid access, exit through the door is delayed for 15 seconds while the local horn sounds. At the end of 15 seconds the user may exit. The alarm will continue to sound. Rearming the system requires the use of the digital keypad.

If daytime immediate egress is required, the device may be disarmed using the keypad. The keypad is required to rearm the system.

If the fire alarm is triggered, the local horn sounds and immediate egress is available. The fire alarm must be cleared before the digital keypad can be used to reset the system.

**Exterior day operation.**   Entering a valid keypad access code will disarm the device and allow entry. The exterior cylinder must be unlocked to permit entry.

**Interior/exterior night operation.**   With the exception of supervisor security codes, all users may be locked out to prevent after-hour access.

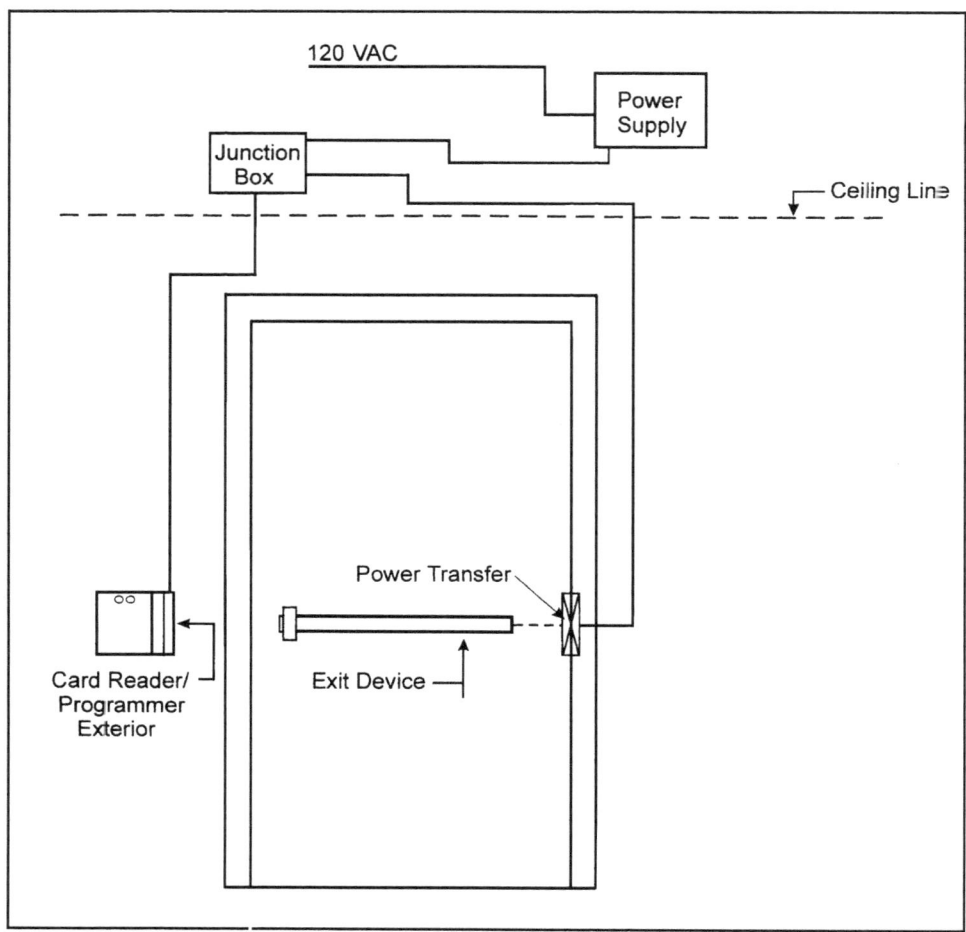

**Figure 13-17**   Exit/entry door.

The electrified exit device combined with an access control system allows authorized personnel access to various areas.

The access control will accept either a card or personal identification number (PIN) code entry. Electrified latch retraction provides access without the need for a conventional key operation.

### System description of operation

**Interior operation.** Free egress is permitted at all times. There is no alarm; the door may be held open for an indefinite time.

**Exterior day operation.** A valid card or PIN is required to open the door which may be propped open for an indefinite time. Upon power loss, a key is required to mechanically open the door.

**Exterior night operation.** For a higher level of security, the unit may be set to accept access cards only.

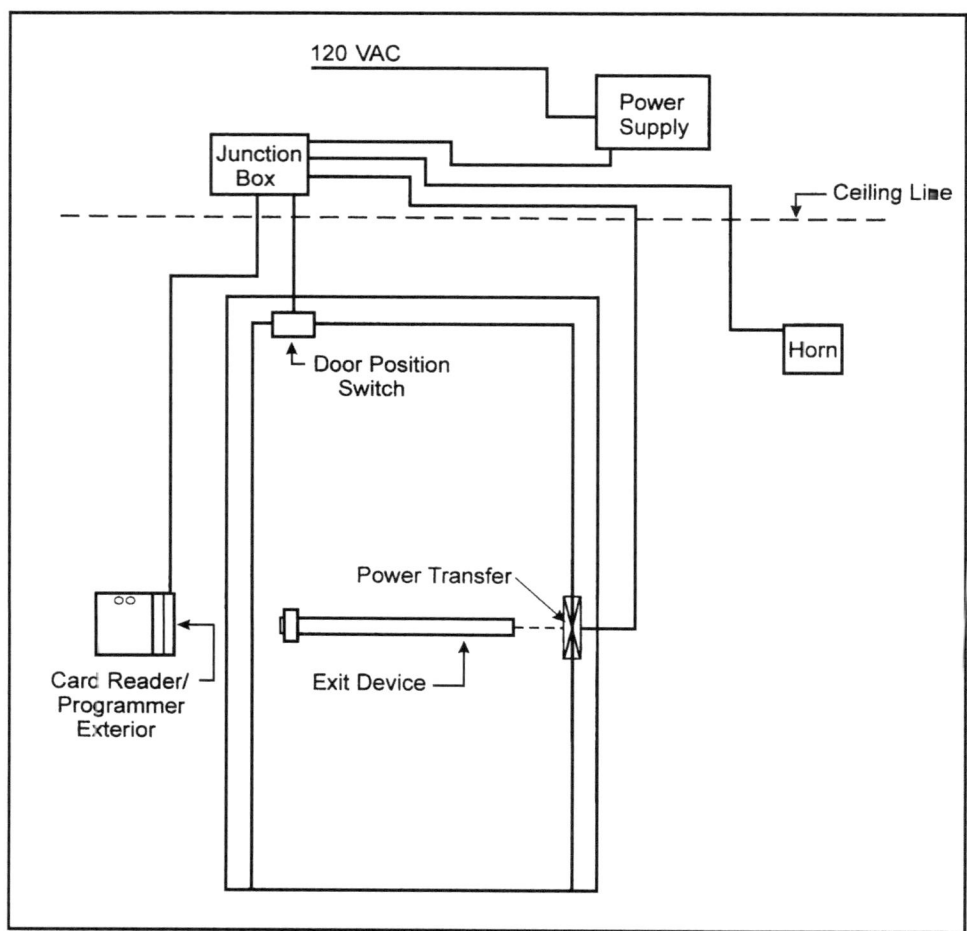

**Figure 13-18** Exit/entry door—with alarm.

The electrified exit device combined with an access control system allows authorized personnel access to various areas.

The access control will accept either a card or PIN entry. Electrified latch retraction provides access without the need for conventional key operation.

The alarm circuit is activated when the door has been propped or forced open. To deactivate the alarm from the exterior, a valid card or PIN must be presented to the reader. To deactivate the alarm from the interior, the user simply exits through the door; which automatically disarms the alarm.

## System description of operation

**Interior operation.**  Free egress is permitted at all times. The exit device will automatically disarm the alarm.

**Exterior day operation.**  Presenting a valid card or PIN to the reader will retract the latch. The latch will reextend automatically, allowing the door to relock when closed. If the door is held or forced open, the alarm will sound and must be deactivated by a valid card or PIN.

**Exterior night operation.**  For a higher level of security, the unit may be set to accept access cards only.

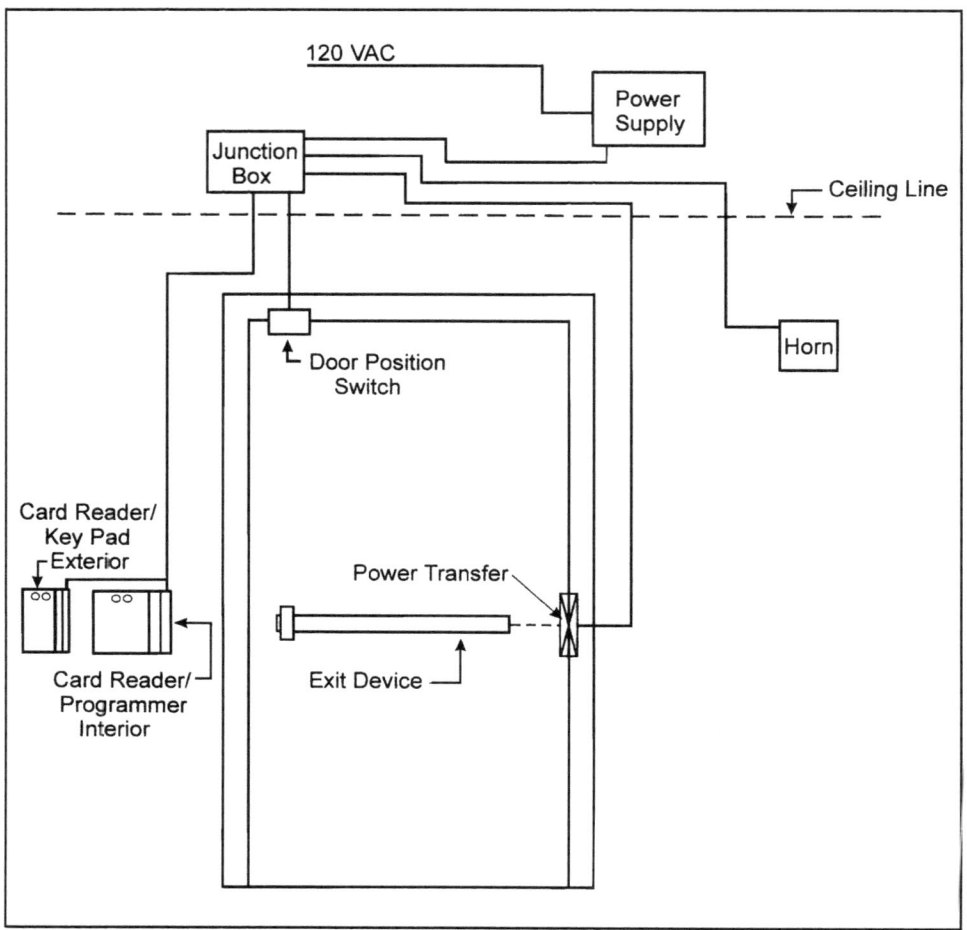

**Figure 13-19**   Exit/entry door—with alarm and egress/access control.

The electrified exit device combined with an access control system allows both entry and exit control with an alarm system.

### System description of operation

**Interior day operation.**   Presenting a valid card or PIN to the reader will retract the latch and inhibit the alarm, allowing the door to be opened. If the door is not reclosed and relatched in a preset time, the alarm will sound. If the door is opened without deactivating the access system, the alarm will sound. A valid access or PIN is required to reset the alarm. An optional printer enables printing out the user activity. The access unit may also be set to keep the door unlocked for an extended period for unrestricted use.

**Interior night operation.**   Presenting a valid card or PIN to the reader will retract the latch and inhibit the alarm, allowing the door to be opened.

**Exterior day operation.**   To gain entry, an authorized user presents a valid card or PIN to the reader. A valid entry will cause the exit device latch to retract, allowing the door to be pulled open with the nonfunctional trim.

**Exterior night operation.**   For a higher level of evening and weekend security, the unit may be set to accept access cards, PIN, or card and PIN, or to allow no access at all. On a valid access the door may be pulled open. The latch will reextend automatically, allowing the door to relock when closed. If the door is held or forced open, the alarm will sound and must be deactivated by a valid card or PIN.

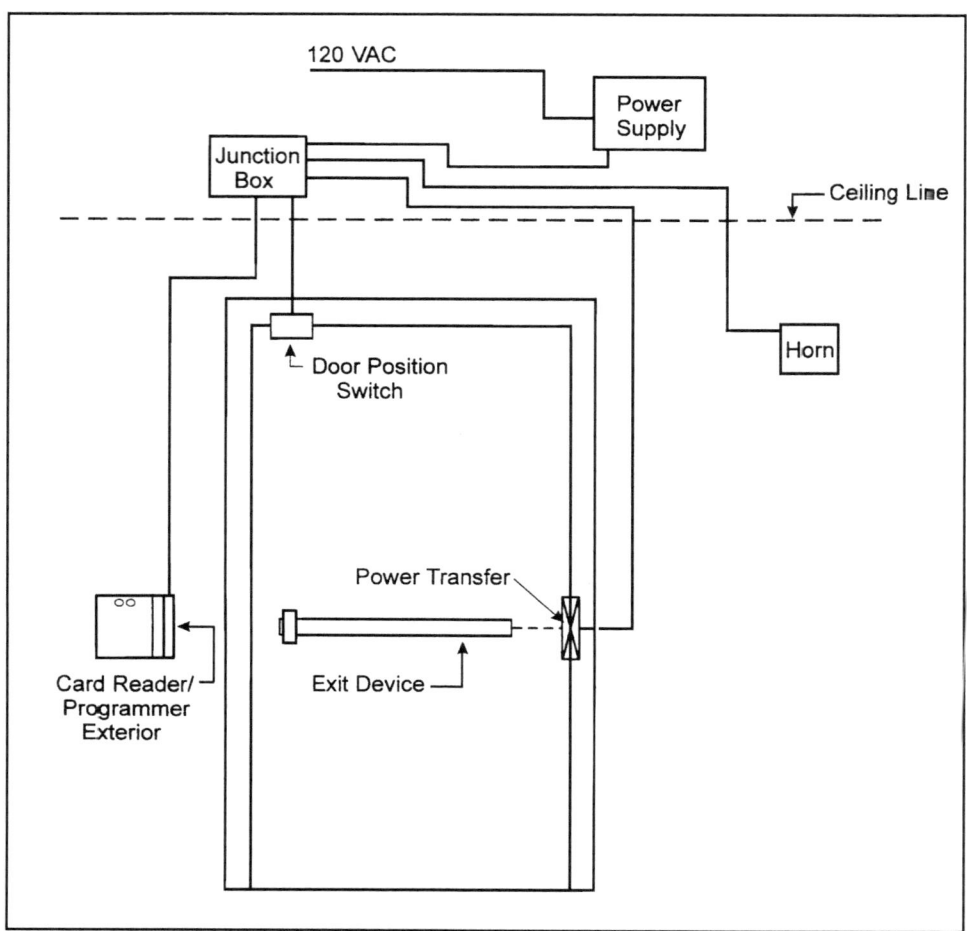

**Figure 13-20**   Employee entry door—with alarm and access control.

Employee entry control provides control and accountability of use by employees. The card reader access control limits the number of times employees may enter and records in memory when they use the access control system.

Valid access allows the door to be opened from the exterior. The door can be set to be unlocked from 1 to 60 seconds by the access control system. The system's memory records authorized entries and unauthorized attempts as well. From the interior the exit device provides free egress at all times.

### System description of operation

**Interior operation.**   Depressing the touchpad will inhibit the alarm, allowing the user to exit. When the door is closed, it will immediately latch and relock and the device will rearm itself. If the hold-open timer elapses, the alarm will sound. A valid access or PIN is required to reset the alarm.

**Exterior day operation.**   To gain entry an authorized user presents a valid card or PIN to the reader, causing the exit device lever to unlock. The card system has the capability, based on time and day, to automatically change the access requirements from a card, to PIN, to unlocked. If the door is held or forced open, the alarm will sound.

**Exterior night operation.**   For a higher level of evening and weekend security, the unit may be set to accept access cards or PIN or to prohibit any access. On a valid access the door may be opened. When the door is closed, the device will relock automatically.

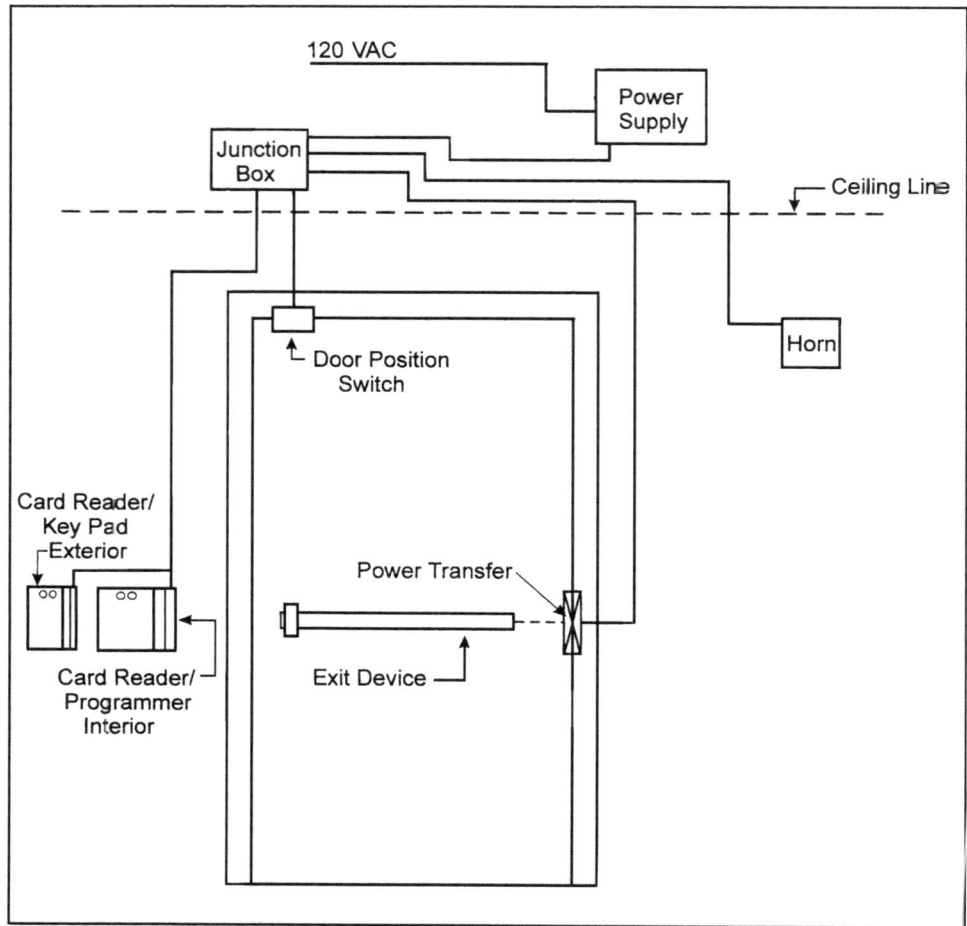

**Figure 13-21**   Employee exit/entry door—with alarm and access/egress control.

Employee entry and exit control provides control and accountability of use by employees. The card reader access control limits the number of times employees may enter and records in memory when they use the access control system. Valid access allows the door to be opened from the exterior. The door can be set to unlock from 1 to 60 seconds by the access control system.

In addition, the system's memory records current transactions. This keeps track of not only authorized entries but unauthorized attempts as well.

The access control system also controls, monitors, and records activity through the exit. While the exit device provides free egress at all times, use without first using a valid card or PIN number will cause the alarm to sound.

### System description of operation

**Interior day operation.**   Presenting a valid card or PIN to the reader will inhibit the alarm and allow the door to be opened. If the door is not reclosed and relatched, the alarm will sound. Someone who opens the door without deactivating the access system will be able to exit, but the alarm will sound. The access unit may also be set to keep the door unlocked for an extended period for unrestricted use.

**Exterior day operation.**   To gain entry, an authorized user presents a card or PIN to the reader. A valid entry will cause the exit device lever to unlock, allowing the door to be opened.

**Exterior night operation.**   For a higher level of evening and weekend security, the unit may be set to accept access cards or PIN or to prohibit any access. On a valid access the door may be opened. Once the door is opened, the lever will relock automatically, allowing the door to relock when closed.

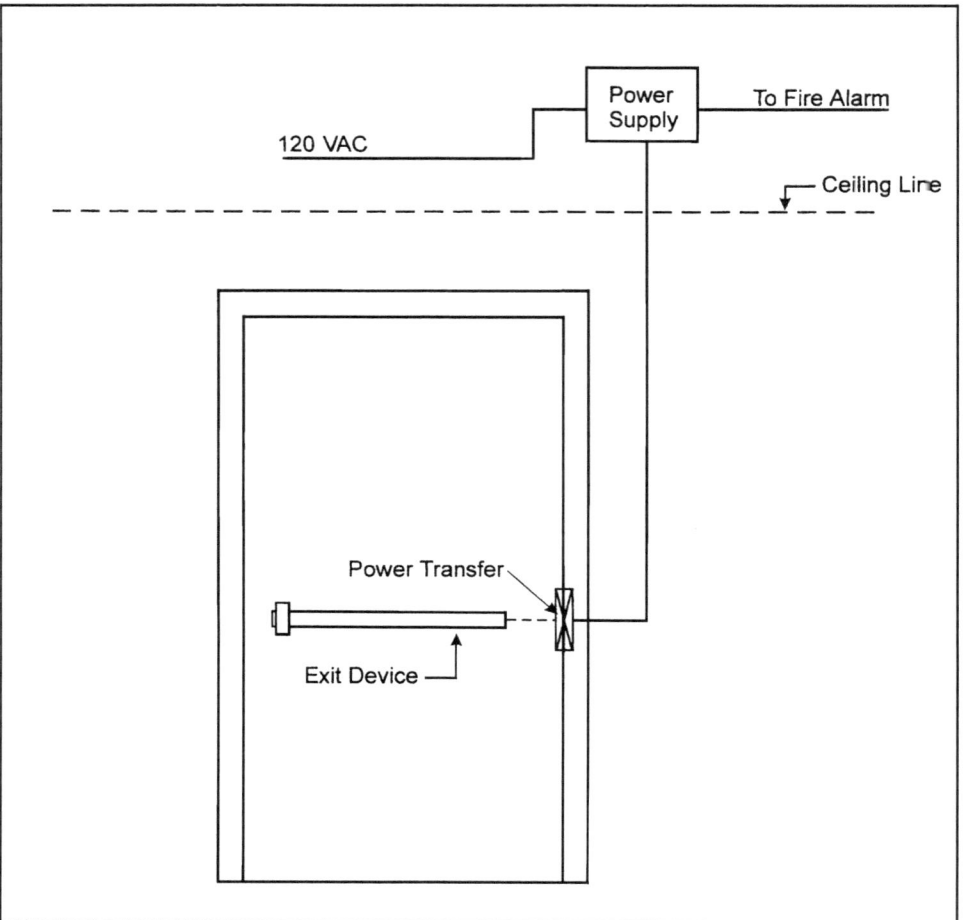

**Figure 13-22**   High-rise stairwell door.

Many life safety codes require reentry through the emergency exit on the stairwell side of the building. In most codes this may be mandated only when there is a fire alarm or loss of electrical power. The electrified exit device allows for reentry to the floor. The device must be fail-safe to ensure the trim will unlock in the event of power loss, allowing free reentry to the floor. The system requires the exit device to be tied into the fire alarm.

### System description of operation

**Floor side operation.**   Free egress is permitted at all times.

**Stairwell side operation.**   The door is locked at all times, preventing entry from the stairwell to a floor. If the fire alarm is actuated, reentry from the stairwell is permitted to the floor. On power failure, reentry is permitted from the stairwell to the floor.

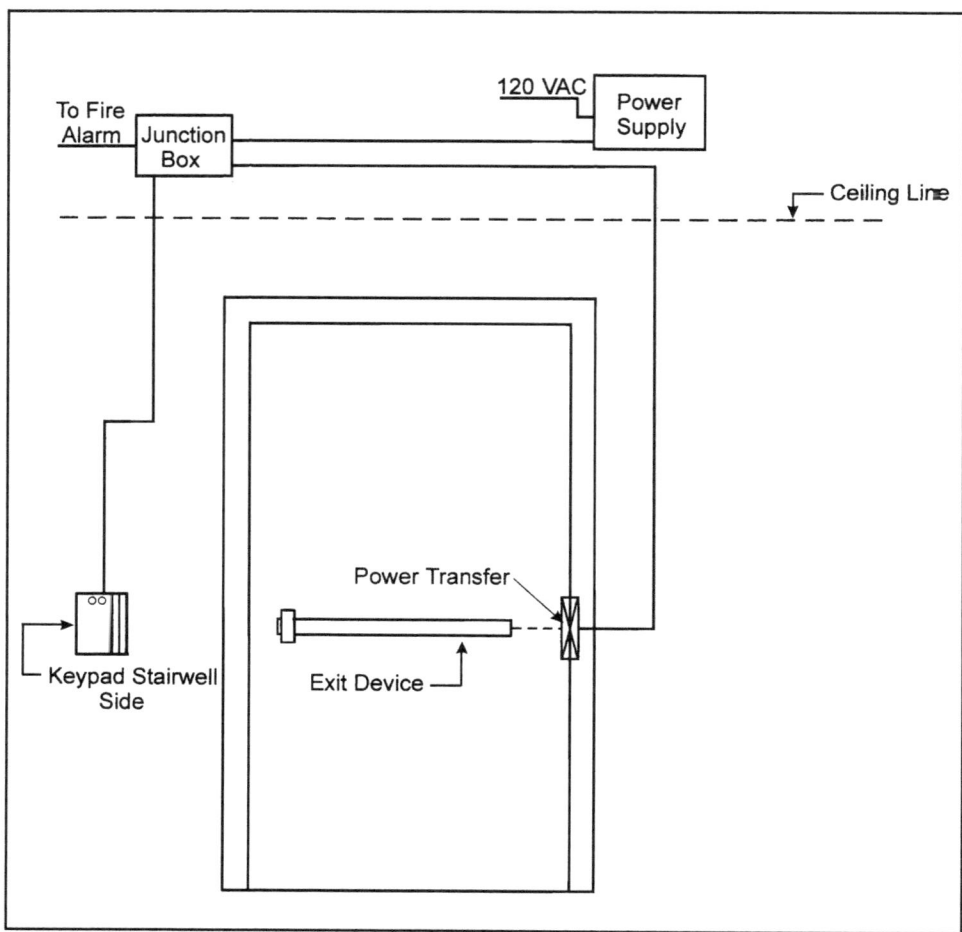

**Figure 13-23**    High-rise stairwell door—with keypad reentry.

Many life safety codes require reentry through the emergency exit on the stairwell side of the door. In most codes this may be mandated only when there is a fire alarm or loss of electrical power. The electrified exit device allows for reentry to the floor. The device must be fail-safe to ensure the trim will unlock in the event of power loss, allowing free reentry to the floor. The system requires the exit device to be tied into the fire alarm. The digital keypad mounted in the stairwell permits authorized users to reenter the floor from the stairwell.

**System description of operation**

**Floor side operation.**  Free egress is permitted at all times.

**Stairwell side operation.**  Entering a valid code into the digital keypad will unlock the trim, allowing it to retract the latch and permit the door to be pulled open. If the fire alarm is actuated, reentry from the stairwell is permitted to the floor.

On power failure the trim will unlock, allowing reentry from the stairwell to the floor.

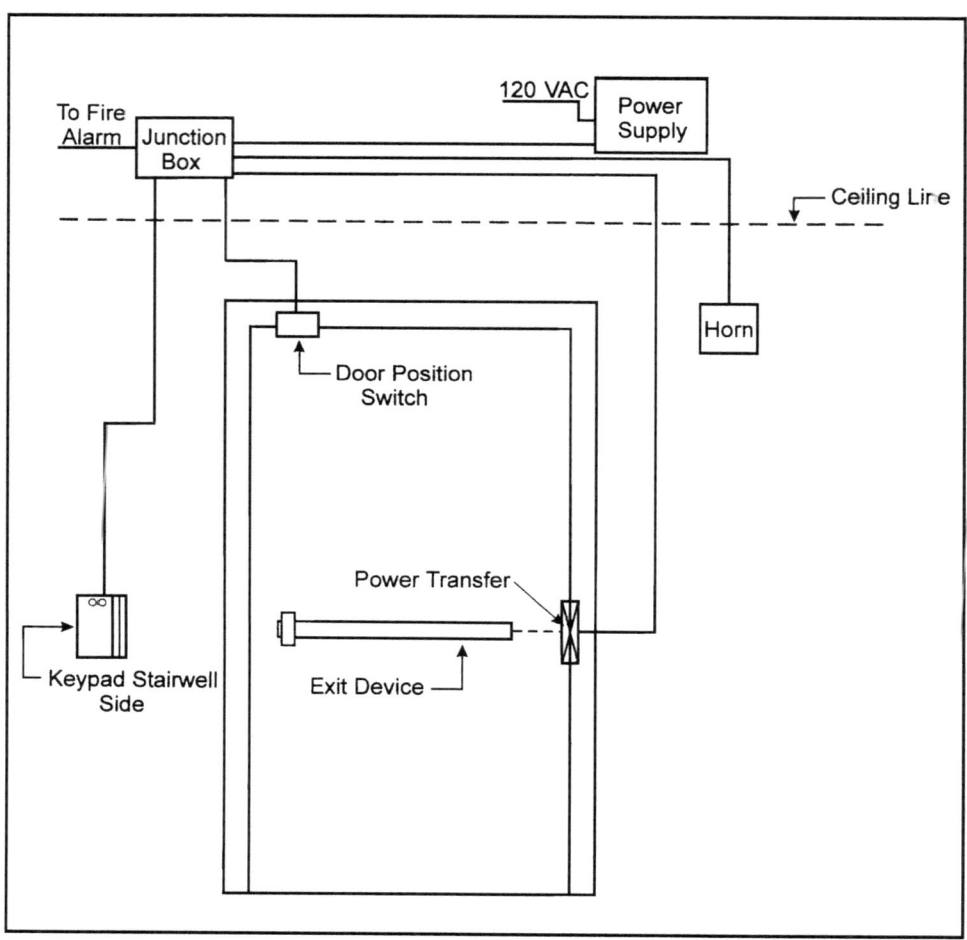

**Figure 13-24**   High-rise stairwell door—with alarm and access control.

Many life safety codes require reentry through the emergency exit on the stairwell side of the door. In most codes this may be mandated only when there is a fire alarm or loss of electrical power. The electrified exit device allows for reentry to the floor. The device must be fail-safe to ensure the trim will unlock in the event of power loss, allowing free reentry to the floor. The system requires the exit device to be tied into the fire alarm. The digital keypad mounted in the stairwell permits authorized users to reenter the floor from the stairwell. The alarm system will provide an alarm if the door is held open.

## System description of operation

**Floor side operation.**   Free egress is permitted at all times. Pushing the touchpad will inhibit the alarm, allowing exit without alarm. If the door is not closed and resecured within the keypad time setting, the alarm will sound. Pushing the touchpad or entering a valid code will silence the alarm and reset the timer.

**Stairwell side operation.**   Entering a valid code into the digital keypad will unlock the trim, allowing the lever to open the door. If the door is held or propped open after entry, or the door is forced open, the alarm will sound. Pushing the touchpad or entering a valid code will silence the alarm.

If the fire alarm is actuated, reentry from the stairwell is permitted to the floor. On power failure the trim will unlock, allowing reentry from the stairwell to the floor.

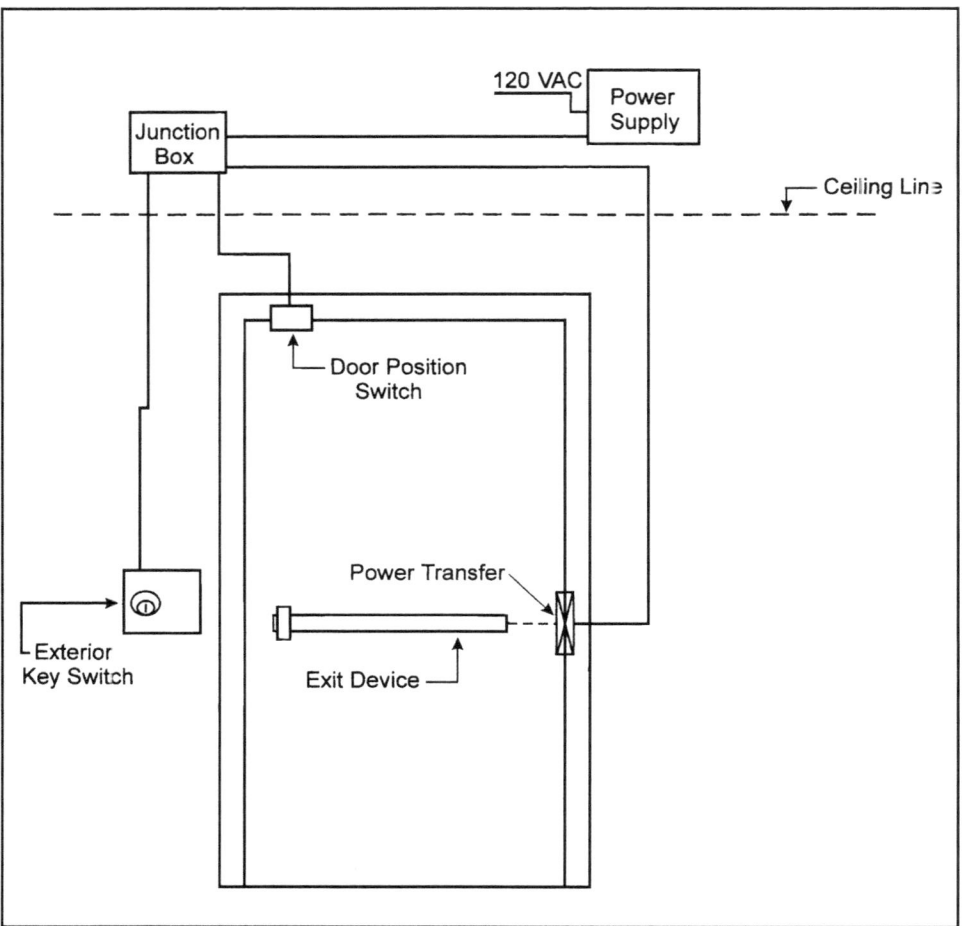

**Figure 13-25**  Perimeter door—with day/night control.

During open hours, input from the key switch will hold the electric latch retraction exit device in the retracted position. This will allow the opening to operate in a push/pull mode. A key is required to electrically dog or undog (latch or unlatch) the device. Multiple door openings with additional electrified exit devices may be linked to this application.

### System description of operation

**Interior operation.**    Free egress is permitted at all times.

**Exterior day operation.**    Using the key switch will retract the exit device latch, allowing the door to be pulled open. The key may be turned in the clockwise direction for momentary unlock or in the counterclockwise direction for maintained unlock.

**Exterior night operation.**    The key may be turned in the clockwise direction to momentarily retract the exit device latch, allowing the door to be pulled open.

On power failure the exterior cylinder may be used to mechanically retract the latch, allowing the door to be pulled open.

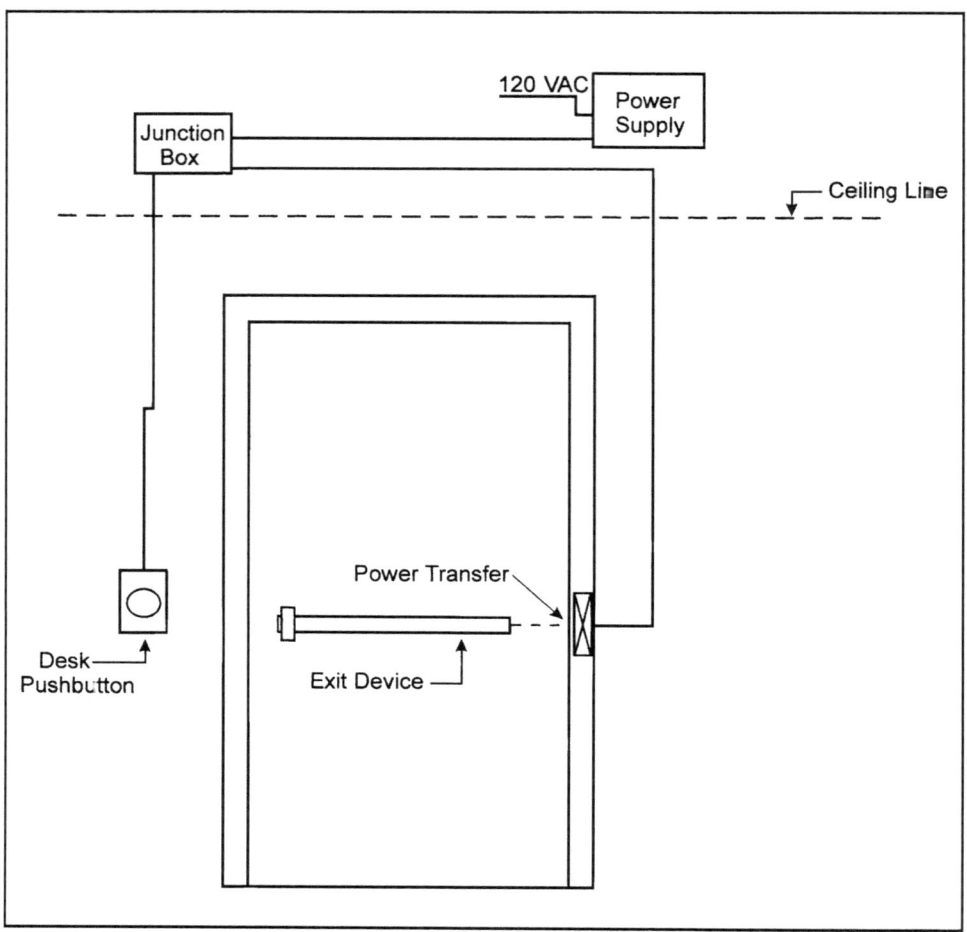

**Figure 13-26**   Perimeter door—with remote entry control.

A pushbutton may be used to momentarily release the electrified exit device and allow entry from the exterior.

Unlatched time can be adjusted from 5 to 60 seconds. This will ensure that the door will relock from the exterior while allowing free egress for life safety. The pushbutton may be located on a wall, behind a counter, or on a desk within view of the entry under control.

### System description of operation

**Interior operation.**   Free egress is permitted at all times.

**Exterior day operation.**   On the interior, a desk- or wall-mounted switch is in view of the opening. When a person on the interior recognizes an authorized person approaching the door for entry, he or she simply depresses the pushbutton, which retracts the latch, allowing the door to be pulled open from the exterior.

**Exterior night operation.**   If no one is inside to assist with entry, a mechanical key may be used to retract the latch from the exterior to allow entry.

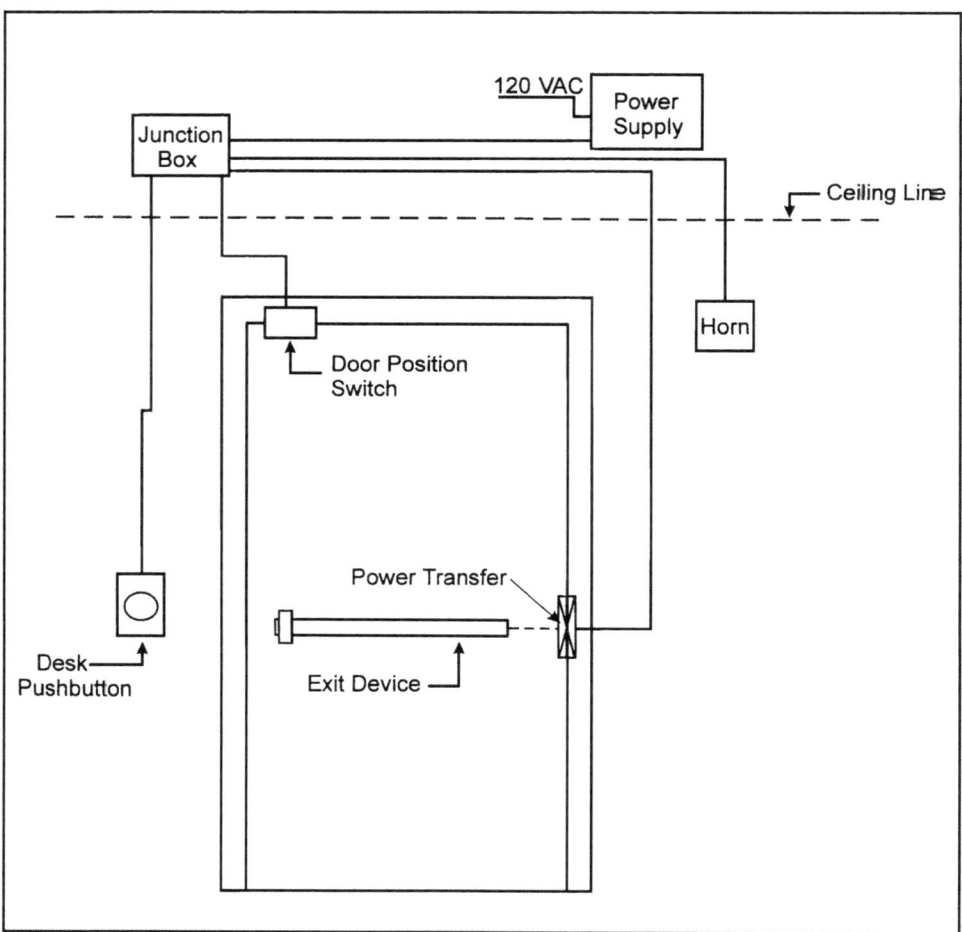

**Figure 13-27** Perimeter door—with alarm and remote entry control.

An electrified exit device and remote pushbutton allow remote control of the normally locked entry with the added security of alarm monitoring. The pushbutton provides remote control of entry. The alarm circuit indicates that the door has been forced or propped open for an extended period, warning of a security violation.

## System description of operation

**Interior operation.**  When the pushbutton is depressed, the device will stay retracted for a period of 5 to 60 seconds. The opening is equipped with a door position switch. If a user exits without first depressing the pushbutton, the device will sound an alarm. Additionally, if the door is held or propped open beyond the time set on the timer, the alarm will sound. Depressing the pushbutton will silence the alarm.

**Exterior day operation.**  On the interior, a desk- or wall-mounted switch is in view of the opening. When a person on the interior recognizes an authorized person approaching the door for entry, he or she simply depresses the pushbutton, which retracts the latch, allowing the door to be pulled open from the exterior.

**Exterior night operation.**  Forcing the door open or opening it with a key will cause an alarm condition.

## Automatic door applications

Many automatic door systems require the added security of locking devices. The locking devices must be electric to allow coordination with the power operator. Multidoor automatic door systems also require electrically controlled coordination with each door in the paths of egress and ingress. Multidoor systems also may require interlocking with one another. At first all this may seem complicated, but technical advances in power operators can actually make designing these systems quite easy.

The systems presented in this section may tempt you to design a wiring diagram utilizing timers and relays. And this is good; it is what you have learned to do throughout this book. It will also sharpen your skills in understanding the logic required to make these systems work. But I would be remiss in not informing you that there are "ready-built" electronic packages available. These electronics may be built into the power operator or available as separate controllers. They can often provide the coordination required by simple inputs and outputs between system components. I would check first with the manufacturer of the power operator to see what is available. It could make designing the system much easier.

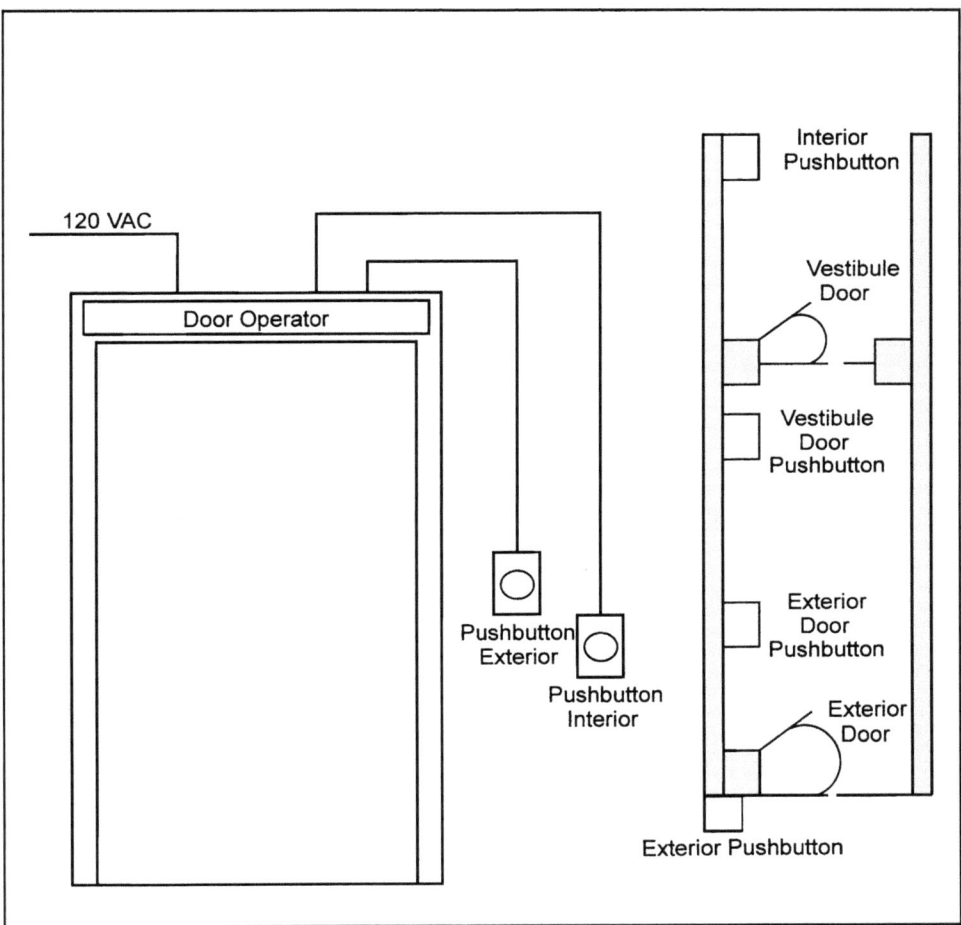

**Figure 13-28** Automatic doors—bidirectional vestibule system.

The system shown in Figure 13-28 permits two power-operated doors to sequentially open when activated by either the interior or the exterior pushbutton. Directional time delay is provided between door openings. Additional vestibule pushbuttons are provided for activating each door if necessary.

### System description of operation

Doors within the vestibule or exterior application are normally in a closed position. Manually depressing the interior pushbutton will activate the vestibule door operator. The exterior door operator will remain inactive for a predetermined time. Upon expiration of time delay, the exterior door operator will automatically activate as the vestibule door begins to close. The reverse sequence of operation occurs upon manual depressing of the exterior pushbutton. Additional pushbuttons are mounted within the vestibule for the activation of each operator if necessary.

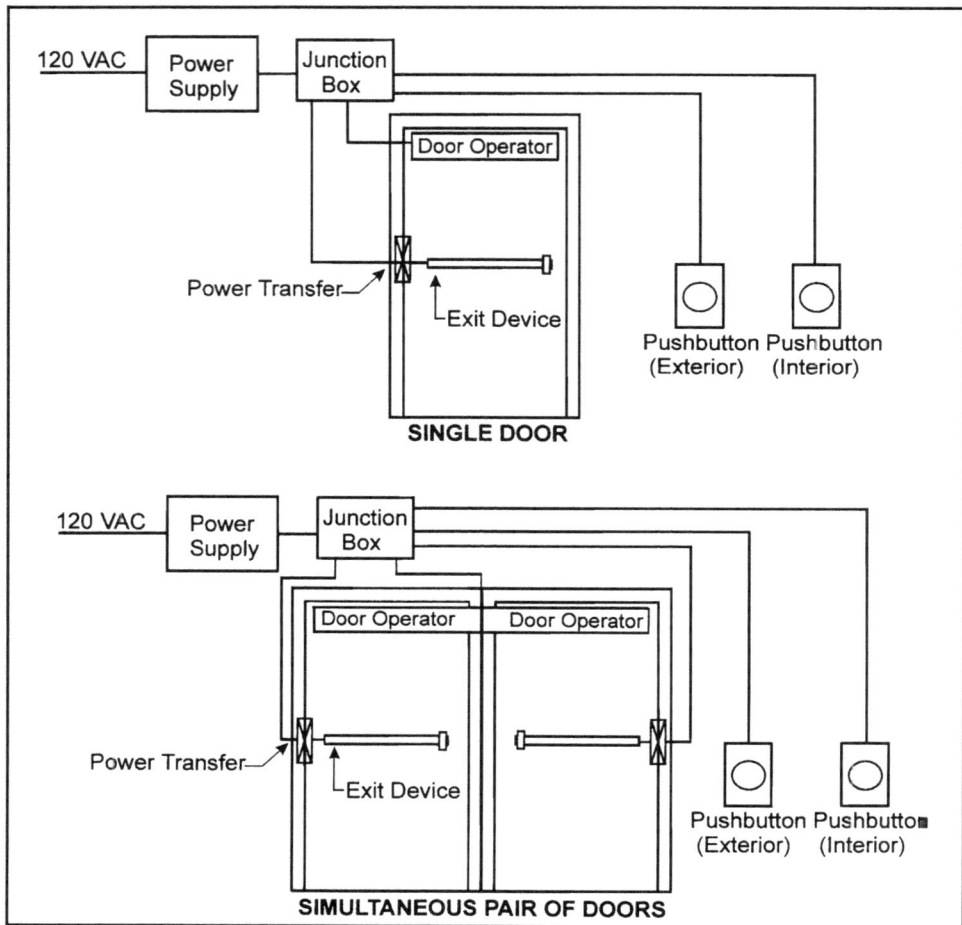

**Figure 13-29**  Automatic doors—with electric latch retraction exit devices.

The system shown in Figure 13-29 coordinates the unlocking of the electric exit devices with the activation of the power operators. The time delay between latch retraction and power operator activation is normally only a few seconds.

## System description of operation

Doors with electric latch retraction exit devices are normally in a closed and latched position. Manually depressing the interior or exterior pushbutton or depressing the exit device touchpad will activate the exit devices, retracting the latch bolt. Operators will remain inactive for a predetermined time. Upon expiration of the time delay, the door operator will activate. The exit device latch bolt will return to its normally extended position.

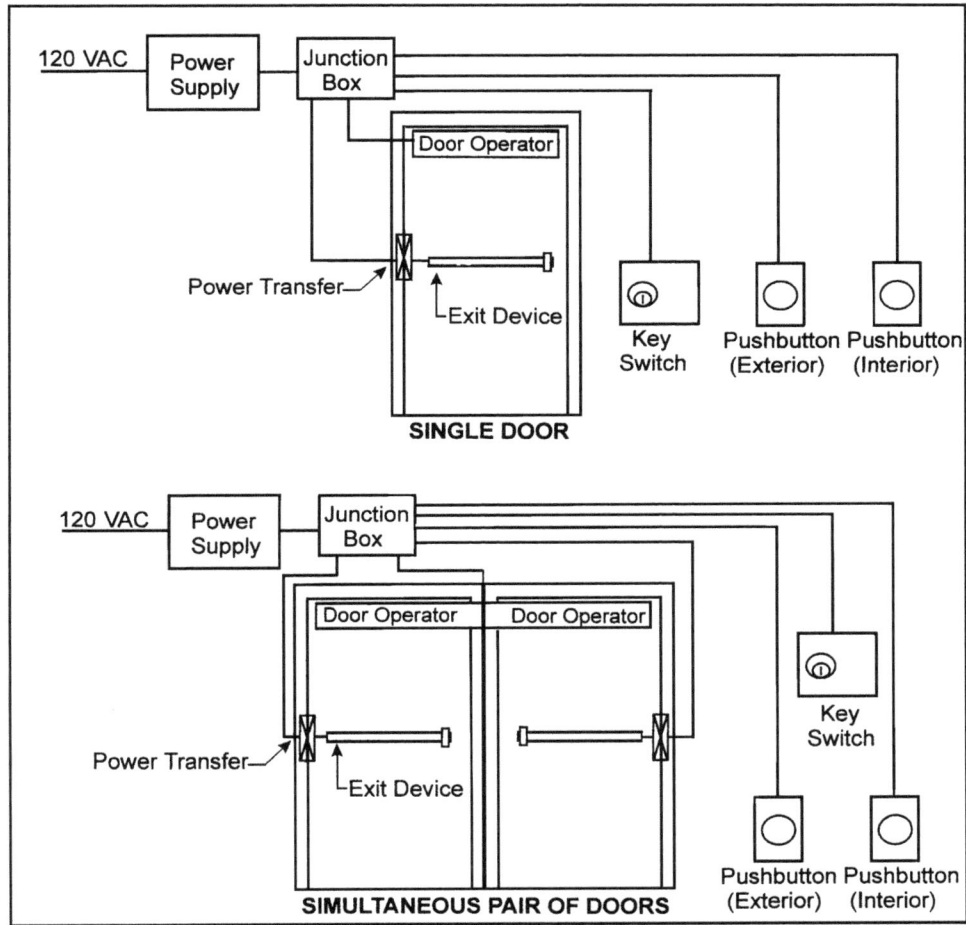

**Figure 13-30** Automatic doors—with daytime electric dogging exit devices.

The system shown in Figure 13-30 allows continuous use of automatic doors by electrically dogging the exit devices. Dogging may be released by remote key switch or fire panel tie-in.

## System description of operation

Doors with electrically dogged exit devices are normally in a closed position. Exit devices are dogged or locked by means of a key switch at a remote location. Manually depressing the interior or exterior pushbutton will activate the door operator. Doors will act as a push/pull assembly when electrically dogged. The key switch can relock the exit devices for manual nighttime operation. Fire-rated exit devices must be interfaced with the fire alarm system. Activation of the fire alarm system will release the dogging, latching the doors.

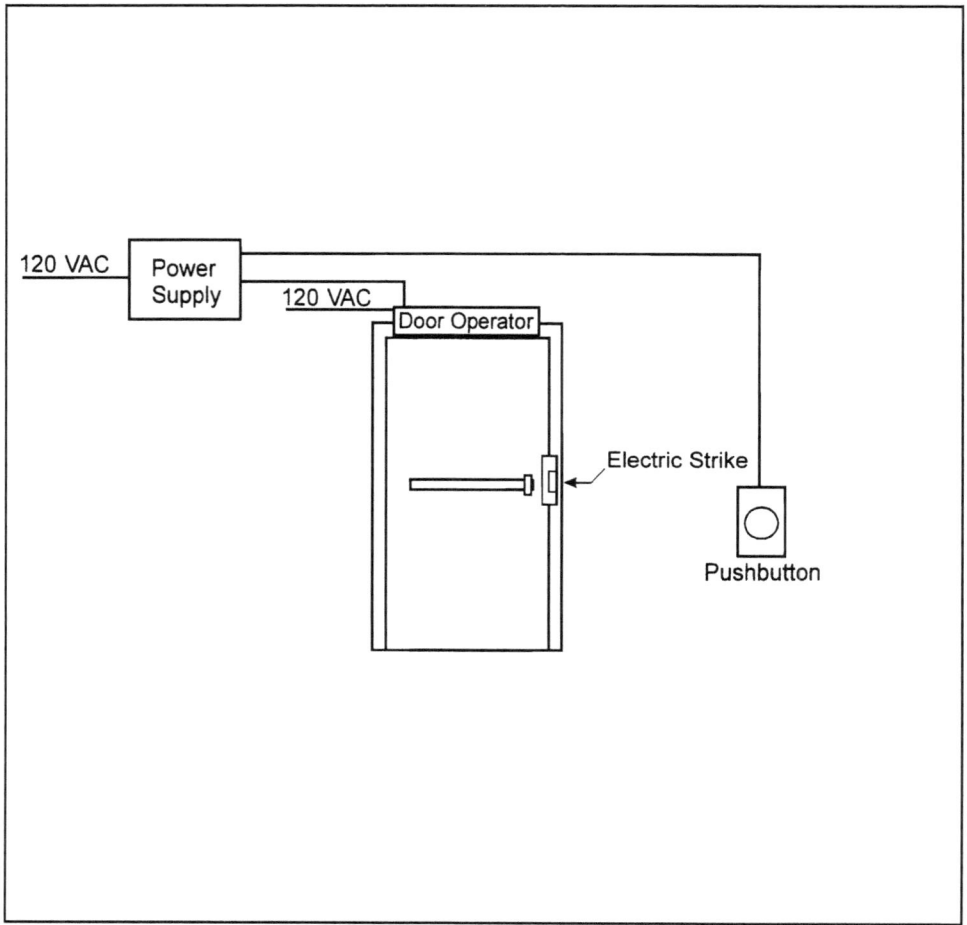

**Figure 13-31**   Automatic door—with electric strike.

The system shown in Figure 13-31 coordinates the release of the electric strike with activation of the power operator. The power supply commonly contains time delay and relay circuitry to provide proper sequencing. Control circuitry may also be available built into the power operator.

## System description of operation

Door with electric strike is in the normally closed and locked position. Manually depressing the interior pushbutton will release the electric strike. The door operator will remain inactive for a predetermined time. Upon expiration of the time delay, the door operator will activate. The electric strike will relock after a predetermined time.

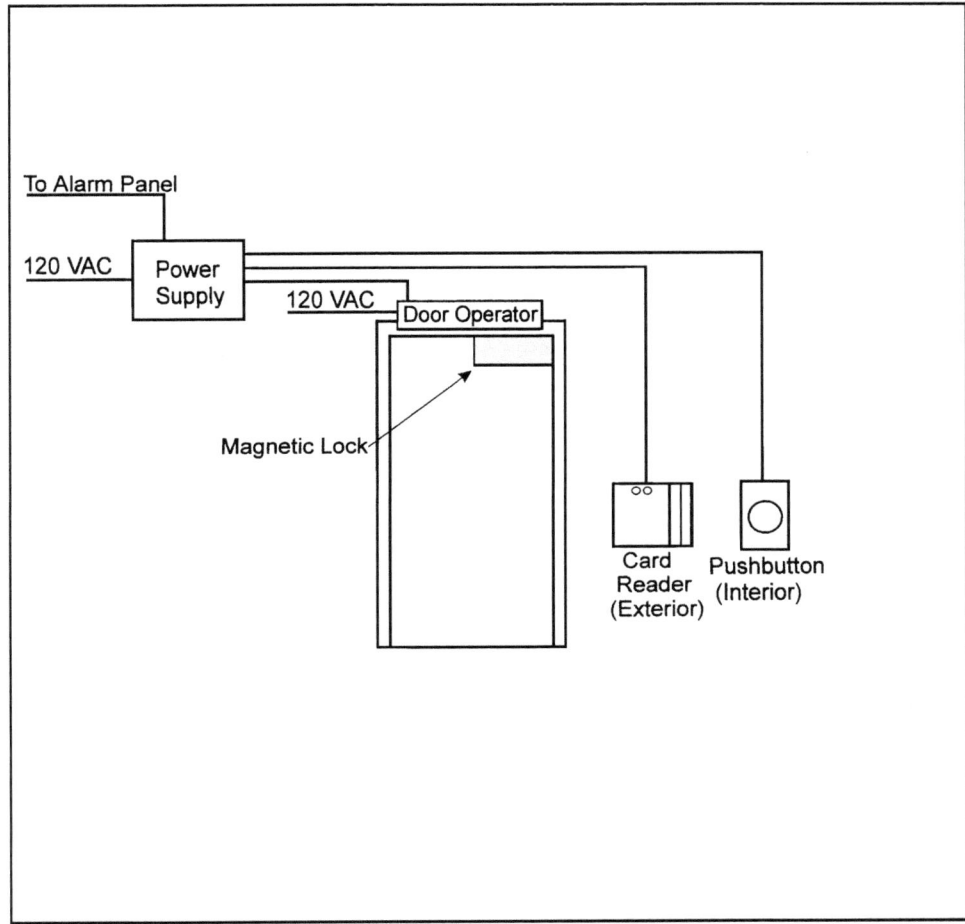

**Figure 13-32** Automatic door—with magnetic lock.

The system shown in Figure 13-32 coordinates the release of the magnetic lock with the activation of the power operator. Simple control circuitry is normally provided built into the power operator.

### System description of operation

Door with magnetic lock is in the normally closed and locked position. Manually depressing the interior pushbutton or use of the exterior card reader will deactivate the magnetic lock. The door operator will remain inactive for a predetermined time. Upon expiration of the time delay, the door operator will activate. The magnetic lock will relock upon closing of the door. The system must be interfaced with a fire alarm system.

# Solution of Chapter 5 Test Exercise

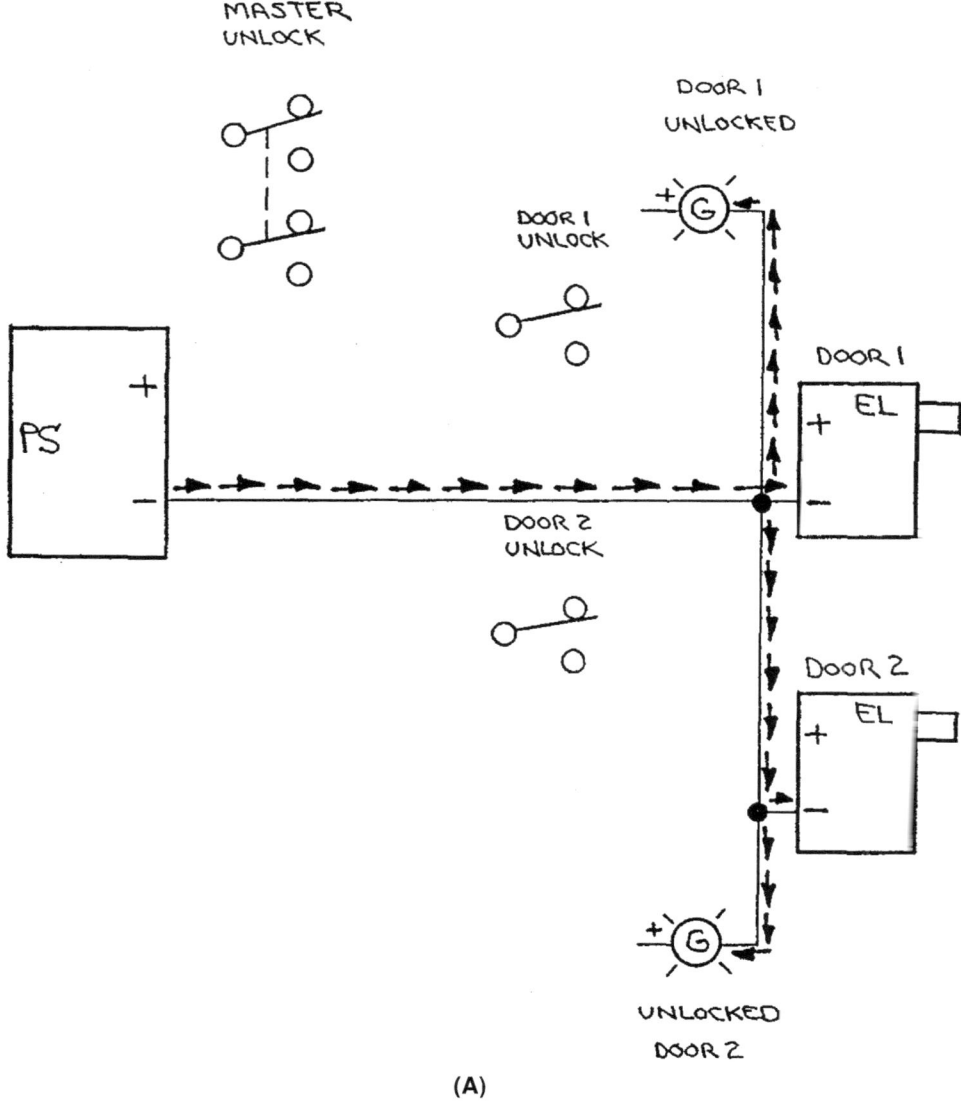

**(A)**

**Figure A-1A.** You may find it helps to "highlight" your wire runs as you check your drawing to the solution diagrams. The easiest way to start is to wire the power supply (−) to all the loads (−).

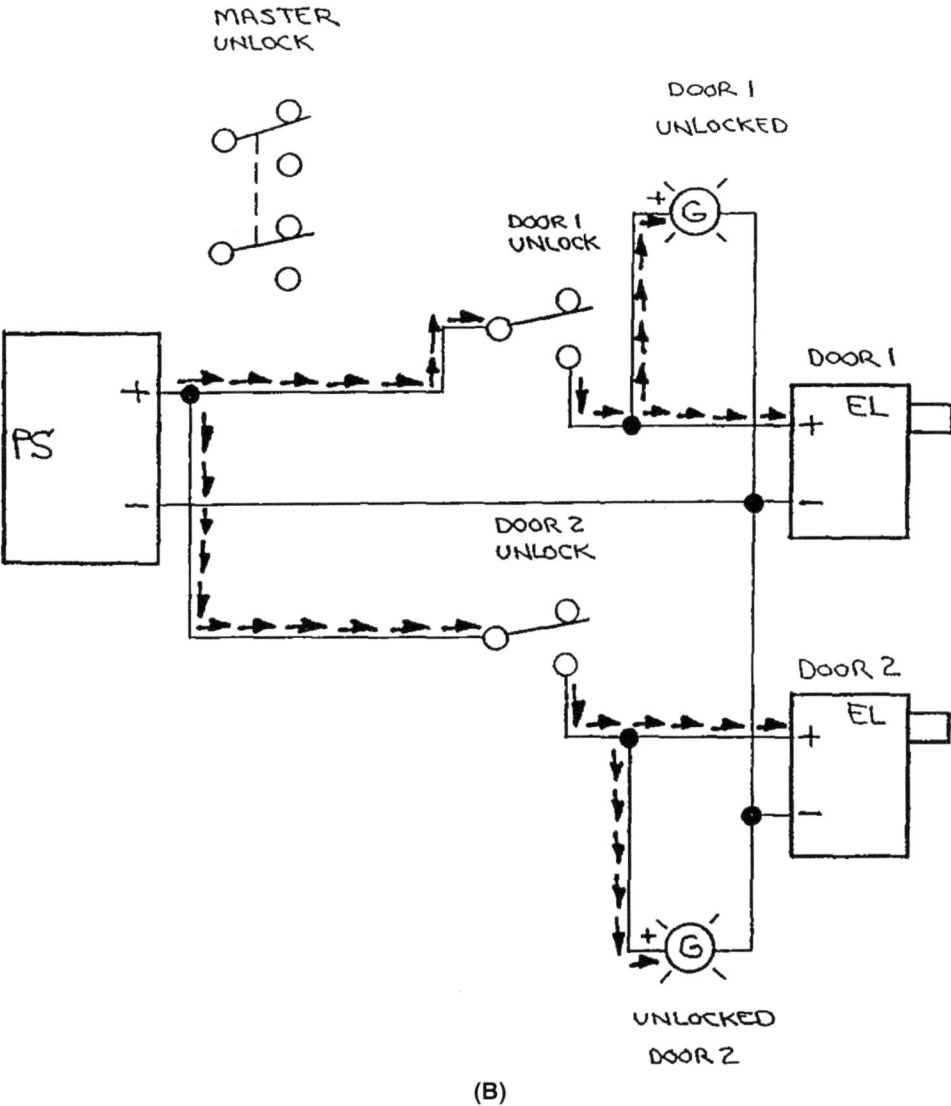

**(B)**

**Figure A-1B.**   The next step is to wire the power supply (+) to the loads. I chose to start with the individual switches. Notice the (+) wire run travels to door 1 switch. The wire run then continues from the *open* switch contact to the lock and indicator. When the switch closes (+) travels to the lock, releasing it, and to the green light, illuminating it. The same route is run for door 2.

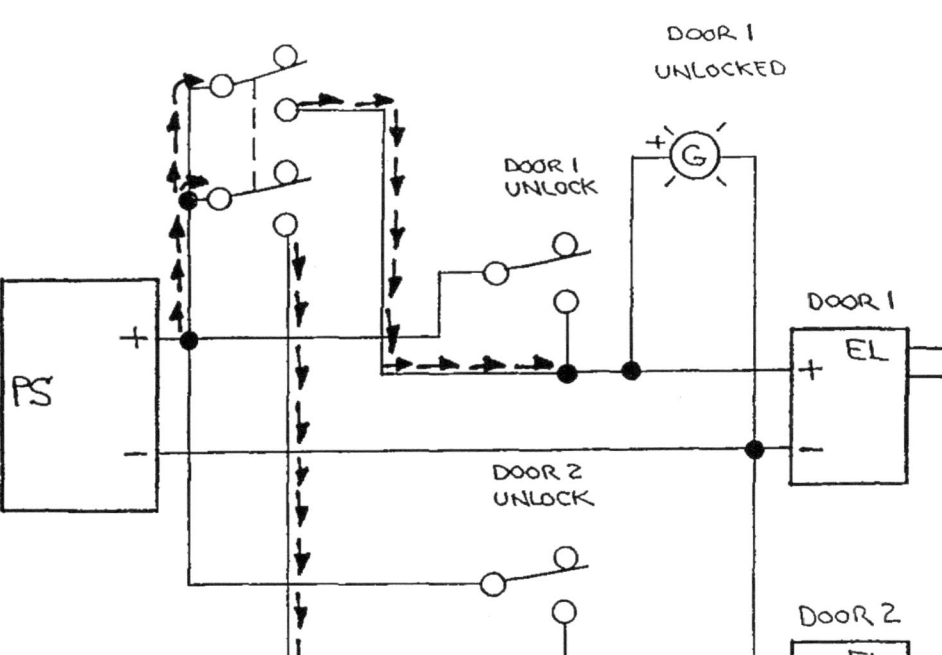

**(C)**

**Figure A-1C.**  The final step is to wire (+) through the master switch. First wire (+) to the common of the two contacts sets. Then wire the open contact of one contact set, following a route that ensures (+) will reach the lock and indicator when the contact closes. Notice that you can hook up with an existing wire run that follows a path directly to the lock and indicator. Repeat this wiring for the second contact set.

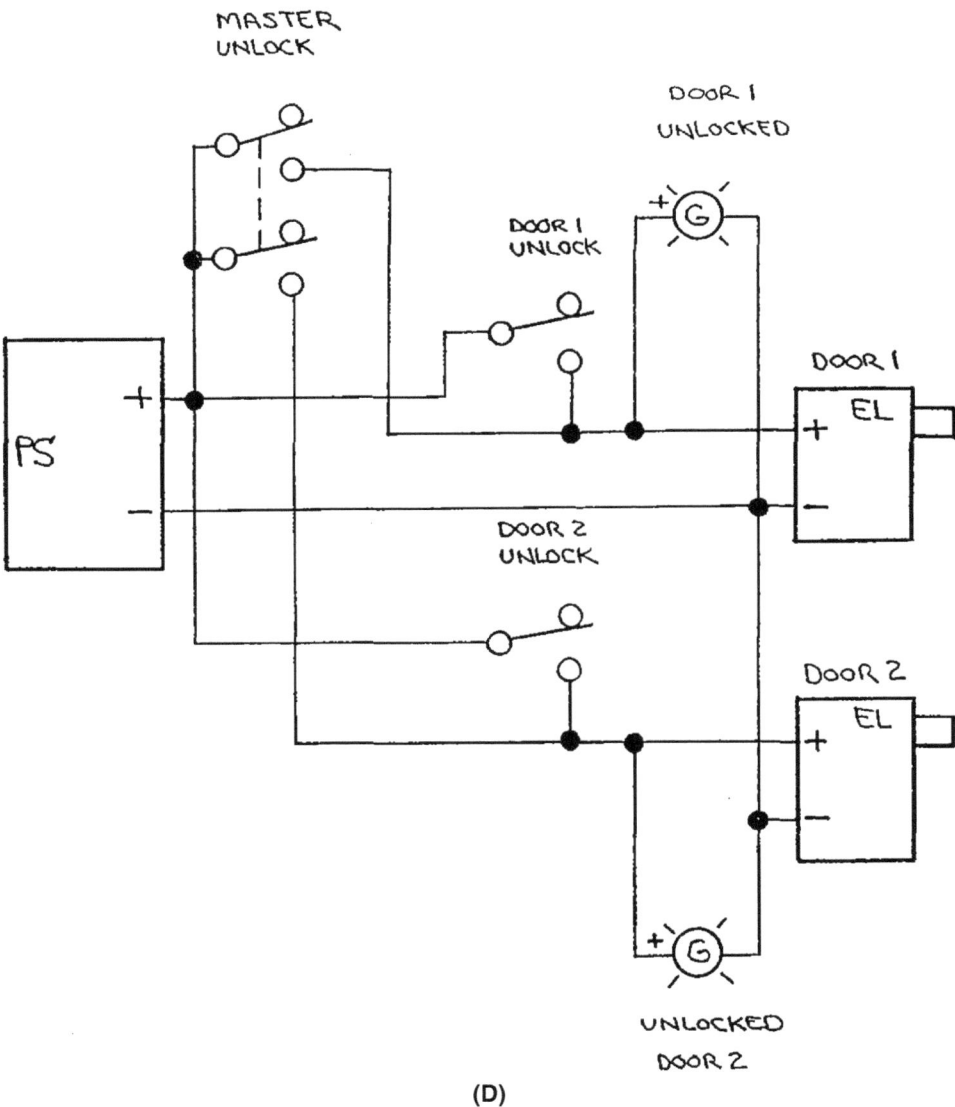

**(D)**

**Figure A-1 D.** The wiring is now complete. The finishing touches for this drawing would be terminal identifications, wire colors, voltage, and current notations, and any other pertinent information. A final touch would be to rewrite the original system notes as a "description of operation" to include on the wiring diagram.

# Test Answers for Chapters 6 and 7

1. Junction box

2. *b*

3. *a*

4. F (check to see if it is required).

5. *b*

6. Door status and lock status

7. Audible or visual

8. T

9. Interlocks

10. DSS  = door status switch
    BPS  = bolt position switch
    LCM  = latch (lock) cam monitor
    DPS  = door position switch
    LED  = light-emitting diode
    dB   = decibel

# Solution of Chapter 9 Test Exercise

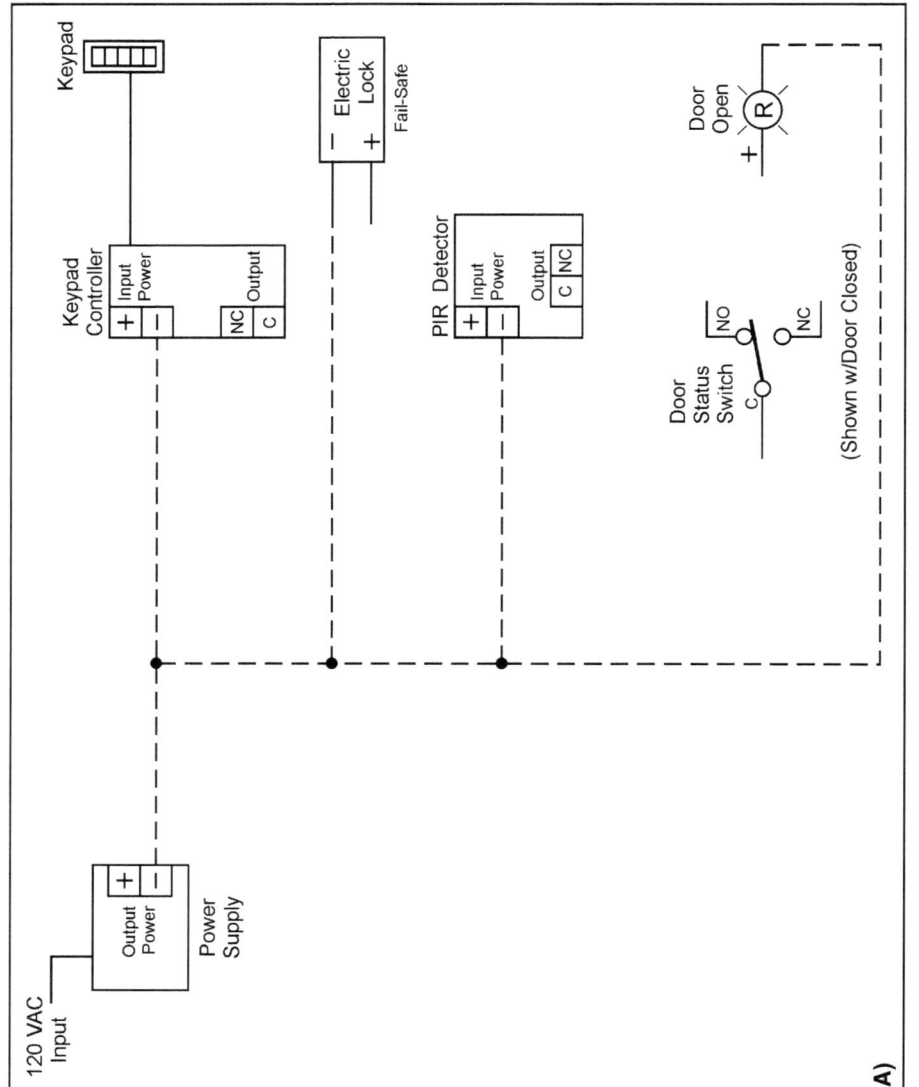

**Figure C-1A.** Hookup (−) from the power supply to all the loads in the system.

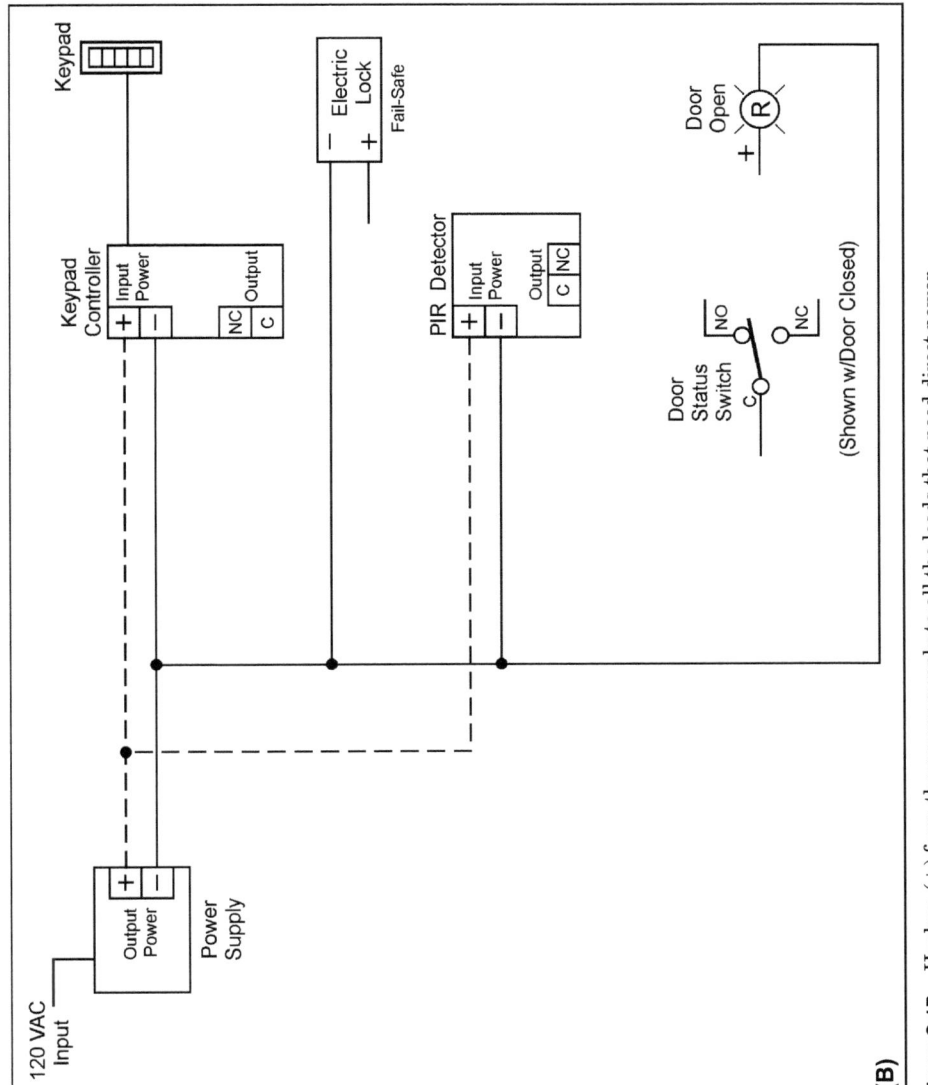

**Figure C-1B.** Hookup (+) from the power supply to all the loads that need direct power (not switched).

249

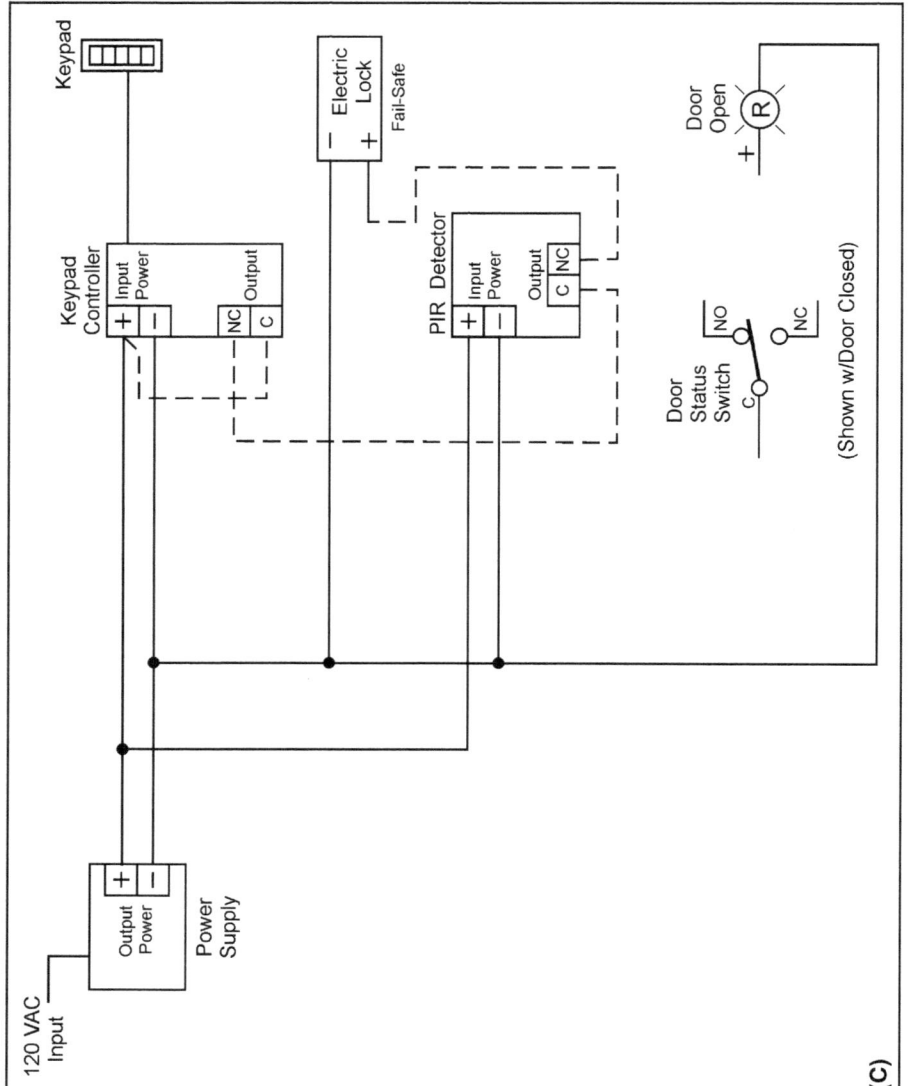

**Figure C-1C.** Hookup (+) from the power supply to the electric lock. Note that (+) must be controlled by the keypad controller and the PIR detector. The (+) run can be picked up anywhere along a (+) run from the power supply provided it is before any switch in the run. A handy place here is at the (+) terminal of the keypad controller since ( | ) has to run through its control contacts anyway. Note that (+) runs through two NC contacts before reaching the lock. Opening either of the contacts releases the lock.

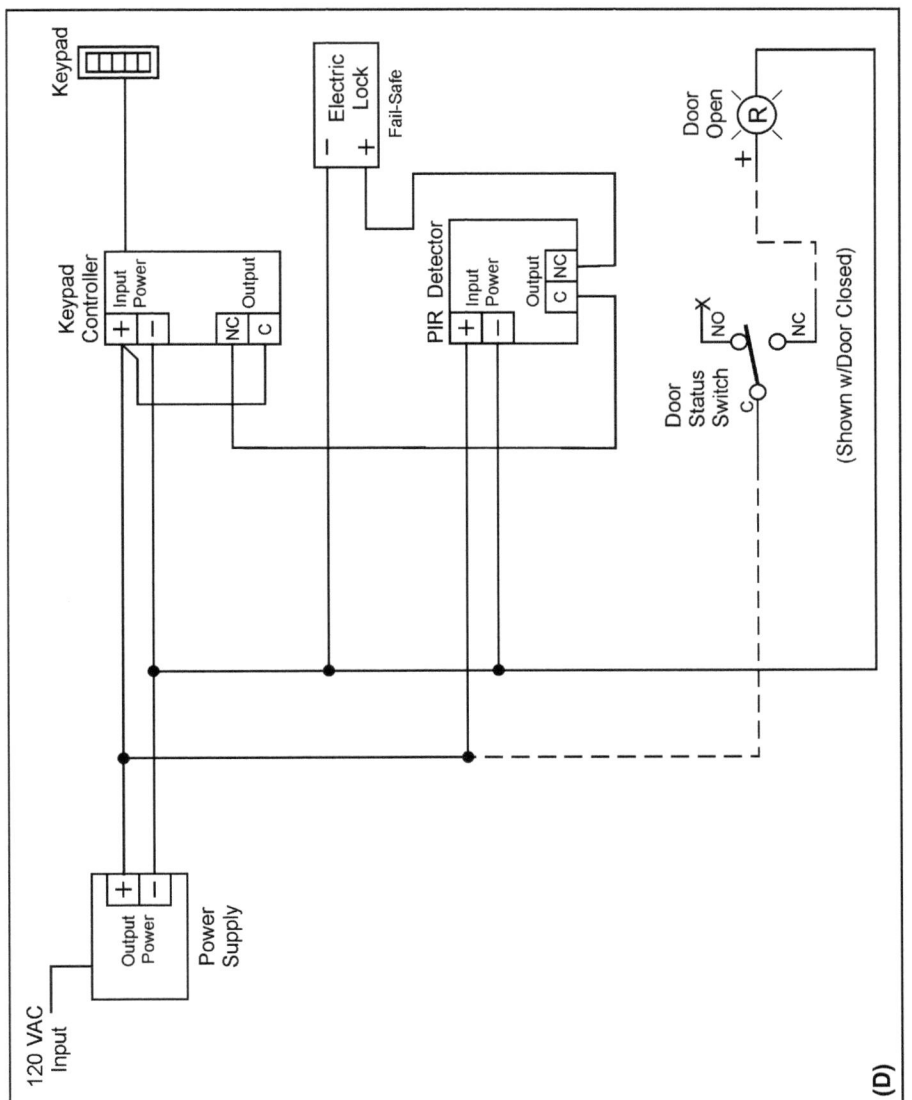

**Figure C-1D.** Hookup (+) from the power supply to the door open light. Once again (+) can be picked up on any unbroken (+) line that is convenient. Note that we wire the normally open held closed side of the door status switch. Opening the door causes the switch to "fall" closed illuminating the light. (Terminate the unused wire from the NO side of the switch by marking with an X.)

251

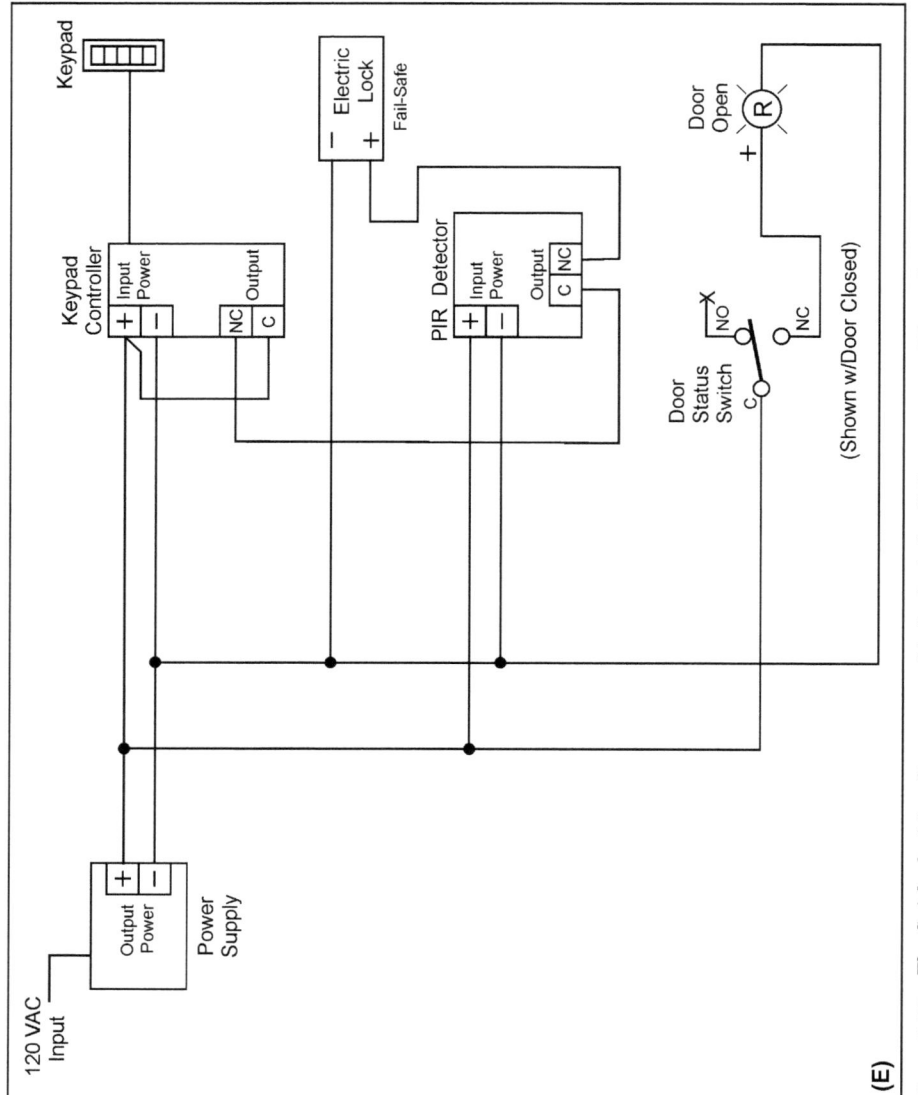

**Figure C-1E.** The finished wiring diagram would also include all the component IDs and any other pertinent information. The wire runs drawn are not necessarily the way the system would be wired. The drawing is meant to be neat and easily understood. A professional installer will use his expertise to efficiently route the wiring.

# Glossary

**access code** Numeric or alphanumeric data which, when correctly entered into a keypad, allows authorized entry into a controlled area without causing an alarm condition.

**access control** The control of persons or vehicles through entrances and exits of a protected area.

**access control card** A card containing coded information which, when read by a card reader, grants access to a protected area if the coded information is deemed valid by the controlling electronics.

**access level** See **authorization level.**

**access mode** The operation of an alarm system such that no alarm signal is given when the protected area is entered.

**access parameters** Programmed information defining system variables such as access codes, entry times, and authorization level.

**access point** Each means of entry into a controlled security area, consisting of an access control device, monitor switches, and door-locking devices.

**access status** The existing conditions which specify a person's ability to enter or leave a protected area.

**acquired data** Data on a transaction or event gathered in real time from remote data collection devices such as card readers, alarm monitors, and input to a central processor for analysis or decision making.

**active card** A type of access control card, powered by the card reader into which it is inserted, that transmits a readable signal read by specialized scanning sensors.

**actuator** The mechanism of a switch that operates the contacts.

**alarm** A device used to indicate an emergency or other specific condition.

**alarm disable** The ability to turn off an alarm electronically or physically.

**alarm enable** The ability to turn on an alarm electronically or physically.

**alarm monitor** Usually a centrally controlled computer that monitors a system via a system of interconnection wires, cables, and radio-frequency networks. It is used to poll devices on the lines or to collect data from other monitors for control and storage of data.

**alarm report** An alarm message formatted for listing at a printer or on a message screen and stored in an alarm system file for historical recall.

**alphanumeric** Combinations of numbers and letters typically used in an access code.

**alternating current (AC)** Electric current that reverses its direction regularly and continually. The voltage alternates its polarity and direction of current flow from negative to positive. Alternating current flows back and forth in the conductor and is expressed in cycles per second, or hertz (Hz).

**ambient temperature** The temperature of the air immediately surrounding a specific area.

**American wire gauge (AWG)** A standard gauge for designating the size of solid wires.

**ampere (A)** The unit of measurement for the rate of electrical current flow, characterized by the symbol $I$ (in Ohm's law formulas) and abbreviated as A. One ampere is the current flowing through one ohm of resistance at one volt potential.

**ampere-hour**  A measurement of a battery's capacity. One ampere of current flowing for one hour equals one ampere-hour.

**annunciator**  A visual or audible device which indicates an existing condition or the former condition of each circuit being monitored.

**ANSI**  The abbreviation for the American National Standards Institute.

**antipassback**  A feature of an access control system which prevents successive use of one card to pass through any door, turnstile, or portal in the same direction. To attain this type of protection, a separate reader is required at each entry and exit. Antipassback prevents a card from being passed back to another person for the purpose of unauthorized access.

**arc**  An electrical current through air or across the surface of an insulator associated with high voltage. An arc can occur when a contact is opened, deenergizing an inductive load. Arcing will limit the life of switch contacts.

**armor**  A metal jacket surrounding wires for mechanical protection.

**attended ID station**  A place where a guard or someone in authority supervises identification of people attempting to gain access to a protected area.

**audit trail**  A historical record sequentially accounting for all activities with an access control system. Such a record allows reconstruction and analysis of events during a given time period.

**authorization level**  A predefined security rating which must be met before access to a protected area is granted. Authorization level is sometimes referred to as access level or status level.

**authorized person**  A person who has been cleared to enter secure areas.

**authorized release device**  A device that allows authorized persons to enter or exit monitored and controlled openings without triggering an alarm. The authorized release should be a restricted device, for example, a key switch, card reader, or keypad.

**automatic time switch**  A timing device that turns locking devices on and off at preset times.

**auxiliary code**  In some access control and intrusion alarm systems, a secondary code, often used as a temporary code, issued for granting access or operating an alarm system without revealing the primary code.

**badge reader**  A sensor used to scan and interpret data encoded in a personnel identification badge.

**bar code card**  A series of lines printed on a card in a certain format which forms a code that is optically read by a reader.

**bar code reader**  A device that scans bar code information and transmits it to a central computer or control unit. These units can take the form of readers, wands, or gun-shaped scanners.

**barium ferrite card**  A card in which magnetic material is embedded that is encoded to produce individual codes. This card is sometimes referred to as a sandwich card.

**barium ferrite reader**  A device that reads the information magnetically encoded in a barium ferrite card. These readers are typically an insertion type, or some similar variation depending on the manufacturer.

**batch programming**    A method of processing data in which transactions are collected and prepared for input to the controller for processing as a single unit. In a central control system, the ability to program, alter, or cancel groups of cards with same access parameters in a single high-speed operation.

**battery**    A source of electrical energy produced by a controlled chemical reaction. Primary batteries are those that are discarded when discharged; secondary batteries are rechargeable.

**battery backup**    A standby energy source which provides DC power in the event of a power failure. Battery backups can be used to maintain data in memory, keep a system clock updated, or operate card readers, door strikes, deadbolts, and magnetic locks if a power failure occurs. The backup system may power the whole system for a time, but it normally falls into a degraded mode, powering only essential parts of a system.

**biometric access control**    A method of access verification in which persons seeking entry into a protected area are identified by their biometric parameters (fingerprints, hand geometry, etc.).

**biometrics**    The technology of personnel verification by measuring unique biological characteristics such as fingerprints, hand geometry, voice analysis, and the eye's retinal pattern.

**block diagram**    A drawing that shows the relationship of equipment in a system. Blocks are used to represent each piece of equipment and are arranged into a system diagram to show their physical or operational relation to one another.

**bolt**    The projectable member of a lock or latch mechanism that engages the doorframe strike plate.

**bolt position switch (BPS)**    A switch used in a locking device to monitor whether the locking bolt is in the locked (projected) or unlocked (retracted) position.

**break**    To open an electric circuit.

**break-before-make**    The action of a type of switch in which one contact opens before another closes.

**breakdown voltage**    The voltage at which the insulation between two conductors is destroyed.

**brownout**    Low line voltage which can cause misoperation of and possible damage to electrical equipment.

**bus**    Solid metal or wire connector, uninsulated, sometimes called a busbar. In a computer or data transmission system, it is the principal channels connecting major elements (for example, address bus, control bus, data bus).

**cable**    A group of insulated conductors in a common jacket.

**cable clamp**    A device used to give mechanical support to wire or cable.

**cable tie**    A beltlike plastic strip that loops around bundles of cables or insulated wires to hold them together.

**card access**    A term denoting a type of access control system that uses encoded cards and card readers to allow access to protected areas.

**card encoder**    A device used to encode access information on cards used for purposes of gaining entry into protected areas.

**card reader**   A device that scans or reads encoded information on access control cards.

**chip**   A microminiature electronic circuit on a tiny silicon wafer or other conductive material.

**chip-in-card**   An identification or access control card with a built-in integrated circuit (chip), giving the card-coded memory or microprocessor intelligence to record and store data. Also known as a smart card.

**circuit**   The path through which electrical energy flows.

**circuit breaker**   An automatic switch that opens an electric circuit if abnormal current conditions (such as an overload) occur.

**circuit, closed**   An electric circuit in which current normally flows until interrupted by the opening of a switch or a switch-type electronic component.

**circuit, open**   An electric circuit in which current does not flow until permitted by the closing of a switch or a switch-type electronic component.

**closure**   The point at which two contacts meet to complete a circuit.

**code, digital**   An access code or signal transmission containing data in digital or numerical form.

**coded card**   A plastic card (usually polyvinyl chloride) that has a combination secreted in its design to permit authorized entry and exit.

**coil, electric**   Successive turns of insulated wire that create a magnetic field when an electric current is passed through them.

**conductivity**   The capability of a material to carry electric current.

**conductor**   A material that will allow electricity to pass readily through it. Most metals are conductors, silver being the best.

**conduit**   A tube or trough for protecting electrical wires or cables.

**connector**   Generally, any device used to provide rapid connect/disconnect service for electrical cable and wire terminations.

**contact**   A magnetically or mechanically operated, electrically conductive point or set of points that opens or closes to interrupt or permit the flow of current.

**contact rating**   Load rating of a switch. Ratings are stated at their maximum voltage or current.

**continuity**   The state of being complete and uninterrupted, for example, a normally closed circuit.

**continuity check**   A test performed on a length of wire or cable to determine whether the electric current flows continuously throughout the length.

**continuous-duty**   Refers to a device that is designed to operate continuously with no off or rest periods.

**continuous-duty locking unit**   An electric lock equipped with a heavy-duty solenoid that can be energized indefinitely.

**control center**   A center in a facility where the access and alarm subsystems are supervised and where personnel are maintained continuously to record and investigate alarms and trouble signals.

**control point** An entry or exit point, such as a door, turnstile, or portal, where access is controlled and subject to verification.

**controlled access area** Any clearly demarcated area to which entry and exit are monitored and controlled.

**crimp** To compress (deform) a connector barrel around a wire to make an electrical connection.

**current** The flow of electrons through an electrical conductor. Current is measured in amperes.

**current-carrying capacity** The maximum current an insulated conductor can safely carry without exceeding its insulation temperature limitations.

**cycle (frequency)** The number of times per second that alternating current reverses its direction of flow. The standard commercial current in the United States is 60 cycles (60 Hz).

**deadbolt** A bolt operated manually and not actuated by springs. When blocked, the bolt cannot be forced back. A deadbolt is projected and retracted by a key cylinder or lever handle.

**deadlatch** A latch in which the latchbolt is positively held in the projected position by an auxiliary mechanism.

**decibel (dB)** An increment of measurement used to compare measured levels of sound energy (intensity) to the apparent level detected by the human ear. A sound that has 10 times the energy of another sound is said to be 10 decibels louder; 100 times the energy is 20 decibels louder; 1000 times the energy is 30 decibels louder; and so on. Decibel levels are correctly expressed as the number of decibels at a measured distance from the source of sound (for example, 125 decibels at 10 feet).

**dedicated line** A telephone line connecting two points, such as a protected premises and a central station, for alarm signaling. Also called lease line, direct wire, and direct connect.

**deenergize** To remove power.

**degausser** A device that erases data from magnetically encoded media, such as magnetic stripe access control cards.

**degraded mode** A mode of operation that creates a minimal authorization level in the event of a central processing unit failure.

**digital multimeter (DMM)** A multifunction meter typically capable of measuring voltage, current, and resistance with a digital readout.

**DIP switch** A miniature switch used to program, set, or change circuit functions. DIP is an abbreviation for the dual-in-line package, which houses the switch.

**direct current (DC)** Electric current that travels in only one direction and has negative (−) and positive (+) polarity.

**distributed card access control** A system in which all access control decisions are made at the control point, independent of the central processing unit.

**distributed system** An access control system where the devices make their own access decisions, uploading event messages periodically to the central processing unit for storage.

**door status switch (DSS)**   A switch used to monitor whether a door is in an opened or closed position. Also referred to as door position switch (DPS).

**double-pole, double-throw (DPDT)**   A term used to describe a switch or relay output contact form in which two separate switches are operating simultaneously, each with a normally open and normally closed contact and a common connection. This form is used to make and break two separate circuits.

**download**   The act of sending information from a central processing unit to a peripheral device such as a card reader or other slave processor in the system.

**dry contact**   Metallic points making (shorting) or breaking (opening) a circuit. The switched circuit must have its own source of power and is merely routed through the dry contacts.

**duress**   In access control terms, descriptive of a situation in which a person is forced to gain access to a protected area against his or her wishes.

**duress alarm**   A device which produces either a silent alarm or an alarm signal under a condition of personal distress. This device is normally manually operated and may be fixed or portable. Duress alarms can also be built into the operation of a combination keypad/card reader device.

**duress code**   A special code which, when entered into a keypad, will alert the system to a duress condition.

**duty cycle**   The percentage of "on" time or operating time of a device. For example, a device that is on for 1 minute and off for 9 minutes is operating at a 10 percent duty cycle.

**earth ground**   Connection to the earth. Standard earth ground is a copper rod at least $1/2$ inch in diameter and 8 feet long, driven into moist earth. For many purposes, underground water pipes can be used. This is known as cold water pipe (CWP) ground.

**electric strike**   An electric door-locking device (usually solenoid-operated) that will unlock the door when electrical power is applied to it. A fail-safe configuration will operate in the reverse condition (normally locked when power is applied and unlocked when power is interrupted).

**electromagnet**   A coil of wire, usually wound on an iron core, that produces a strong magnetic field when current is sent through the coil.

**electromagnetic**   Pertaining to combined electric and magnetic fields associated with movements of electrons through conductors.

**electromagnetic lock**   A locking device that uses magnetism to hold the door securely closed. An electromagnetic lock is typically used as a secondary means of securing the door. See **magnetic lock.**

**electromotive force (emf)**   Pressure or voltage; the force that causes current to flow in a circuit.

**embossed card**   A type of access control card that uses a raised pattern as a means of encoding.

**emergency release**   An optional release device to release an electric lock in an emergency.

**encapsulant**   A material, usually epoxy, used to encase and seal all components in an electronic circuit.

**encoding**   The act of writing data to a card.

**end-of-line (EOL) resistor**   Resistance in a supervised circuit, usually at the farthest point from the alarm control unit, restricting the flow of current to a known value which is monitored. Shorting the circuit in an attempt to bypass protective devices in the loop (say, door contacts) will create increased flow of current and cause an alarm. Opening (breaking) the circuit also triggers an alarm if the system is armed, or a supervisory signal if the system is disarmed.

**energize**   To apply power.

**entrance code**   See **access code.**

**entrance delay**   The time between actuation of a sensor on an entrance door or gate and the sounding of a local alarm or transmission of an alarm signal by the control unit. This delay is used if the card access reader is located within the protected area and permits a person with an access card or key to enter without causing an alarm. The delay is provided by a timer within the control unit.

**EPROM**   Erasable programmable read-only memory.

**exit alarm**   An electrically operated device indicating either audibly or silently the unauthorized opening of a secured door.

**exit reader**   A reader that controls egress from a controlled area.

**exit switch**   A switch operating an electronically controlled door, allowing egress from a protected area.

**explosionproof device**   Any device, such as a contact switch, that is enclosed in an explosionproof housing to prevent possible sparking in potentially volatile environments.

**facility code**   A code typically used in security and access control systems that identifies the customer or location of the system.

**factory calibration**   Factory setting of a control circuit by the manufacturer to bring the circuit into specification.

**factory-fixed**   Refers to adjustment made by the manufacturer and not changeable by the user.

**fail-safe**   In alarm systems, descriptive of an operation that will trigger an alarm or trouble condition in case of equipment failure or power loss. In access control systems, this operation will unlock all controlled doors in case of equipment failure or power loss.

**fail-safe lock**   An electric lock that automatically unlocks with any power interruption.

**fail-secure**   In an access control system, a condition whereby controlled doors automatically lock in the case of equipment failure or power loss.

**fail-secure lock**   An electric lock that requires power to unlock.

**false alarm**   An alarm signal transmitted in the absence of an alarm condition.

**fast-on terminal**   A solderless, easy-to-use, female/male push-on terminal that comes in various sizes.

**Federal Communications Commission (FCC)**   The U.S. government agency that regulates communications by telephone, telegraph, radio, and television.

**fingerprint pattern area**   The part of a fingerprint's loops, whorls, and arches in which identifying characteristics appear.

**fingerprint reader**   A high-security biometric access control device that identifies a person by fingerprints. After the finger is placed on a light-sensitive plate, the print is read and compared to images stored in a computer's memory. If the print matches one stored in memory, access is granted.

**flasher**   A control in which the output to the load (normally a lamp) is turned on and off repeatedly at a given rate of operation or flashes per minute (FPM).

**flux**   The lines of force that make up an electrostatic field. The rate of flow of energy across or through a surface. A substance used to remove oxides from surfaces to be joined by soldering or welding.

**form C contact**   A switch mechanism that contains three contacts (normally open, common, and normally closed).

**frequency**   The number of complete operations or cycles that take place within a given period of time (normally 1 second), as in the AC line frequency of 60 Hz (60 cycles per second).

**full-wave**   A term used for both AC and DC voltages, suggesting that both halves of the sine wave are utilized (for example, full-wave AC and full-wave rectified AC or unfiltered full-wave DC).

**function key**   A key on a keyboard or keypad dedicated for a specific use. Pressing the key causes a specific predetermined response, such as bypassing an intrusion alarm sensor.

**fuse**   A protective device placed in a circuit as a safeguard, containing a strip of easily melted metal. When the current flow becomes too great, the metal melts, breaking the circuit.

**ground**   The electrical connection with the ground, which remains essentially at a same potential point, assumed to be zero for reference to other potential points. Electrical connection to a common point, which is the zero-voltage reference for the circuit, not necessarily at ground potential or connected to it. Also known as chassis ground or earth ground.

**half-wave**   Refers to the passing or the use of only one-half of the AC sine wave. The result is half-wave rectified AC, or unfiltered half-wave DC.

**hand geometry**   An access control technology that verifies a person's identity by comparing relative variations in finger lengths and thicknesses.

**handwriting dynamics**   See **signature verification.**

**hard-wired**   Refers to groups of connections that require the use of wire conductors.

**hertz (Hz)**   The international unit of frequency equal to one cycle per second; named after the German physicist Heinrich Rudolph Hertz (1857–1894).

**hi-pot**   A test designed to determine the highest potential that can be applied to a conductor without breaking through the insulation.

**historical recording**   A chronological record of events in an alarm or access control system. Also called historical logging.

**Hollerith card**   A type of access card having small holes which can be read by a light source or contact brushes. A Hollerith card is easily duplicated and is not suitable for high-security applications.

**hookup wire**   Insulated wire used for low-current, low-voltage applications internaly within enclosed electronic equipment.

**host computer**   The main controlling entity in a system with multiple processing units.

**hot**   Connected, alive, energized.

**ID station**   A location where a machine alone performs the identification function.

**identification**   The act of recognizing a person as unique from all others. This ac may be done by another person on sight aided by proper documentation, by a device using card or keypad data, or by using biometric techniques.

**induced AC**   A condition caused when low-voltage wiring is placed near high-voltage wiring. The higher-voltage line may interfere with the lower-voltage line and may interfere with or damage microprocessor-based equipment.

**induction**   An influence exerted by a charged body or by a magnetic field on neighboring bodies without apparent communication; electrifying, magnetizing, or inducing of voltage by exposure to a field.

**inductive load**   An electric device made of wire, wound or coiled, to create a magnetic field to produce mechanical work when energized. Components such as motors, solenoids, and relay coils are all inductive loads by nature. An inductive load can exhibit an inrush current of up to 5 times its normal running or steady-state current when energized. When deenergized, the magnetic field collapses and a high-voltage transient is generated, which can cause arcing across contacts or a malfunction of and/or damage to electronic circuits. When transients are present, they should be suppressed. (See **transient.**)

**infrared**   Light waves that are too low in frequency to be seen by the unaided human eye.

**infrared card**   A card technology that uses infrared light waves to read a code pattern hidden with an opaque barrier. There are two basic types: reflective and transmissive. With reflective infrared, both the infrared emitter and detector of the reader are on the same physical side of the card. Infrared light is reflected off the code pattern and read by the detector. With transmissive technology the infrared emitter and detector of the reader are on opposite sides. Infrared light goes through the card where it is picked up by the detector on the other side.

**infrared reader**   A card reader that uses an infrared light source to read information encoded in an access control card. This reader is an optical technology and is based on an optical density principle.

**inhibit**   A temporary prevention of an alarm during the authorized opening of a protected door or device.

**input voltage**   The designed power source requirement needed by equipment in order to operate properly.

**inrush**   The initial surge of current through a load when power is first applied. Lamp loads, inductive motors, solenoids, and capacitive load types all have inrush or surge currents higher than the normal running or steady-state currents. Resistive loads, such as heater elements, have no inrush.

**insertion card**   In card-based access control, a card that is inserted into a reader, rather than swiped through or placed near.

**insulation** A material that provides high electric resistance, making it suitable for covering components, terminals, and wires to prevent possible contact of adjacent conductors, resulting in a short circuit.

**intelligent terminal or device** A microprocesor-based input/output device with free-standing logic capabilities. Distributed access control device that makes its own access decisions, uploading event messages periodically to the central processing unit for storage. These devices can also communicate with the central processing unit, and new operating instructions can be downloaded from the central processing unit to the intelligent device.

**interlock** A system of multiple doors with controlled interaction. Interlocks are also known as lighttraps, airtraps, mantraps, and sallyports. (See **safety interlock and security interlock.**)

**intermittent-duty** Descriptive of a solenoid designed to be energized for short periods of time. Continuous operation may damage an intermittent-duty solenoid.

**interval** A period of time from one event to another. An interval timer controls the time at which a load is energized or deenergized.

**isolation** No electrical connection between two or more circuits.

**jacket** Pertaining to wire and cable, the outer sheath that protects it against the environment and may also provide additional insulation.

**jumper** A connection between two points. A plug-in device that makes a connection between points as on a circuit board or backplane.

**junction** A point in a circuit where two or more wires are connected.

**junction box** A protective enclosure for connecting circuit wires.

**keyless access control** An entry control system using a means other than a key. This term usually refers to a digital keypad or card reader used in conjunction with an electrically controlled locking device.

**keypad** A device for inputting information into a computer-controlled system for the purposes of arming and disarming an alarm system, or operating an access control system.

**key switch** A lockable switch that is operated by a key.

**kilohm** 1000 ohms.

**labeled** Descriptive of equipment or materials that have a label, symbol, or other identifying mark of an organization that is approved by the authority having jurisdiction over product evaluation. The label indicates compliance by the manufacturer with appropriate equipment or performance standards.

**latching relay** A relay that latches in either the on or off condition until reset manually or by a signal.

**life** The number of performance hours, days, years, or actual operations for which an item is designed.

**light-emitting diode (LED)** A solid-state device that gives off virtually heatless colored light when electric current is passed through it. LEDs are very efficient and long-lasting and are often used for digital readouts and annunciators. Common colors include red, green, and amber.

**lighttrap or airtrap**   A room with two or more doors controlled to prevent more than one door being opened at one time.

**line cord**   A cord, terminating in a plug at one end, that is used to connect equipment or appliances to a power outlet.

**line drop**   A voltage loss occurring between any two points in a power or transmission line. Such loss, or drop, is due to the resistance, reactance, or leakage of the line.

**line supervision**   The electrical supervision of a wire run to detect tampering (a cut or shorted wire). Line supervision usually requires a terminating element at the end of the monitored wire loop. (See **end-of-line resistor.**)

**line voltage**   The voltage existing in a main cable or circuit, such as at a wall outlet.

**listed**   Descriptive of equipment or materials included in a list published by an authorizing organization. The listing states that the equipment or material meets appropriate standards or has been tested for and is suited to a specific application.

**load**   Any device that consumes electrical power; the amount of power required for operation of a circuit or device.

**load rating**   A control specification outlining the type of load, the minimum and the maximum currents, and the voltage.

**local alarm**   A visual or audible signaling device located at a monitored door, window, or other opening.

**lock**   A device for securing a door in the closed position against unauthorized or forced entry.

**logging**   The process of creating a permanent record or log. For example, some access control and alarm systems keep a record of all system events which are immediately printed or stored for later printing.

**magnetic keycard**   A plastic card containing thousands of magnetic bits or particles which can be arranged to match the required pattern set up in a card reader.

**magnetic lock**   A type of door lock consisting of an electromagnet and strike plate. The electromagnet is mounted to the door frame, opposite the strike plate, which is mounted to the door. When power is applied, the strength of the magnet holds the door locked.

**magnetic spot card**   A type of access control card with a barium ferrite core. Codes are determined by the magnetic polarity of dots embedded in the card.

**magnetic stripe card**   A type of access control card with a data-encoded strip of magnetic material.

**magnetic switch**   A switch which consists of two separate units: a magnetically actuated switch and a magnet. The switch is usually mounted in a fixed position (door jamb or window frame) opposing the magnet, which is fastened to a door or window. When the movable section is opened, the magnet moves with it, activating the switch.

**maintained switch**   A switch designed for applications requiring sustained contact, but with provision for resetting (say, a light switch).

**make**   To close or establish an electric circuit.

**make-before-break**   The action of a type of switch in which one contact closes before another opens.

**mantrap**  An arrangement of doors, usually forming a small corridor or booth, that allows a person to enter and be identified before proceeding into a controlled area. Most mantraps are engineered so that both doors are locked as soon as a person enters, effectively trapping the individual until identification is made. (Also see **interlock.**)

**manual override**  A feature in some access control systems which allows for manual shutdown and operation during power failure and emergency conditions.

**master code card**  A special access control card containing a code that grants entry and exit at all card readers in a system.

**maximum rating**  The absolute maximum condition in which a device is designed to operate. Voltage, frequency, current, temperature, humidity, shock, and other parameters can be specified as maximum.

**megohm**  1 million ohms.

**memory**  The section of a computer that stores data and instructions in binary code.

**microprocessor controller**  In an access control system, a device that drives the program stored in its read-only memory (ROM) and random-access memory (RAM). This program typically contains access authorization data such as areas allowed for entry and time frames for entry.

**mil**  One one-thousandth (0.001) of an inch; a unit used in measuring the diameter of wire and the thickness of insulation over a conductor.

**milliampere (mA)**  One one-thousandth (0.001) ampere.

**millisecond (ms)**  One one-thousandth (0.001) second.

**mode of operation**  The specified operational condition of a switch, lock, door system, and so forth.

**modem**  A device that converts the computer system's digital information into analog information and transmits it over a telephone line. Another modem must be used when the information is received to convert the information back from analog to digital.

**momentary switch**  A switch designed for applications requiring a momentary signal. When pressure is removed, the switch contacts automatically return to their original state (for example, doorbell).

**monitoring loop**  A continuous loop of wire starting at a control panel and running through switches in a system to indicate a breach of security through an open switch or a cut wire.

**motherboard**  A master printed-circuit board used to interface the activities of individual printed-circuit boards and the devices being controlled or monitored. The motherboard is usually located at the back of a control panel assembly; individual printed-circuit boards plug into it.

**multiconductor cable**  A cable consisting of two or more conductors, either cabled or laid in a flat parallel construction, with or without a common overall covering.

**multiplex**  Descriptive of a system of transmitting several messages simultaneously on the same circuit or channel. Multiplex equipment greatly reduces the number of wire cables needed in a system.

**National Electrical Code (NEC)**  The code used throughout the United States to control wiring and other electrical installations in the interest of public safety. It is published by the National Fire Protection Association (NFPA).

**National Electrical Manufacturers Association (NEMA)**   The organization that establishes standards in the manufacture of electronic components.

**noise**   Unwanted and/or unintelligible signals picked up on a cable circuit.

**normally closed (NC)**   The state or position of a contact prior to initiation or energization—in this case, a closed condition.

**normally open (NO)**   The state or position of a contact prior to initiation or energization—in this case, an open condition.

**octal plug**   An eight-pin male connector with a locating key for proper orientation.

**offline card reader**   A stand-alone access control card reader that contains its own intelligence for granting and denying access.

**ohm**   A unit of measurement for resistance $R$ and impedance $Z$.

**Ohm's law**   One of the most widely used principles of electricity. It expresses the relationship between voltage $E$, current $I$, and resistance $R$ according to the following equations: $E = IR$, $I = E/R$, and $R = E/I$.

**online access control system**   A system of card readers, microprocessor controllers, or other sensing devices connected to a centralized decision-making computer.

**operating temperature**   A temperature range over which a device will perform within its specified design tolerances; may be stated in degrees Fahrenheit (°F) or degrees Celsius (°C).

**operating voltage**   The voltage by which a system operates; a nominal voltage with a specified tolerance applied; the design voltage range necessary to remain within the operating tolerances. For example, for a system specified 120 volts ±10 percent of nominal, 120 volts is the nominal voltage and the design voltage range is 108 to 132 volts AC.

**optical card**   A type of access control card technology which employs several rows of spots with varying transparency. The relative passage of light through the pattern of spots forms a readable code.

**output voltage**   The designed power source produced by a power supply to operate equipment.

**palm geometry reader**   A biometric access control device that scans and reads the size and shape characteristics of a person's palm as criteria for granting and denying access.

**panic bar**   A quick-release door-mounted exit bar permitting fast opening in case of fire or emergency situation. Also called a crash bar and exit device.

**parallel**   A method of connecting an electric circuit whereby each element is connected across the other. The addition of all currents through each element equals the total current of the circuit.

**personal identification number (PIN)**   A unique numerical code used alone or in conjunction with other access control technologies to gain access to a protected area.

**piggybacking**   See **tailgating.**

**PIN**   Abbreviation for personal identification number.

**polarity**   The positive or negative orientation of a signal or power source.

**potentiometer (pot)**   Variable resistor.

**potting**   The process in which the space between a component and its case is filled with an insulatory compound which hardens to provide an airtight, moistureproof, insulating seal.

**power transfer**   Means to transfer power from the door to the frame.

**primary**   The transformer winding that receives the energy from a supply circuit.

**printed-circuit (PC) board**   A means of making electrical interconnections without using insulated wires. Printed-circuit boards provide a supporting and insulating medium for components and conductors in a form that is readily adaptable to machine assembly.

**programmable card**   An access control card onto which data may be encoded after the card has been manufactured.

**programmable stand-alone card reader**   An access control card reader with its own intelligence for granting and denying access.

**protected area**   An area monitored and controlled by a staffed or electronic security system, or enclosed by barriers. A secured area.

**proximity card**   A radio-frequency-based card technology that utilizes a microcircuit which, when presented to a proximity reader, activates the card's circuitry, thus transmitting the data stored in the card.

**punch hole card**   An access control technology based on the Hollerith principle, having a specific pattern of punched holes.

**rack-mounted**   Descriptive of a method of housing many control and security panels. Nineteen-inch rack mounting is a standard for the electrical equipment trades. Rack mounting allows equipment of several different manufacturers, different types of communications, fire/smoke alarm and security equipment to be used in the same area without taking up a large amount of space. It also achieves a more uniform and organized appearance.

**random access memory (RAM)**   Read/write memory, either volatile or nonvolatile, whose contents may be altered at will or read out without alteration and which may be randomly addressed.

**rated voltage**   The maximum voltage at which an electric component can operate for extended periods without undue degradation or safety hazard.

**reactance**   Opposition offered to the flow of alternating current by inductance or capacitance of a component or circuit.

**read-only memory (ROM)**   Nonvolatile memory whose contents are mask-programmed during microelectronic fabrication; consequently they cannot be altered. This type of memory is used to store microinstructions, reference data, and microprocessor programs.

**real-time command**   An operational command that has no built-in delay. The operation takes place as soon as the command is entered.

**rectifier**   A solid-state electrical device that will allow current to flow in one direction only. It is designed to convert alternating current to direct current.

**regulated power supply**   A power supply that provides a constant output during input voltage variations.

**relay**  A type of switch actuated by current flowing through a coil wound around an iron core, which is thereby magnetized and attracts an armature, opening or closing the switch contacts. Removal of the current returns the contacts to their resting condition (normally open or normally closed), except in the case of latching relays, which must be reset.

**remote alarm**  A visual or audible signaling device used to signal violations at locations removed from the central control station or monitored openings. For example, a remote alarm may be placed on a roof, in a stair tower, or at guard stations outside a building.

**remote reset**  A switch located at a monitored opening. If a violation occurs, the alarm at the main control console cannot be turned off until the door is secured and the remote reset is activated. Its purpose is to ensure the inspection of an opening that has been violated or left open.

**remote terminal**  A device for communicating with a host computer system from a location that is apart from the central computer facility.

**reset time**  The time required to return the output to its original condition.

**resistance**  The opposition to the flow of an electric current (measured in ohms); the reciprocal of conductance.

**resistor**  A circuit element whose chief purpose is to oppose the flow of current.

**resolution**  The degree of setability.

**restricted area**  A room or other area requiring controlled access due to the sensitive nature of materials contained there.

**retina reader**  A biometric-based access control device that scans the blood vessel pattern in a person's retina and compares it with stored images before granting or denying access.

**reverse-polarity-protected**  Descriptive of DC controls where if the polarity of the input were reversed, there would be no damage.

**rigid conduit**  A metal piping for housing the insulated wires of an electric circuit.

**riser diagram**  A document which explains the number of conductors to be run between equipment in a system.

**safety interlock**  A multidoor system in which all doors are normally closed and unlocked; opening any door locks all other doors.

**sallyport**  See **interlock.**

**secondary**  The transformer winding that receives energy by electromagnetic induction from the primary.

**secure mode**  The condition of an alarm system in which all sensors and control units are ready to respond to an event.

**secured area**  The user-defined subdivision of a facility; each security area usually has its own time zones, access points, and/or alarm points.

**security interlock**  A multidoor system in which all doors are normally closed and locked; releasing one door disables the releases for all other doors until the first door is closed and relocked.

**self-contained card reader**  A card-reading device containing built-in intelligence to grant or deny access. A stand-alone reader not connected to a central computer for control. Also called an offline reader.

**semiconductor**   A material that has a resistance between those of insulators and conductors.

**sequential card reader**   A card/code access control system that requires the use of a card plus the input of a code on a keypad.

**series circuit**   An electric circuit in which all the receptive devices are arranged in succession, as distinguished from a parallel circuit. The same current flows through each part of the circuit in sequence.

**shielding**   The act of enclosing a circuit, conductor, or component in a grounded, low-resistance metal container, or placing a grounded metal plate between two components, that provides isolation by short-circuiting the lines of force of any electric field within or without the shielded area.

**short circuit**   The connection, often unintentional, that provides a low-resistance path between two points in a circuit or to ground. Depending upon the circumstances, a short circuit may stop the circuit operation, alter it, or have no effect. If it results in excessive current flow, damage to the circuit may result.

**shunt**   The deliberate shorting out of a portion of an electric circuit.

**signature verification**   A biometric-based access control technology that records and compares signature dynamics (writing speed, pen pressure, etc.) against those characteristics stored for that person in memory as criteria for granting and denying access.

**single-entry access system**   A system designed to prevent the passback of an access control card. A card must be read by an exit reader before it is valid again in an entry reader.

**single pole, single throw (SPST)**   A switch with only one moving and one stationary contact, available either normally open (NO) or normally closed (NC).

**sink**   A component or connection into which current flows. A means of dissipating heat; heat sink.

**slave**   Any device that operates on command from another device, but does not do so independently.

**smart card**   An identification card containing an integrated circuit, allowing it to receive and store data, which gives it limited microprocessor intelligence.

**smart terminal**   A hardware device that processes access requests from card readers or report requests to a system processor that can accommodate multiple access control panels. Additionally, it provides the connection points and controls for electronic related items for its access points, such as door contacts and door strikes.

**soldering**   A method of making an electrical connection. The two components to be connected are physically placed together and heated. Solder, a conductive metallic alloy with a low melting point, is then placed on the heated components. It melts and flows around the components to make a permanent connection.

**solenoid**   An electromechanical device that operates a lockbolt. When electricity is applied, a mechanical motion is obtained that moves the bolt.

**spike**   A voltage pulse of short duration, but high amplitude. Also called transient or surge. Spikes can damage electronic equipment.

**splice**   A connection of two or more conductors to provide good mechanical strength as well as good conductivity.

**stand-alone card access system**   An access control system in which the card reader is connected to the control electronics. All required components are contained in one enclosure, and there is no reliance on an external or host computer.

**status level**   See **authorization level.**

**steady state**   A term used to specify the current through a load or electric circuit after the inrush current is complete; a stable run condition.

**storage temperature**   The maximum temperature that any one material in a system can withstand without sustaining damage; a nonworking condition.

**stranded conductor**   A conductor composed of several single solid wires twisted together.

**strike**   A plate mortised into or mounted on the door jamb to accept and restrain a bolt when the door is closed. In some metal installations of a deadlock, the strike may simply be an opening cut into the jamb.

**supervised circuit**   A circuit that will indicate alarm and trouble conditions.

**surface-mounted switch**   A mechanically or magnetically based contact point mounted on the surface of a protected door, window, or gate in such a way that it is visible.

**swipe reader**   An access control card reader through which a card is passed (swiped) instead of inserted.

**switch, maintained**   A switch that, when activated, maintains its activated position until it is reactivated.

**switch, momentary**   A switch that, when activated, automatically returns to its original position when influence is removed.

**switch, normally closed**   A switch that, when not energized, is closed to form a path for current.

**switch, normally open**   A switch that, when not energized, is open and does not let current flow.

**switches**   Devices that make or break connections in an electric or electronic circuit.

**tailgating**   In access control, the act of one or more individuals entering a controlled area by using a single card. Also the act of following an authorized person into a controlled area. Also known as piggybacking.

**tap**   A special lead brought out from an intermediate point of a coil or winding.

**telephone entry system**   An access control system using a telephone located outside the protected area. An individual desiring access dials a coded sequence of numbers to connect with an operator who grants or denies access.

**terminal block**   A device that provides a place for safe and convenient interconnection of current-carrying conductors.

**terminals**   Metal wire termination devices designed to handle one or more conductors and to be attached to a board, bus, or block with mechanical fasteners or clipped on. Common types are ring tongue, spade, flag, hook, blade, quick-connect, offset, and flanged.

**terminating element**   An electric device connected at the end of a pair of electrical conductors that provides the means of supervising those conductors. (See **line supervision.**)

**throughput rate**   The rate at which people or vehicles pass through an access point.

**time delay**    An electronically controlled delay period designed into a component that will either send a prolonged signal or delay transmitting a signal.

**time-delay relay**    A relay with contacts that are delayed on make or break after a short, fixed time interval.

**time zone**    The time periods for each security area as defined by the user. The time zone specifies during what time periods a security area can be entered. Each employee or group of employees is also assigned a time zone. A time zone may contain multiple time intervals, such as holidays and times of day for cleaning activities.

**tinned copper**    Copper with a tin coating added to aid in soldering and to inhibit corrosion.

**tolerance**    Normally stated as a percentage, the maximum allowable deviation of electrical, environmental, or dimensional parameters.

**touchpad**    See **keypad.**

**transformer**    An electric device that changes voltage in direct proportion to currents and in inverse proportion to the ratio of the number of turns of its primary and secondary windings. The input side of a transformer is called the primary side; the output or low-voltage side is called the secondary side.

**transient**    Any increase or decrease in the excursion of voltage, current, power, heat, and so forth above or below a nominal value that is not normal to the source. (See **transient voltage.**)

**transient voltage**    Refers to several parameters of a transient: the peak or maximum voltage reached, the rate of rise of the transient, and the duration of the transient. Transient voltages are generated when inductive loads such as solenoids, motors, relays, and so forth are deenergized. Although some devices have excellent protection against these sometimes damaging excursions, when a transient is known to be present, it should be suppressed at the source. Diodes and metal-oxide varistors (MOVs) are commonly used as suppressors.

**transponder**    An electronic device that automatically transmits a signal in response to the receipt of a signal.

**trickle charge**    A low-powered electric energy source provided to keep standby batteries fully charged.

**twisted-pair**    A type of cable composed of two small insulated conductors twisted together without a common covering. The two conductors of a twisted-pair are usually substantially insulated, so the combination is a special type of a cord.

**type A error**    An error in an access control system in which a person who should be granted access is denied it. Also called a type 1 or type I error.

**type B error**    An error in an access control system in which a person who should be denied access is granted it. Also called a type 2 or type II error.

**Underwriters Laboratories (UL)**    UL-listed or -labeled signifies that production samples of the product have been found to comply with established Underwriters Laboratories requirements and that the manufacturer is authorized to use the UL listing marks on its products.

**verification**    The act of verifying an individual after being provided with adequate information such as the PIN number. Also known as authentication.

**verification error rate**   The rate of errors in an access control system that are attributable to the granting/denying of access which should have been denied/granted.

**voice analysis**   Personal identification by individual voice frequency.

**voice recognition system**   An access control device used to verify a person's identity by comparing previously recorded keywords and phrases digitally stored in a computer's memory against those spoken when access is sought.

**volt (V)**   A unit of electromotive force. It is the difference of potential required to make a current of one ampere flow through a resistance of one ohm.

**volt-amp (VA) rating**   The product of rated input voltage and the rated current. This establishes the "apparent energy" available to accomplish work.

**voltage**   The term most often used (in place of electromotive force, potential, or potential difference) to designate electrical pressure that exists between two points and is capable of producing a flow of current when a closed circuit is connected between the two points.

**voltage drop**   Voltage loss experienced by electric circuits due to two principal factors: wire size and length of wire runs.

**watt**   The common unit of electrical power. One watt is dissipated by a resistance of one ohm through which one ampere flows.

**Watt's law**   $P = I \times E$.

**Wiegand card**   An access control card based on the Wiegand effect, whereby small lengths of specially processed wire are permanently embedded in a vinyl card. These wires are placed in an array that emulates a binary code, which is sensed as the card is swiped through, or inserted into, a Wiegand reader.

**Wiegand effect**   A pulse-generating phenomenon in a special alloy wire that is processed in such a way as to create two distinct magnetic regions in the same homogeneous piece of wire, referred to as a shell and a core. These two magnetic regions react differently to any applied magnetic field. The shell requires a strong magnetic field to reverse its magnetic polarity, whereas the core will reverse under weaker field conditions. It is at the point where the shell and core change to different-polarity orientations that the Wiegand pulse is generated, sensed by a pickup coil (the reader).

**wire**   A slender rod or filament of drawn metal.

**wire nut**   A connector used to make and insulate an electrical connection. Wire ends are stripped and placed into a caplike connector (wire nut), and the wire nut is then twisted to secure the wire ends together. The cap design serves to insulate the connection.

**zone**   A specific area of protection; a portion of a large protected area.

# Index

## ABOUT THE AUTHOR

**John L. Schum** is the author of *Learn About... Electronic Locking Devices* and *Learn About... Basic Electricity*, published by Locksmith Publishing Corporation. He writes technical articles for trade magazines and has given numerous seminars throughout the United States and Canada for various trade groups. He also has been a program speaker at several security industry conventions.

John is a certified instructor for the Associated Locksmiths of America and has instructed for Yankee Security Conference since 1983. He has also taught courses on electromechanical hardware for the Door and Hardware Institute since 1985. He is currently head instructor for the DHI Electronic Hardware II course.

John has been involved in the design and application of electronic security hardware since 1979. He has served as engineering manager, customer service manager, and technical services manager for several manufacturers. John is currently vice president of sales for DynaLock Corporation and resides in Thomaston, Conn.